THE ECOLOGY
OF SOCIAL BEHAVIOR

Paul SH Hart .

Academic Press Rapid Manuscript Reproduction

THE ECOLOGY
OF SOCIAL BEHAVIOR

Edited by

C. N. Slobodchikoff

Department of Biology
Northern Arizona University
Flagstaff, Arizona

ACADEMIC PRESS, INC.
Harcourt Brace Jovanovich, Publishers

San Diego   New York   Berkeley   Boston
London   Sydney   Tokyo   Toronto

ACADEMIC PRESS, INC.
1250 Sixth Avenue
San Diego, California 92101

United Kingdom Edition published by
ACADEMIC PRESS INC. (LONDON) LTD.
24-28 Oval Road, London NW1 7DX

Library of Congress Cataloging-in-Publication Data

The Ecology of social behavior.

Includes index.
1. Social behavior in animals. 2. Animal ecology.
I. Slobodchikoff, C. N.
QL775.E26   1987   591.52'46   87-35148
ISBN   0-12-648780-4   (hardcover)(alk. paper)
ISBN   0-12-648781-2   (paperback)(alk. paper)

PRINTED IN THE UNITED STATES OF AMERICA
88 89 90 91   9 8 7 6 5 4 3 2 1

CONTENTS

PREFACE

The goal of this book is to explore the relationships between ecology and the origins and maintenance of social behavior. In preparing the chapters, the authors were asked to try to discuss empirical data from the standpoint of a theoretical framework. Consequently, some discussions may seem highly speculative. We tend to forget all too often, however, that speculation is where science begins. Without speculation, we have a set of random observations, that, outside the framework of a paradigm, may not even be called facts. Scientists often get buried in data and observations, and are subsequently unable to see the forest for the trees. This book is an attempt to delineate the forest more closely.

The chapters in this book suggest that a consideration of ecological factors is necessary to any paradigm that tries to explain the origins and maintenance of social behavior. This is not to deny, however, the importance of other factors, such as the genetic aspects of kinship or the phyletic aspects of evolutionary lineages, in the genesis of some social systems. Most of the chapters in this book suggest that there are some trade-offs between ecology, genetics, and phylogeny in the development and persistence of specific social systems.

Many people contributed to the production of this book. Henry Hooper, Associate Vice President of Academic Affairs at Northern Arizona University (NAU), and Richard Foust, Director of the Bilby Center at NAU, greatly encouraged the project and subsidized many of the costs of assembling these chapters. Luella Holter and Evelyn Wong of the Bilby Center staff patiently typed and retyped the entire manuscript, a herculean task for which I am very grateful. Cheryl Fischer, Kitty Gehring, Judy Kiriazis, Tony Robinson, Anne Slobodchikoff, and Tad Theimer patiently proofread, edited, and assembled materials. The Academic Press staff handled all my questions with understanding. I thank all of these people for their efforts, which were often beyond the call of duty. Last but not least, I thank the authors for their enthusiastic response to the theoretical nature of this book, and I am grateful to reviewers for taking time out of busy schedules to offer criticism and constructive advice.

C. N. Slobodchikoff

THE ECOLOGY
OF SOCIAL BEHAVIOR

I. INTRODUCTION

Chapter 1

ECOLOGICAL TRADE-OFFS AND SOCIAL BEHAVIOR

C.N. Slobodchikoff

Department of Biology, Northern Arizona University
Flagstaff, Arizona 86011

William M. Shields

Department of Environmental and Forest Biology,
State University of New York, College of
Environmental Science and Forestry
Syracuse, New York 13210

I.   COST-BENEFIT ANALYSES AND THE ORIGIN AND MAINTE-
     NANCE OF SOCIAL BEHAVIOR
II. ECOLOGICAL TRADEOFFS AND SOCIAL BEHAVIOR
REFERENCES

I.   COST-BENEFIT ANALYSES AND THE ORIGIN AND MAINTE-
     NANCE OF SOCIAL BEHAVIOR

Recent discussions regarding the origin and maintenance of social behavior are often based on conflicting, or at least confusing, collections of hypotheses and data about the evolution of: (1) groups (colonies or societies); (2) behavior within groups (social behavior); or (3), both (1) and (2). We define a group, with Wilson (1975, p. 585), as "any set of organisms, belonging to the same species, that remain together for a period of time interacting with one another to a distinctly greater degree than with other conspecifics." We define a society as an interacting group consisting of more individuals than parents and their immediate dependent offspring. In this context, social behavior includes any behavioral interactions among members of social groups. Social behavior, then, includes any selfish, spiteful, cooperative, or altruistic behavior directed at group members and need not be limited to social species (Hamilton 1964).

At first glance, such discussions of social behavior offer a plethora of potential explanations at odds with themselves. Thus, some treatments have emphasized the importance of ecological factors in the adaptive origin and maintenance of group living and social behavior (e.g., Crook 1965, 1970; Lack 1968). Other treatments have emphasized the importance of kinship in the adaptive shaping of social behavior (e.g., Hamilton 1964; Alexander 1974; Trivers 1985). Still others have emphasized that the groups or social behaviors we currently see may be the result of phylogenetic inertia, i.e., patterns developed in evolutionary time within the phylogeny of the animals in question (Wilson 1975).

These different emphases can be categorized into three classes of hypothesis. The classes are not mutually exclusive and each could have an independent and important influence on explaining the origin or maintenance of sociality in any particular species, population, or group. *Genetic* hypotheses assume that high levels of kinship among group members are necessary and sufficient to explain the kind of society being investigated. *Ecological* hypotheses assume that nongenetic environmental factors are necessary and sufficient to explain an observed social system. Finally, a *phylogenetic* hypothesis assumes that a social system evolved in the context of conditions occurring earlier in the evolutionary history of the lineage in question and does not represent an adaptation to current conditions.

The recent emphasis on genetics stems from the attention generated by Hamilton's (1964) quantification of inclusive fitness, which was intended to explore the effects of kinship on the evolution of altruistic behavior. Hamilton's rule states that altruistic behavior will evolve only when the gain for the beneficiary of a behavior (B, where the beneficiary is a different individul from actor also called donor), multiplied by the probability (r) that the beneficiary and the actor have genes in common, exceeds the cost of that behavior to the actor or donor (C). Costs and benefits here are both positive terms. When the probability of having genes in common (i.e., relatedness) between the actor and the beneficiary is zero, the cost to the actor is often greater than the benefit to the beneficiary. This translates into the classic statement that only when $Br > C$ will altruistic social behavior evolve. This has also been translated into an expectation that relatedness will have profound effects on the evolution of sociality.

For example, Hamilton (1964) originally argued that the social system of honeybees might have arisen, in part, because of haplodiploidy, i.e., because worker bees may be more closely related to each other than they would be to their own progeny. The eusocial Hymenoptera, however, make up a relatively small proportion of species. Most Hymenoptera are not social, although all non-social bee and wasp females have the same probability as the honeybees of being more closely related to sibs than to progeny. Although the actual relatedness among worker honeybees within a hive remains an empirical question and is probably less than that assumed by Hamilton (Page 1986), the different levels of sociality observed in the Hymenoptera imply that relatedness alone is not a sufficient condition for the evolution of eusociality.

Ecological hypotheses also assume that cost-benefit functions are driving factors in the evolution of sociality, although usually less explicitly than the genetic arguments. As Alexander (1974) noted, there are always costs and potential benefits associated with living and interacting in a group. The implication is that a particular kind of group, or social behavior within a group, will occur only when the benefits to actor (B) exceed the costs to the same actor (C), that is when B > C. Note that relatedness is not explicit in this equation. Here the actor and beneficiary are the same individual, and the relatedness in question (the r in Br) is always one (i.e., r = 1) and so can be ignored during the calculations.

Most hypotheses of this type suggest that groups originate in order to exploit resources that cannot be exploited as efficiently by solitary individuals (e.g., Crook 1972; Slobodchikoff 1984; Wittenberger and Hunt 1985). They imply that the presence of conspecifics generates a direct and net benefit for an individual joining, or interacting within, a group (e.g., Alexander 1974; Lott 1984; Wittenberger and Hunt 1985). Here, resources are taken to mean any of a variety of factors that can affect fitness: food, space, mates, nonmate social partners, refugia nest sites, climate, and time, to name a few. Since resources can include other animals, social groups also possess the potential benefit of group defense against other animals, particularly predators. Thus, the testing of ecological hypotheses also entails cost-benefit analyses of social groups versus solitary individuals, but without the constraint of relatedness. Unlike the genetic arguments, which have involved extensive theoretical modeling and analysis of kinship, the ecological hypotheses have been explored primarily through studies using the comparative method (e.g., Crook 1965, 1970, 1980; Lack 1968).

Phylogenetic hypotheses are usually proposed as a last resort when extensive estimation of costs and benefits has failed to demonstrate a net benefit in a particular case, or when comparative analysis or experimentation have failed to document the expected type of sociality in a particular case (e.g., Berger this volume). In order to deduce phylogenetic patterns of sociality, the comparative approach has been used, with correlations between social systems and morphological types that are believed to reflect actual phylogeny (e.g., Lin and Michener 1974; Brown 1975). Such phylogenetic arguments suggest that any genetic or ecological factors influencing the adaptive value of sociality have been operative over evolutionary time scales. According to these arguments, extant species are unable to respond to immediate conditions in an adaptive manner. Adaptive arguments about the origin of social systems may be valid, but arguments about the maintenance of sociality do not require adaptative explanations.

To explore the origin and maintenance of groups or sociality within groups, one should investigate all of the factors that can influence the costs and benefits of group living or of particular behaviors within groups. The underlying hypothesis is that only when total benefits exceed total costs (i.e., when B > C), on a per capita basis, will groups or specific social behaviors evolve. For a complete analysis, however, the total benefits and costs need to be partitioned into their direct and indirect components (e.g., Hamilton 1964; Shields 1980; Brown and Brown 1981). Thus, B includes both direct (b_d) and indirect benefits ($b_i r$, where r is the relatedness between actor and those affected positively by his act not including himself) and C includes direct (c_d) and indirect costs ($c_i r$, where r is the relatedness between actor and others which are affected negatively by the same act). Group living or specific types of social behavior are expected to arise only when the per capita combined direct and indirect benefits exceed the per capita combined direct and indirect costs (i.e., when ($b_d + b_i r$) > ($c_d + c_i r$)).

Under most conditions, selfish, spiteful, and aggressive behaviors are expected to limit group cohesion by increasing the direct and indirect costs of grouping, while cooperative and altruistic behaviors should favor grouping and group cohesion by increasing the direct and indirect benefits associated with group living. Insofar as all indirect effects are influenced by patterns of kinship within groups, kinship will be critical to understanding genetically based sociality. Insofar as most ecological factors disproportionately influence the direct costs and benefits of group living, kinship may be less important, or even irrelevant, in exploring ecologically based sociality.

It is easy to perceive the genetic and ecological arguments as alternative hypotheses. We would caution, however, that it is incorrect to assume that they cannot have operated independently or jointly during the evolution of any social system. For example, it is not difficult to imagine ecological circumstances that originally favored animals living in groups regardless of the relatedness of group members (e.g., a shortage of nest sites). Once members of such a species were regularly living together in groups, however, kinship patterns could be expected to influence the kinds of social behavior displayed among group members according to Hamilton's rule. The end result is that investigators faced with the many *fait accompli's* provided by different extant species (for a review, see Rubenstein and Wrangham 1986) would do best to keep all of the alternatives in mind when interpreting the behavior of their species. It seems most likely to us that genetic, ecological, and phylogenetic factors have all contributed to the origin and maintenance of sociality in specific animal groups. The mix of causative factors at any given time for any given species probably represents a specific solution to a specific set of selection pressures given the historical constraints of its phylogeny.

II. ECOLOGICAL TRADEOFFS AND SOCIAL BEHAVIOR

The chapters in this volume consider some of the ecological tradeoffs that might lead to sociality. While the primary emphasis of these chapters is on a consideration of ecological mechanisms, the chapters also illustrate the interplay between ecology, genetics, and phylogeny.

Several papers treat the general questions of origins and maintenance. Slobodchikoff and Schulz (Ch. 2) suggest that groups will form to collectively exploit resources that cannot be efficiently exploited by solitary individuals. They propose that a proximate mechanism for regulating group size is aggression, and that aggression mediates the quantity and distribution of a particular resource that is available to each individual within the social group. They also suggest that although group size can increase to a point where many individuals within the group do not breed, such an increase can occur in an evolutionary sense only if all the individuals in the group have some expectation or probability of breeding. Giraldeau (Ch. 3) predicts that groups will not be of optimal size, but rather of a size that makes it unattractive for solitary individuals to join the group. This size, which Giraldeau calls the stable group size, varies with kinship patterns and is somewhat less than the maximum group size, where the net benefit in terms of resource extraction per individual in a social group equals the net benefit to a solitary individual. Barlow (Ch. 4) discusses the general terms and conditions necessary for resources to influence social systems. He suggests that mating systems are an important component of resource-determined sociality. Barlow points out that monogamy, perhaps the most simple social system, fits resource models in ways that are similar to the more complex social systems. However, he points out that monogamy in different species has evolved in various ways in response to different selective pressures. Rodman (Ch. 5) addresses a variety of different ecological hypotheses, particularly those that have been used to explain sociality in primates. He points out that there is a need for assessing different alternative hypotheses and assessing the relative complexity of the factors that could be influencing the size of social groups.

Several papers address the relationships between resource abundance, resource distribution, and social behavior. Uetz (Ch. 15) shows that social spiders form groups where prey abundance is high and variance in prey abundance is low. Under conditions where prey are patchily distributed and the variance in prey abundance is high, the daily intake levels of spiders do not support group living, and the spiders living in such habitats are solitary. Using a comparative approach, Gautier et al. (Ch. 14) show that cockroach species living in forest habitats, where food is not abundant and very patchily distributed tend to be non-social. Other cockroach species dwelling in caves, where food is abundant and clumped, tend to have variable social systems, depending on the specific abundance and distribution of the food. These general patterns are found among phylogenetically unrelated species, suggesting that in cockroaches, phylogeny has played a secondary role to ecology in the development of sociality. Heinrich (Ch. 12) shows how sociality is utilized by ravens, which feed on carcasses in the winter. A carcass represents a superabundant and temporary food source, and no single raven can use all the food that is available at any one time. Under these short-term, ephemeral conditions in ecological time, the ravens share information about food sources by calling in other ravens and forming a feeding flock.

Resource abundance and distribution of food and habitat sites can be modified by another resource--climate. Marzluff and Balda (Ch. 11) show how climate affects the abundance of pinyon nut crops. The abundance and distribution of pinyon nut crops in turn influence the social systems of two species of corvids, the pinyon jay and the scrub jay. Both species have to deal with the same set of resources, but the morphological adaptations derived from the phylogeny of the two species mandate different solutions to dealing with the gathering and utilization of pinyon nuts. These different solutions translate into different social systems. Ligon and Ligon (Ch. 10) show that woodhoopoes have a physiological intolerance to low temperatures, and that this intolerance necessitates roosting in groups within cavities for warmth. However, the number of good territories containing good roost cavities that are relatively free from predation is limited, so the birds form alliances of either related or unrelated individuals for the purpose of territory acquisition and maintenance.

Inheritance of resources can be an important feature of the ecological constraints that social animals face. Shields et al. (Ch. 9) show that swallows, faced with environments that vary cryptically in such a way that habitats cannot be immediately assessed, use traditional breeding grounds as an indicator of habitat quality. Shields et al. argue that the decision rule in such cases is: where others have been able to live and breed succesfully in the past, so can I. Waser (Ch. 6) points out that banner-tailed kangaroo rats need mounds as refuges, and the supply of mounds is very limited. The chance of successful dispersal by young is low, and the cost of sharing a mound with young is also low. Waser suggests that when resources are non-depreciable, philopatry is likely to occur. Philopatry may be an important precursor to the development of kin-related social behavior in some animal groups. Myles (Ch. 16) argues that in a variety of animal groups it is not the relatedness of individuals that is important, but the tradeoffs involved in dispersal versus the inheritance of some necessary resources. Since in practice kin often inherit resources, many social groups are going to be kin groups. However, Myles argues that inheritance need not be only among kin, and that kin arguments are neither necessary nor sufficient for explaining the formation and maintenance of social groups.

Within a social group, there may be competition for resources. Such social competition is discussed by Pollock and Rissing (Ch. 13), who suggest that subordinates aid dominants in order to remain close to resources that would otherwise be unavailable to them. They argue that such social competition has been an important feature in the origins of a number of social groups, particularly the social Hymenoptera. A game-theoretic model of such social competition is explored by both Pollock and Rissing and Slobodchikoff and Schulz (Ch. 2). A similar view of social competition is suggested by Armitage (Ch. 7), who suggests that among the ground squirrels sociality is both cooperative and competitive, and is aimed at maximizing direct fitness rather than indirect fitness.

Finally, what happens when an investigator manipulates the resources available to a group of genetically unrelated animals and finds that the same social system is maintained? Berger (Ch. 8) shows that some ungulates maintain the same social system, regardless of the distribution and abundance of resources. Berger ascribes this to phylogenetic inertia, the canalizing effect of evolution on the social system of these animals. He also carefully considers, however, all of the difficulties involved in actually determining and measuring the resources that are relevant for manipulation, and shows how difficult it can be to determine that phylogenetic factors have actually insulated a social system from ecological or genetic considerations.

REFERENCES

Alexander, R.D. 1974. The evolution of social behavior. *Annual Review of Ecology and Systematics* 5:325-383.

Brown, J.L. 1975. *The evolution of behavior*. New York:Norton.

Brown, J.L. and E.R. Brown. 1981. "Kin selection and individual selection in Babblers." pp. 244-256. In: R.D. Alexander and D.W. Tinkle, eds. *Natural selection and social behavior*. New York:Chiron Press.

Crook, J.H. 1965. The adaptive significance of avian social organizations. *Symposia of the Zoological Society of London* 18:237-258.

Crook, J.H. 1970. Ed. *Social behaviour in birds and mammals: Essays on the social ethology of animals and man*. London:Academic Press.

Crook, J.H. 1972. "Sexual selection, dimorphism and social organization in the primates." pp. 231-281. In: B. Campbell, ed. *Sexual selection and the descent of man*. Chicago:Aldine.

Crook, J.H. 1980. *The evolution of human consciousness*. Oxford:Clarendon Press.

Hamilton, W.D. 1964. The evolution of social behavior. *Journal of Theoretical Biology* 7:1-52.

Lack, D. 1968. *Ecological adaptations for breeding in birds*. London: Methuen.

Lin, N. and C.D. Michener. 1972. Evolution of sociality in insects. *Quarterly Review of Biology* 47:131-159.

Lott, D.F. 1984. Intraspecific variation in the social systems of wild vertebrates. *Behaviour* 88:266-325.

Page, R.E. 1986. Sperm utilization in social insects. *Annual Review of Entomology* 31:297-320.

Rubenstein, D.I. and R.W. Wrangham. 1986. Eds. *Ecological aspects of social evolution: Birds and mammals*. Princeton:Princeton University Press.

Shields, W.M. 1980. Ground squirrel alarm calls: Nepotism or parental care? *American Naturalist* 116:599-603.

Slobodchikoff, C.N. 1984. "Resources and the evolution of social behavior." pp. 227-251. In: P.W. Price, C.N. Slobodchikoff and W.S. Gaud, eds. *A new ecology: Novel approaches to interactive systems*. New York:Wiley Interscience.

Trivers, R. 1985. *Social evolution*. Menlo Park, CA:Benjamin/Cummings
    Publishing Co.
Wilson, E.O. 1975. *Sociobiology*. Cambridge, MA:Belknap Press.
Wittenberger, J.F. and J.L. Hunt, Jr. 1985. "The adaptive significance of
    coloniality in birds." pp. 1-75. In: D.S. Farner, J.R. King and K.C.
    Parkes, eds. *Avian biology*, v. 3. New York:Academic Press.

II. THEORY OF ECOLOGICAL INTERACTIONS

Chapter 2

COOPERATION, AGGRESSION, AND THE EVOLUTION OF SOCIAL BEHAVIOR

C.N. Slobodchikoff

Department of Biology
Northern Arizona University, Flagstaff 86011

William C. Schulz

Department of Mathematics
Northern Arizona University, Flagstaff 86011

I.   INTRODUCTION
II.   SOCIAL DYNAMICS: A SOCIAL EQUILIBRIUM MODEL
III. SOCIAL DOMINANCE
IV. DISPERSAL AND REPRODUCTIVE DOMINANCE
V.   MAXIMUM GROUP SIZE
VI. SOCIAL DYNAMICS: A DYNAMIC EQUILIBRIUM MODEL
VII. A GAME THEORETIC MODEL OF SOCIALITY
VIII. REPRODUCTIVE DOMINANCE AND REPRODUCTIVE EXPECTATION
IX. SUMMARY: SOCIAL DYNAMICS AND EVOLUTION
REFERENCES

I.   INTRODUCTION

Until recently, the relationship between resources and sociality in animals has been poorly understood. Although kin-related models of sociality have been explored extensively (e.g., Alexander 1974; Orlove 1975; Milinski 1978; Sherman 1980; Weigel 1981; Michener 1983), since Hamilton's (1964) seminal work describing the theoretical relationship between social Hymenoptera workers in a haplodiploid system, ecological factors contributing to the development and maintenance of social

13

groups have been less extensively modeled. In the mid-1960s to the mid-1970s, Crook and his associates (Crook 1965; Crook and Gartlan 1966; Crook 1970, 1972; Crook et al. 1976) attempted to describe relationships between social systems, resource abundances, and resource distributions for social birds, social mammals, and social and solitary primates. Similar studies have been done with some other animals, such as African antelopes (Jarman 1974), and primates. These types of studies have been sometimes termed socioecology (Crook 1970; Gautier 1982; Terborgh and Janson 1986), because they deal with the ecology of social groups. From an ecological perspective, the primates have been studied most extensively. A series of papers has attempted to document the relationship between food type, mating system, home range, body morphology, and social behavior (Eisenberg et al. 1972; Jorde and Spuhler 1974; Milton and May 1976; Clutton-Brock and Harvey 1977a,b).

The resources hypothesis of sociality suggests that social groups form and are maintained in response to the cooperative benefits that can be obtained by extracting certain types of resource distributions (Crook 1965; Slobodchikoff 1984). These groups may consist of either kin or non-kin. Slobodchikoff (1984) suggested that while kin-related assemblages may be a necessary condition for predisposing some animal groups toward social behavior, such kin-related associations may not be a sufficient condition for sociality to be maintained. Following and extending Crook's (1965, 1970, 1972) arguments, Slobodchikoff suggested that ecological factors such as resource abundances and distributions play a role in determining whether or not a particular group of animals is going to be social.

For example, Hamilton (1964) argued that sociality in the Hymenoptera arises because of the high degree of relatedness that is potentially possible between worker sisters as a result of haplodiploidy. However, while the ants and some bees are social, most species of Hymenoptera are not, and all have the same system of haplodiploidy as the honeybees used in Hamilton's arguments for sociality. The parasitic Hymenoptera constitute an extremely large number of species, yet they are all non-social (Askew 1971). Two major differences between the solitary parasitic Hymenoptera and the social Hymenoptera are the abundance and the distribution of food resources available to the larvae of each group.

The relationship between resources and sociality is expressed in several different models (Crook 1965, 1972; Wrangham 1980, 1983; Terborgh 1983; Van Schaik and Van Hooff 1983; Slobodchikoff 1984). These models each make some similar predictions, and some predictions that are different. Crook (1972) suggested that (1) group size will be high where patchiness of resources is low; (2) group size will be lowest at intermediate levels of patchiness; and (3) group size will increase again as patchiness begins to increase beyond this intermediate level. The logic of this argument, as suggested by Crook, is that at uniform resource distributions a social group can easily defend a particular territory containing necessary resources against other groups or other individuals. As patchiness increases, the members of the group are forced into more direct competition with each other for utilization of each patch, and the more subordinate individuals will leave the group, causing the group size to drop. As patchiness continues to increase, a larger group size is again more favored, since more individuals would then be available for locating rich patches. Crook also suggested that there is a relationship between resource abundance and group size: the more abundant the resource, the larger the group size of the social group utilizing that resource.

Wrangham (1980, 1983) and Slobodchikoff (1984) make somewhat different predictions concerning resources and sociality. Wrangham suggested that related females group together to defend resources, while males join groups to have access to females. In this model, the size of the patches determines the group size. Smaller patches lead to monogamy, since the food resources will not support a larger group size. Larger patches lead to polygyny, either with a single male or with multiple males. Slobodchikoff (1984) suggested that the animals did not have to be related in order to band together to defend patchy resources. He further suggested that monogamy would occur primarily when resources were poor in quality and were uniformly distributed, while polygyny would occur when resources were rich in quality, supporting a larger group size, and when resources were patchily distributed.

Terborgh (1983) and Van Schaik and Van Hooff (1983) made similar predictions about the relationship between group size and resource levels, but also suggested that predation would play a role. They suggested that another important component other than food is the protection that is derived from having other members of a group watch for predators. In a cost-benefit formulation of this relationship, Terborgh (1983) and Terborgh and Janson (1986) suggested that costs increase with the food in a patch, while benefits associated with antipredator behavior decrease along some exponential function. The difference between the benefit curve and the cost curve represents the optimal group size at different levels of resource abundance.

Essentially, a common thread through the models is that (1) group size should increase with increasing resource abundance; and (2) at some level of resource abundance, there should be a switch from solitary individuals to monogamous associations, and then a switch from monogamous associations to polygynous ones, first with single male polygyny and then with multiple male polygyny. Differences are that (1) Crook's (1972) model predicts that the lowest group size will be at the intermediate levels of resource abundance and patchiness; (2) Wrangham's (1980, 1983) model predicts that the females are all related while the males are not necessarily related, and that resources have to be defensible by the females; (3) Slobodchikoff's (1984) model predicts that neither the males nor the females are necessarily related, that both sexes would participate in group defense, and that the dispersion of the resources, as well as the abundance, would determine the type of social system present; and (4) Terborgh's (1983) model and Van Schaik and Van Hooff's (1983) model predict that predator protection will modify the effect of food resources on group size.

All the models assume some degree of resource defense, or some sort of aggression between individuals; however, none of the models explicitly deals with the conditions by which such aggression would affect the utilization of the resources and the size of the social group. In this chapter we develop two models that explore the relationship between resources and group size as a function of aggression. We show how aggressive dominance can affect both the size of the social group and the decision to stay and participate in the social group. Finally, we address the question of why animals might stay in a social group without reproducing, and occupy a subordinate status, when they could instead leave the group and breed as solitary individuals.

II.   SOCIAL DYNAMICS: A SOCIAL EQUILIBRIUM MODEL

Social behavior can be viewed as a way for animals to cooperatively exploit resources that they would not be able to exploit as individuals. As such, a relationship can be modeled between resources and the size of the social group. Part of this relationship involves aggression. Aggression can provide a mechanism for controlling the size of the group, for defining the distribution of resources within the group, and for defining the reproductive relationships of the group. In this view, aggression acts as a governor on the social group, translating the available resources into a group size that is able to cooperatively exploit the resources.

The relationship between resources, group size, and aggression can be stated simply. Let us first define these variables:

R_T = total resource
(gs) = group size
n   = the amount of resource required by a single animal to breed
a   = the coefficient measuring the proportion of resource lost as a result of aggression

The social equilibrium (SE) can then be defined as

$$SE = \frac{R_T}{a\ n\ (gs)} = 1 . \tag{1}$$

Let us look at this formula in some depth. The total resource that is available to the animals is R_T. This resource can be food, habitat, mates or other animals, or a combination of these resources. The availability of this resource is modified by a, such that $\frac{R_T}{a}$ represents the fraction of the resource that is actually available to the animals as a result of aggression. A considerable amount of aggression between the animals means that a is large, and the available resource is correspondingly small, while little aggression means that a is small, and the available resource is large.

Another way of looking at the relationship between $\underline{a}$ and R_T is as follows: If we define R_L as the resource lost due to aggression, and R_A as the resource available after aggression, then

$$R_T = R_L + R_A \tag{2}$$

and $\underline{a}$ can be defined as

$$a = \frac{R_T}{R_A} \tag{3}$$

or

$$R_T = a \, R_A \tag{4}$$

Thus, $\underline{a}$ represents the proportion of the resource that is lost through the effects of aggression. The assumption that is made here is that aggression entails a cost in terms of time, energy, and risk of injury, and that the effort expended on aggression can be translated directly into a loss of resource.

Given this assumption, the proportion $\underline{a}$ can be subdivided into resource loss through intra-group and inter-group aggression. If R_g is the resource lost through intra-group aggression, and R_i is the resource lost through inter-group aggression, then

$$R_L = R_g + R_i \tag{5}$$

Substituting equation (4) into equation (2), we have

$$R_L = R_A \, (a - 1).$$

With a substitution of R_g and R_i from equation (5), we have

$$a = \frac{R_g + R_i}{R_A} + 1 \tag{6}$$

Group size (gs) can be seen to depend on the interaction of the resources and aggression in the system. A rearrangement of the social equilibrium formula shows that group size is expressed as

$$(gs) = \frac{R_T}{a \, n} \tag{7}$$

This means that for a constant ratio of $\frac{R_T}{n}$, as (gs) increases, $\underline{a}$ decreases, and as $\underline{a}$ increases, (gs) decreases (Fig. 1). At first glance, this may seem counterintuitive. It seems to say that there is less aggression in a big group than there is in a small group. The argument is not quite that simple. The argument says that a smaller group will lose more of the available resource, whether through intra-group or inter-group aggression, than a larger group. This includes the cost of social interactions between the group members as well as the costs of defending the resource against non-group members. Even given that, however, the logical extension is that the costs of aggression decrease with increasing group size.

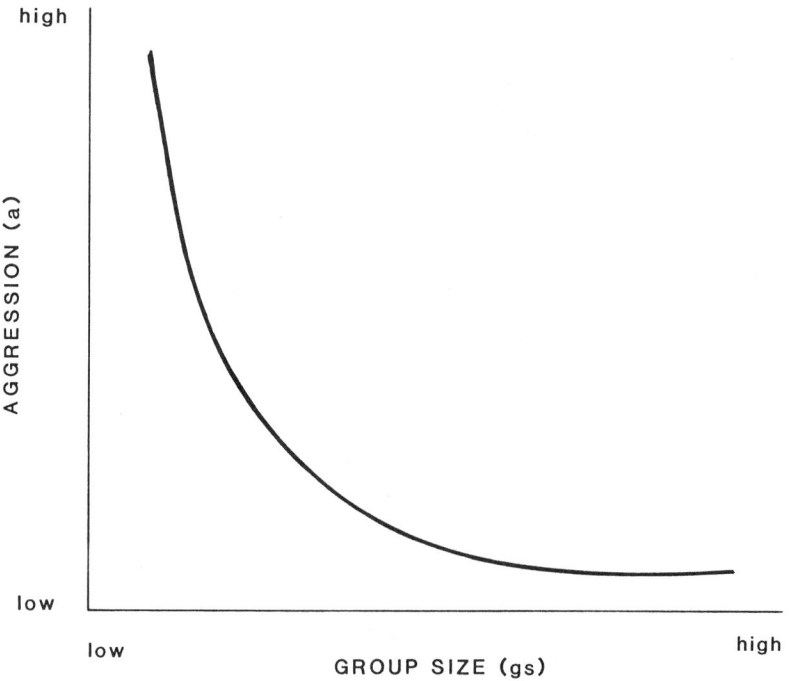

Fig. 1. The relationship between group size and aggression.

III.   SOCIAL DOMINANCE

Many social groups set up dominance hierarchies even as the group is forming. Examples of dominance hierarchies in animals such as the primates have proven to be ambiguous as to whether the dominant animals have significantly more access to resources than subordinate ones. By viewing dominance hierarchies, however, as a way of limiting aggression within the social group, and thus decreasing a̲ while increasing the fraction of R_T that becomes available to the group, it can be seen that both dominant and subordinate animals can stand to benefit by limiting the costs of aggression.

The social equilibrium model may be modified to express this relationship. Under conditions of resources being more abundant than the social equilibrium formula predicts for a given group size, dominant animals may have more access to resources than subordinate ones. In this situation, if k_i is the proportion of resource garnered by a single individual, and $\sum^{(gs)} k_i = 1$, with (gs) being the number of animals in the social group, an unequal distribution of resources can be obtained between dominants and subordinates under the condition of the inequality

$$\frac{\Sigma \, k_i \, R_T}{a \, n \, (gs)} > 1 \tag{8}$$

In this circumstance, $\underline{a}$ is expected to be high, and as (gs) increases with a corresponding decrease in $\underline{a}$, an unequal distribution of resources through social dominance is expected to disappear (Table I).

IV. DISPERSAL AND REPRODUCTIVE DOMINANCE

So far, we have seen what happens when the sociality equilibrium equals 1 or exceeds 1. But what happens when the equilibrium drops below 1? As group size increases, $\underline{a}$ continues to decrease. However, the critical ratio of $\dfrac{R_T}{n}$ becomes too small for some animals to have enough resources available to be able to breed. Although the predicted strength of cooperation in the social group should increase (as $\underline{a}$ decreases), the choice for some animals becomes (1) to stay in the social group and not breed, or (2) to leave the social group and try to find another group that has not yet approached its sociality equilibrium of SE = 1.

The choice of whether to stay or to leave should depend on two probabilities. One is the probability, P_d, of surviving as a solitary individual and finding another group where breeding is possible. The other is the probability, P_b, of breeding eventually by staying in the social group. If $P_d > P_b$, the animal should leave, and if $P_b > P_d$, the animal should stay.

If the animal stays, its chances of breeding decrease with increasing group size once SE < 1. The probability of breeding can be approximated

Table I. Characteristics of Social Systems under Different Conditions of the Sociality Equilibrium

SE > 1	SE = 1	SE < 1
Social dominance	No dominance	Reproductive dominance
All animals breed	All animals breed	Some animals breed
High aggression	Low aggression	Low aggression
Low cooperation	High cooperation	High cooperation
Unequal resource distribution	Equal resource distribution	Unequal resource distribution

through the sociality equilibrium formula, by the approximate relationship, when SE < 1, or as follows:

$$P_b = \frac{R_T}{n\ (gs)} \qquad (9)$$

This probability function is shown in Fig. 2. The consequence of this declining probability of reproduction is reproductive dominance (Table I).

A constraint here is that the animal's share of the resource must be enough to allow the animal to survive, if the resource is food, space, or other animals exluding mates. If r is the minimum resource needed to support one non-breeding animal, then the animals' share of the resource cannot fall below $\underline{r}$. If it does, then the animal must leave the group and disperse.

V.   MAXIMUM GROUP SIZE

The social equilibrium equation (1) can be used to predict the maximum group size that should be possible for a given amount of a resource. When $\underline{n}$ represents the amount of resource needed by each and every $\overline{animal}$ in the group (i.e., when all the animals in the group breed), the maximum group size is predicted by equation (7), or

$$(gs)_{max} = \frac{R_T}{a\ n}$$

But what if there is reproductive dominance, and not all animals in the group need or use resources in the same way? For example, in honey bee colonies, there is usually only one reproductive female per colony, while ant colonies have one to several reproductive females (Wilson 1975).

We can define this situation with a few additional terms. If $\underline{n}$ is the minimum amount of resource needed to support one breeding animal, then let us say that $\underline{r}$ is the minimum amount of resource needed to support one non-breeding animal. Then, if only one animal breeds in the group, the maximum group size will be

$$(gs)_{max} = \frac{R_T - n}{r} + 1 \qquad (10)$$

This can be expanded to a more general case of $\underline{x}$ number of breeding animals and $\underline{y}$ number of non-breeders. The maximum group size under these conditions becomes

$$(gs)_{max} = \frac{R_T - xn}{r} + x \qquad (11)$$

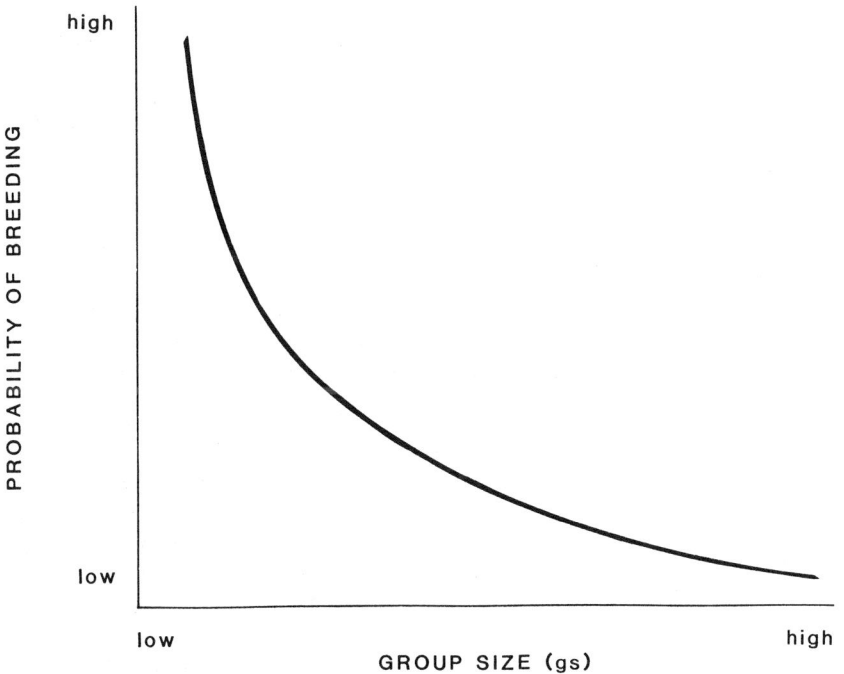

Fig. 2. The relationship between group size and the probability of reproduction in the social group.

If breeding males and females have different resource requirements and $\underline{n}$ is the minimum amount of resource needed to support one breeding female, while $\underline{q}$ is the minimum amount of resource needed to support one breeding male, and $\underline{x}$ is the number of breeding females, while $\underline{w}$ is the number of breeding males, then the maximum group size becomes

$$(gs)_{max} = \frac{R_T - (xn + wq)}{r} + (x + w) \qquad (12)$$

As the number of breeders increases so that the resource is more equitably distributed among the group members, equation (12) approaches the maximum group size predicted by equation (7).

VI.   SOCIAL DYNAMICS: A DYNAMIC EQUILIBRIUM MODEL

So far, in the social equilibrium model, we have not taken into account what happens to group size with small and continuous changes in aggression. This can be modeled with a system of linear differential equations

$$\frac{dB}{da} = kU + 2l\,(N\text{-}B) \tag{13}$$

$$\frac{dU}{da} = 2l\,(N\text{-}B) - 2mU$$

where B is the number of breeding individuals, U is the number of non-breeding individuals, and N is the number of individuals that the resources can support if all of the individuals are completely solitary. In this system of equations, k, l, and m are numerical coefficients that are determined empirically for different species and different resources. Specifically, l represents the proportionality between the rate of increase of the number of breeding individuals when all animals are solitary and the increase in the number of individuals over N due to decreased aggression. The influence, as aggression decreases, of the increase of non-breeding individuals on the number of breeding individuals is represented by k. The coefficient m, assumed here to be small, represents a possible tendency of the non-breeding individuals to assume dominance of the social group. For most species, m would be equal to 0. The coefficients k, l, and m are assumed to be nonnegative.

In order for the model to correspond to biological reality, certain relationships among the coefficients must hold, namely

$$(1 - m)^2 - 2lk > 0.$$

In addition, to solve for the system of equations, certain boundary conditions must be assumed. Although the number of breeding individuals in a social group seldom is less than one, we may for mathematical convenience set B (0) = 0. The effect of setting B (0) = 1, while more realistic biologically, complicates the mathematics without having any appreciable effect on the solutions. The number of non-breeding individuals in social groups is conditioned by the resources, and this is represented by the parameter u.

With these conditions, elementary methods result in the following solution to equations (13)

$$B(a) = N - \frac{1}{2}\left(N - \frac{ku + N(l-m)}{\beta}\right)e^{-(\alpha - \beta)a} - \frac{1}{2}\left(N + \frac{ku + N(l-m)}{\beta}\right)e^{-(\alpha + \beta)a}$$

$$U(a) = \frac{1}{2}\left(U + \frac{2(U+N)}{\beta}\right)e^{-(\alpha - \beta)a} + \frac{1}{2}\left(U - \frac{2l(U+N)}{\beta}\right)e^{-(\alpha + \beta)a}$$

where, for convenience, we have set $\alpha = l + m$ and $\beta = ((l - m)^2 - 2lk)^{1/2}$. Note that $\beta < \alpha$. The important features of the solution can be determined from the graphs of U (a) and B (a) (Fig. 3).

The graphs of U (a) and B (a) show that as aggression increases the number of breeding individuals approaches N and the number of non-breeding individuals approaches 0. Another feature of the solution is that as aggression decreases (i.e., as cooperation increases), the number

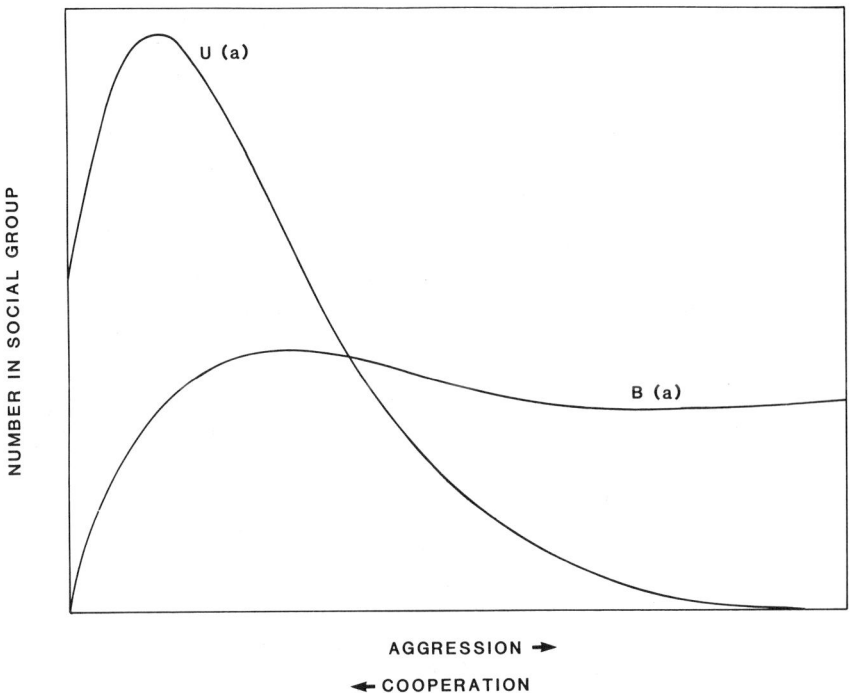

Fig. 3. Graphical representation of the predictions of the dynamic equilibrium model for a given resource level. B represents the number of breeding individuals, while U represents the number of nonbreeding individuals.

of non-breeding individuals rises dramatically, while the number of breeding individuals decreases. This is the same result predicted by the sociality equilibrium model. At the highest levels of cooperation (or the lowest levels of aggression) there are extremely few breeding individuals, with a preponderance of non-breeding individuals. This is similar to the situation that is seen among many of the eusocial Hymenoptera.

VII.   A GAME THEORETIC MODEL OF SOCIALITY

If we assume that resources are unequally distributed among dominants and subordinates, then we can develop a game theoretic model of the resource conditions under which animals should form social groups. This model is a non-zero-sum game that is based on bargaining solutions (Bartos 1967).

In this approach, let us make some assumptions about the social group: (1) both dominants and subordinates can breed; (2) dominants are better at finding, defending, extracting or utilizing resources than are subordinates; (3) dominants can have more offspring in a social group containing subordinates than they can have by themselves as solitary individuals; (4) subordinates may have fewer offspring immediately in the social group but have some expectation of becoming dominants; and (5) subordinates can share access to resources with the dominants in the social group. Thus, in our assumptions, both the dominants and subordinates can derive some benefit from the social group, to different degrees.

Now let us construct a payoff matrix. We will assume that the maximum payoff or loss is +5 or -5. In Case 1, the resources can be shared. We have then four strategies (Table II).

In Strategy A, the dominants (Ds) are willing to share their access to the resource. In the process of sharing, the dominants give up some of the resource for the promise of increased reproductive success. The payoff to the dominants falls to +2. Because the dominants are willing to share, the payoff to the subordinates is the maximum of +5.

In Strategy B, the dominants are willing to share, but the subordinates take some of the resource and leave, not staying to participate in the social group. Here both the dominants and subordinates lose, the subordinates because they give up their access to future resources that the dominants may find, and the dominants because they have shared some of their resource without attracting any subordinates to the social group.

In Strategy C, the dominants do not share their resources, but the subordinates stay anyway. The maximal payoff of +5 goes to the dominants, and the maximal loss of -5 goes to the subordinates.

In Strategy D, the dominants do not share, and the subordinates do not stay. Here there is neither a gain nor a loss to either group, and the payoff to both is 0.

A geometrical analysis of the payoff space (Fig. 4) shows the payoff matrix plotted with respect to the payoffs to both the dominants and subordinates. The equilibrium point V* is a positive payoff for both the dominants and the subordinates. This equilibrium point, however, shows that neither the dominants nor the subordinates gain as much from participation in the social group as predicted by Strategy A. The conse

Table II. Payoff Matrix for Dominants (Ds) and Subordinates (Ss) under Conditions of Resources Sufficient to Share

		Ds	Ss
Strategy A:	Ds share, Ss stay	+2	+5
Strategy B:	Ds share, Ss leave	-1	-3
Strategy C:	Ds not share, Ss stay	+5	-5
Strategy D:	Ds not share, Ss leave	0	0

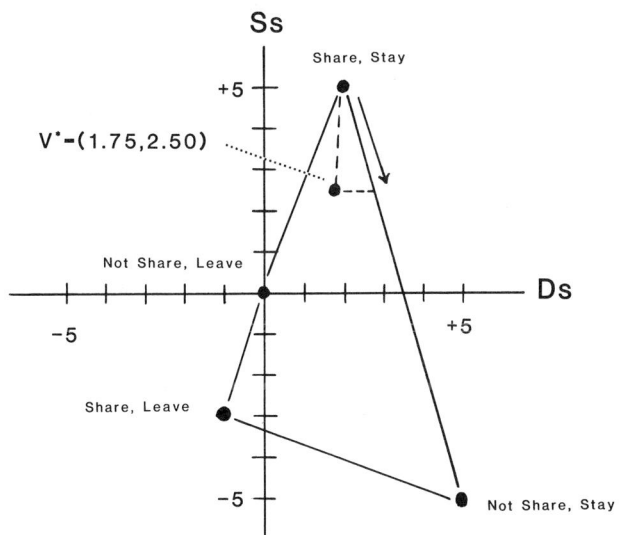

Fig. 4. *Graphical representation of the game theory payoff matrix under the conditions that the resources are sufficient to share.*

quence of this is that there is some latitude for the dominants to withhold sharing the resources without having the subordinates leave the social group (the arrow in Fig. 4 shows the extent to which the dominants can withhold resources without an appreciable change to the equilibrium payoff of the subordinates). The subordinates, on the other hand, can increase their access to the resources to some extent without any serious change in the equilibrium payoff for the dominants. This predicts a constant conflict situation within the social group, where the dominants are trying to skim on sharing with the subordinates, and the subordinates are trying to get more resources out of the dominants.

In Case 2, where resources are insufficient for sharing, we have a different set of predictions. As before we construct a payoff matrix (Table III), where strategies C and D are the

Table III. Payoff Matrix for Dominants (Ds) and Subordinates (Ss) under Conditions of Resources insufficient to Share

		Ds	Ss
Strategy A:	Ds share, Ss stay	-2	+1
Strategy B:	Ds share, Ss leave	-5	+3
Strategy C:	Ds not share, Ss stay	+5	-5
Strategy D:	Ds not share, Ss leave	0	0

same as before, but strategies A and B are now different. In Strategy A, if the dominants share, they lose a part of the resource that they need for maintenance, while the subordinates, by staying, gain a little, although the amount gained is not very much since the resource is not abundant. In Strategy B, the dominants lose the maximum by sharing, since the resource is not abundant.

Under these conditions, the equilibrium point V* is at (0,0), or the point where dominants should not share and the subordinates should leave. However, two interesting results can be seen. One is that both the subordinates and the dominants can improve their payoff by sharing and staying if resources become more abundant (see the direction of the arrow at share, stay, Fig. 5). The other result is that dominants can improve their negative payoff slightly in Strategy B, even if the subordinates leave (see the direction of the arrow in share, leave, Fig. 5). However, the subordinates get a lower payoff by staying at this point rather than leaving. There is, thus, a basic conflict in the initial conditions for sociality, where the dominants may be willing to share but the subordinates may be willing to take but not stay. As can be seen from Case 1, this conflict can be resolved where resources are abundant, but this conflict can prevent the formation of social groups where resources are sparse.

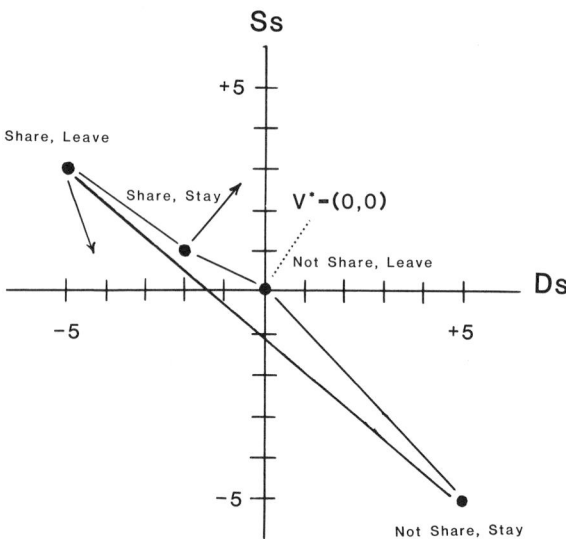

Fig. 5. Graphical representation of the game theory payoff matrix under the conditions that the resources are not sufficient to share.

VIII.   REPRODUCTIVE DOMINANCE AND REPRODUCTIVE EXPECTATION

The sociality equilibrium model shows that once social groups form, they may be maintained even though the resource may not be sufficient for all individuals to breed. We may then ask the question: What prevents animals from leaving the social group under these conditions? To answer this question, we will develop a model of reproductive expectation. This model deals with the expectation of reproducing in a social group versus the expectation of reproducing as a solitary individual.

The assumptions of the reproductive expectation model are: (1) only one female breeds in the social group; (2) if a breeding female dies, another female enters the group in a non-breeding, subordinate position; (3) the mortality rate for all females is the same; and (4) the social group produces more offspring than solitary animals.

We can now define the following terms:

N_s = the number of offspring of a solitary female
N_c = the number of offspring produced by the social group
p = the probability of survival of an individual female
n = the number of animals in the social group

Then the reproductive expectation of a solitary female (E_s) will be

$$E_s = p \, N_s \qquad (14)$$

and the reproductive expectation of a social female (E_c) will be

$$E_c = N_c \frac{1}{n-j} \, p \, (1-p)^j \qquad (15)$$

where j is an index of the number of females. Formula (15) can be approximated by the following:

$$E_c = \frac{N_c}{n} \left[p + \frac{n}{n-1} (1-p) \right] \qquad (16)$$

The results can be seen in Fig. 6, where the reproductive expectations of both the solitary and social group can be seen, with m_c the slope of the social group reproductive expectation, and m_s the slope of the solitary female reproductive expectation. In terms of reproductive expectation, there will be evolutionary pressure to form groups if

$$\frac{E_c}{E_s} > 1 \qquad (17)$$

or if the reproductive expectation of the social group is greater than the reproductive expectation of a solitary individual. If this condition can be satisfied, then a female should stay in the social group, even if she has a low probability of actually breeding.

Let us consider a numerical example. Let us assume that we have a colony of wasps that consists of five females. Of these, only one reproduces at any given time. The probability of survival of an individual wasp, either in the colony or as a solitary individual, is $p = 4/5$. Let us

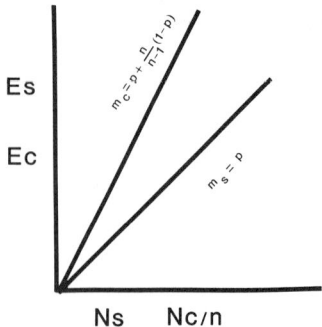

Ns   Nc/n

Fig. 6. Reproductive expectations for a solitary individual and for a
social group of individuals, assuming that only a single individual in the
social group breeds, and all the remaining individuals are nonbreeders.

contrast the reproductive expectation of a wasp in the social group with
a wasp that lives as a solitary individual. Elementary algebra shows that

$$\frac{N_c}{N_s} > \frac{20\ p}{5 - p} = 3.8$$

Thus, a social group of five females must produce at least 3.8 times
more offspring than a solitary female in order to have a reproductive
expectation, per capita, that is higher than the reproductive expectation
of a solitary female. However, if the social group can accomplish this
through increased parental care and offspring survival, then a female
should stay in the social group even if she does not ever reproduce. The
important key is that she should have some probability of becoming a
reproductive female. This appears to be true in most of the social
Hymenoptera and Isoptera, which have large numbers of worker individuals that never reproduce. Each of these worker individuals has some
initial probability of becoming reproductive, and consequently each individual has some reproductive expectation within the social group
(Slobodchikoff 1984).

IX. SUMMARY: SOCIAL DYNAMICS AND EVOLUTION

The dynamics of social behavior can be viewed as an interaction between resources and individuals seeking to exploit those resources.
Through cooperation, the group is able to collectively exploit or defend
resources that each animal individually would not be able to exploit or
defend.
We suggest that this interaction between resources and individuals
can occur independently of the relatedness of individuals in the group
(sensu Hamilton 1964). As we argued earlier, many other species of

Hymenoptera have the same haplodiploid system of sex determination as the honeybees, and so should theoretically have the same levels of genetic relatedness among sisters as worker honeybees, yet most of these species are not social. Electrophoretic analysis of worker ants fails to substantiate a high degree of genetic relatedness among workers in some nests. For example, the Australian desert ant *Rhytidopondera mayri* has relatedness levels of 0.1-0.2 among workers, rather than the 0.75 predicted by Hamilton (Crozier et al. 1984). Similarly, polygyne or multiple queen colonies of the fire ant *Solenopsis invicta* have a relatedness near zero among both the workers and the queens (Ross and Fletcher 1985). The same situation is found among *Polistes exclemans* social wasps, where the relatedness between workers and the brood that they raise is quite low (Strassmann 1985). Even among the honeybees, queens can mate with multiple males, leading to a greatly lowered relatedness among workers (Page and Metcalf 1982).

However, the presence of kin may predispose some animals to form social groups. For example, having offspring stay with their parents during a period of parental care provides a pool of ready individuals who can participate in a social group. In yellow-bellied marmots, kin relationships are important in determining the amount of foraging area shared by individual marmots, and spatial overlap is greatest among close kin, but this relationship is modified by other factors such as individual behavior, reproductive state, age of the animals, and the existence of separate burrow systems (Frase and Armitage 1984). Among the vertebrates, kin groups may be a common initial starting point for the formation of social groups. We argue, however, that kinship is neither a necessary nor a sufficient condition for sociality. We suggest that without appropriate resource levels and without the mediating interactions of aggression, kin groups would not be able to form into social, cooperative groups.

The sociality equilibrium models show how aggression can mediate group size for a given level of resources. Aggression has long been recognized as an important component of social systems (Hall 1964). Aggression often has been viewed as an important part of the socialization process of establishing social bonds and dominance hierarchies, either among kin or among non-kin (Bernstein and Gordon 1974; Bernstein and Ehardt 1986). We suggest that a primary function of aggression in social groups is to mediate the group size with respect to the available resources.

From an evolutionary standpoint, the most stable condition for the group should be when the sociality equilibrium is equal to 1. At this point, all the members of the group can breed and there is an equal access to resources by all the members of the group. When the sociality equilibrium is greater than 1, and the resources are abundant, much of the resources can be lost through aggression and the lack of cooperation, and the group can collect more group members. This collection of group members should occur at the expense of the more dominant individuals who at this point have more access to resources than the subordinates. More individuals will collect in the group, however, because the cost to the dominants of keeping out other potential group members will be greater than the cost that they would face by having competitors within the group use the same resources.

As the sociality equilibrium moves from SE > 1 to SE = 1, the dominants will have to share more of the resources, and the resource then becomes more equitably distributed throughout the group. On the other hand, by sharing more they also do not have to incur the same costs in defending or extracting the resource, since more of the resource defense or extraction is borne by the additional group members. The subordinates at the bottom end of the hierarchy in a social dominance system get access to as much resource when SE = 1 as when SE > 1, so that their access to the resource should be unaffected as SE approaches 1. The dominants can incur a drop in offspring production at this point if their reproductive rate is limited to their increased control of resources, but this loss of offspring production may be balanced by increased cooperation in progeny care within the group as the overall level of cooperation increases. Therefore, actual fitness stays the same or increases for the dominants as SE approaches 1.

As SE drops below 1, reproductive dominance should occur and not all the members of the group will be able to breed. From the standpoint of individual selection, an animal that is unable to breed should not stay in the group unless the probability of finding and breeding in a new group is less than the probability of breeding in the existing group. Under these circumstances, the sociality equilibrium predicts that cooperation should increase and aggression should decrease as the group size increases, and this seems to be the case with the eusocial insects that have an extreme form of reproductive dominance (Wilson 1975). Among these social insects, most species have a small probability at birth of reproducing as adults, and zero probability of joining and breeding in another group. In this respect, some eusocial insect species show incomplete reproductive dominance. Honeybee workers are occasionally able to lay eggs that develop parthenogenetically, because of the haplodiploid system of sex determination, into males. An even more extreme form of reproductive dominance is seen in South African colonies of the ponerine ant, *Opthalmopne berthoudi*, where up to 100 workers in a nest can be inseminated and can produce eggs within the colony. The resulting workers have a low level of relatedness because of the multiple parentage of workers in the colony, but all the workers cooperate in maintaining the colony (Peeters and Crewe 1985). The social equilibrium models predict that this kind of incomplete reproductive dominance is linked to resource levels, and that an increase in resource levels would lead to an increase in the number of reproductive workers.

REFERENCES

Alexander, R.D. 1974. The evolution of social behavior. *Annual Review of Ecology and Systematics* 5:325-383.
Askew, R.R. 1971. *Parasitic insects*. New York:American Elsevier.

Bartos, O.J. 1967. *Simple models of group behavior.* New York:Columbia University Press.
Bernstein, I.S. and T.P. Gordon. 1974. The function of aggression in primate societies. *American Scientist* 62:304-311.
Bernstein, I.S. and C. Ehardt. 1986. The influence of kinship and socialization on aggressive behavior in rhesus monkeys (*Macaca mulatta*). *Animal Behaviour* 34:739-747.
Clutton-Brock, T.H. and P.H. Harvey. 1977a. Primate ecology and social organization. *Journal of Zoology* 183:1-39.
Clutton-Brock, T.H. and P.H. Harvey. 1977b. "Species differences in feeding and ranging behavior in primates." pp. 557-579. In: T.H. Clutton-Brock, ed. *Primate ecology: feeding and ranging behaviour in lemurs, monkeys, and apes.* London:Academic Press.
Crook, J.H. 1965. The adaptive significance of avian social organizations. *Symposium of the Zoological Society of London* 14:181-218.
Crook, J.H. 1970. "The socio-ecology of primates." pp. 103-166. In: J.H. Crook, ed. *Social behaviour in birds and mammals.* London:Academic Press.
Crook, J.H. 1972. "Sexual selection, dimorphism, and social organization in the primates." pp. 231-281. In: B.G. Campbell, ed. *Sexual selection and the descent of man.* Chicago:Aldine.
Crook, J.H., J.E. Ellis and J.D. Goss-Custard. 1976. Mammalian social systems: structure and function. *Animal Behaviour* 24:261-274.
Crook, J.H. and J.S. Gartlan. 1966. Evolution of primate societies. *Nature* 210:1200-1203.
Crozier, R.H., P. Pamilo and Y.C. Crozier. 1984. Relatedness and microgeographic genetic variation in *Rhytidoponera mayri*, an Australian arid-zone ant. *Behavioral Ecology and Sociobiology* 15:143-150.
Eisenberg, J.F., N.A. Muckenhirn and R. Rudran. 1972. The relationship between ecology and social structure in primates. *Science* 176:863-874.
Frase, B.A. and K.B. Armitage. 1984. Foraging patterns of yellow-bellied marmots: role of kinship and individual variability. *Behavioral Ecology and Sociobiology* 16:1-10.
Gautier, J.-Y. 1982. *Socioecologie: l'animal social et son univers.* Toulouse:Bios.
Hall, K.R.L. 1964. "Aggression in monkey and ape societies." pp. 51-64. In: J.D. Carthy and F.J. Ebling, eds. *The natural history of aggression.* London:Academic Press.
Hamilton, W.D. 1964. The genetical evolution of social behaviour. I and II. *Journal of Theoretical Biology* 7:1-52.
Jarman, P.J. 1974. The social organization of antelope in relation to their ecology. *Behaviour* 58:215-267.
Jorde, L.B. and J.N. Spuhler. 1974. A statistical analysis of selected aspects of primate demography, ecology, and social behavior. *Journal of Anthropological Research* 30:199-224.

Michener, G.R. 1983. "Kin identification, matriarchies, and the evolution of sociality in ground dwelling sciurids." pp. 528-572. In: J.F. Eisenberg and D.G. Kleiman, eds. *Advances in the study of mammalian behavior*. American Society of Mammalogists Special Publication No. 7.

Milinksi, M. 1978. Kin selection and reproductive value. *Zeitschrift fur Tierpsychologie* 47:328-329.

Milton, K. and M.L. May. 1976. Body weight, diet and home range area in primates. *Nature* 259:459-462.

Orlove, M.J. 1975. A model of kin selection not invoking coefficients of relationship. *Journal of Theoretical Biology* 49:289-310.

Page, E., Jr. and R.A. Metcalf. 1982. Multiple mating sperm utilization, and social evolution. *American Naturalist* 119:263-281.

Peeters, C. and R. Crewe. 1985. Worker reproduction in the ponerine ant *Opthalmopone berthoudi*: An alternative form of eusocial organization. *Behavioral Ecology and Sociobiology* 18:29-37.

Ross, K.G. and D.J.C. Fletcher. 1985. Comparative study of genetic and social structure in two forms of the fire ant *Solenopsis invicta* (Hymenoptera:Formicidae). *Behavioral Ecology and Sociobiology* 17:349-356.

Sherman, P.W. 1980. The meaning of nepotism. *American Naturalist* 116:604-606.

Slobodchikoff, C.N. 1984. "Resources and the evolution of social behavior." pp. 227-251. In: P.W. Price, C.N. Slobodchikoff and W.S. Gaud, eds. *A new ecology: Novel approaches to interactive systems*. New York:Wiley.

Strassmann, J.E. 1985. Relatedness of workers to brood in the social wasp, *Polistes exclemans* (Hymenoptera:Vespidae). *Zeitschrift fur Tierpsychologie* 69:141-148.

Terborgh, J.W. 1983. *Five New World primates: A study in comparative ecology*. Princeton:Princeton Univ. Press.

Terborgh, J. and C.H. Janson. 1986. The socioecology of primate groups. *Annual Review of Ecology and Systematics* 17:111-135.

Van Schaik, C.P. and J.A.R.A.M. Van Hooff. 1983. On the ultimate causes of primate social systems. *Behaviour* 85:91-117.

Weigel, R.M. 1981. The distribution of altruism among kin: A mathematical model. *American Naturalist* 191-201.

Wilson, E.O. 1975. *Sociobiology*. Cambridge, Mass.:Harvard University Press.

Wrangham, R.W. 1980. An ecological model of female-bonded primate groups. *Behaviour* 75:262-300.

Wrangham, R.W. 1983. "Ultimate factors determining social structure." pp. 255-262. In: R.A. Hinde, ed. *Primate social relationships: An integrated approach.*Sunderland, Mass:Sinauer.

Chapter 3

THE STABLE GROUP AND THE DETERMINANTS OF FORAGING GROUP SIZE[1]

Luc-Alain Giraldeau[2]

Department of Psychology
University of Toronto
Ontario, Canada M5S 1A1

I.   INTRODUCTION

Several hypotheses have been proposed to explain why some animals forage in groups (see Barnard and Thompson 1985 for a review). Most of these hypotheses suggest that individuals are selected to forage in groups because this enhances their foraging returns and reduces their predation risks. Many of these hypotheses have been tested at a qualita-

[1]*This research was supported financially by an NSERC Postdoctoral Fellowship.*
[2]*Present address: Department of Biology, Concordia University, 1455 ouest, boulevard de Maisonneuve, Montréal, Québec, Canada H3G 1M8.*

tive level. At this level the tests consist of demonstrations of some increased foraging success (Horn 1968; Murton et al. 1971; Krebs et al. 1972; Krebs 1973; Feare et al. 1974; Williamson and Gray 1975; Caraco 1979; Howell 1979; Elgar and Catterall 1981; Magurran and Pitcher 1982; Pitcher et al. 1982; Smith 1983; Nentwig 1985; Giraldeau and Lefebvre 1986) or of some reduction in predator success with group size (Morse 1970; Powell 1974; Page and Whitacre 1975; Siegfried and Underhill 1975; Kenward 1978; Lazarus 1979; Caraco et al. 1980; Treherne and Foster 1980; Hoogland 1981; Godin and Morgan 1985; Wolf 1985). A more quantitative test of the selective forces responsible for the evolution of group foraging could predict not only the existence but also the size of the foraging groups. In order to conduct such a test one requires not only the gross benefit function but also the cost (or requirement) function. The benefits are often assumed to increase at a decreasing rate with group size while the foraging requirements increase linearly (Fig. 1). It follows under those conditions that the net benefit function reaches a maximum at some intermediate group size, the optimal group size, and decreases thereafter. Selection is argued to favor individuals that forage in groups of optimal size (Wilson 1975; Caraco and Wolf 1975; Rodman 1981).

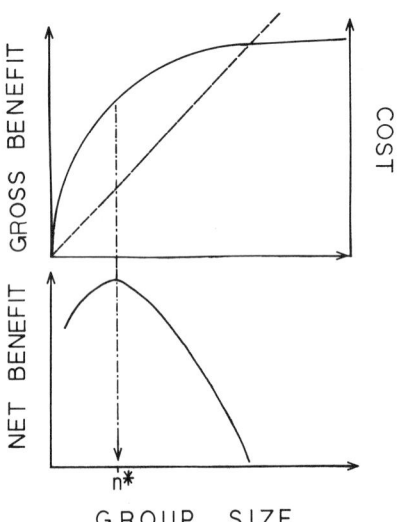

GROUP   SIZE

Figure 1. The top graph depicts the gross benefit (full line) and cost (dashed line) resulting from foraging in a group. The benefit is often thought to increase at a decreasing rate with group size while the food requirements increase linearly. The net benefit function per individual is depicted in the lower graph.

Few studies based on the assumption that animals forage in optimal group sizes have been successful in determining the component of fitness that is maximized by group foraging, that is the determinant of foraging group size (Caraco and Wolf 1975; Hatch 1975; Nudds 1978; Montevecchi 1979; Rodman 1981; Smith 1981). Sibly (1983) and then Clark and Mangel (1984) pointed out that the optimal group size might not be a realistic expectation for selfish animals designed by selection to maximize their individual fitness. They argued that the studies' failures can be attributed to the erroneous expectation of optimal group sizes. Selfish individuals, they predicted, would join groups so long as they would receive some benefit from doing so. The result is a stable group size that is always larger than the optimal group size, exactly what has been reported in most studies (Rodman 1981; Pulliam and Caraco 1984).

In this essay I will review the stable group size models (Sibly 1983; Clark and Mangel 1984) and analyze some of the consequences of incorporating group defense, social dominance and inclusive fitness into the model. Then, using the stable group size model, I will reanalyze results of earlier studies of the determinants of the size of the foraging groups of Inuit seal hunters (*Homo sapiens*) (Smith 1981), ravens (*Corvus corax*) (Montevecchi 1979), and lions (*Panthera leo*) (Caraco and Wolf 1975; Rodman 1981).

II.   THE STABLE GROUP SIZE MODEL

The stable group size model as presented by Sibly (1983) is a game between selfish individuals attempting to maximize their individual fitness (w). For simplicity he assumes that the individuals behave in an "ideal-free" manner, that is, they have equal competitive abilities and are free to join any group of size n that maximizes their w (Fretwell 1972). In addition, the fitness function, w(n), must be increasing in n at least for some range of $1 < n < n^*$, where n^* is the group size that provides the maximum average w for its members, the optimal group size. Additionally the w(n) must be decreasing in n at least for some range of $n > n^*$ as depicted in Figure 2. Sibly (1983) also assumes that animals arrive in the habitat one at a time.

Using the ideal-free assumptions and the fitness function depicted in Figure 2, the game proceeds as follows: The first two individuals to enter the habitat join together because the fitness obtained alone (w(1)) is lower than the fitness obtained in a group of two (w(2)). New arrivals continue to join until w(n + 1) < w(1). The n for which this condition is met is defined as the maximum stable group size $\hat{n}_{(max)}$. For the w(n) depicted in Figure 2, $\hat{n}(max)$ is almost double the size of the optimum. Note that in this model individuals join irrespective of their effect on the group members' w. The only criterion for decision is the recruit's w. Note also that for some fitness functions the optimal group size can meet the condition of stability (w(n* + 1) < w(1)) (Giraldeau and Gillis 1985).

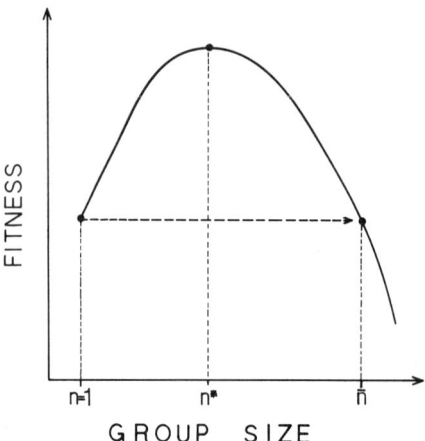

Figure 2. An hypothetical fitness function where fitness increases with group size. The maximum fitness is obtained in the optimal group size (n*). The stable group size (n̄) corresponds to a group size that provides its members and a potential recruit with as much or less fitness than remaining alone.

The outcome of the stability model is rather paradoxical. Gregarious individuals that are argued to live in groups because of increased fitness end up in stable group sizes with no more fitness than solitary individuals. If groups could reach size n̄(max) therefore, sociality would be unlikely to evolve. There are several ways in which groups can be kept from reaching n̄(max). One way follows from the fact that n̄(max), despite being called a stable group size, is technically an unstable equilibrium point (Clark and Mangel 1984; Kramer 1985). If the group happens to be split as the result of an external event, such as the sudden attack of a predator, there would be no tendency for the individuals from either group to rejoin into the original larger group. One possible scenario is that individuals in the larger splinter group (with lower average fitness) would migrate to the smaller splinter group until both subunits were of equal size whereupon a new point of unstable equilibrium would be reached. Any further splitting from external causes would result in even smaller groups up to a minimum of n*. The size of the expected stable group therefore (n̄) depends on the shape of the fitness function which in turn can determine how fissible groups are in relation to external events. A group's propensity to split may vary as a function of the difference between the fitness in the present supra-optimal size group and the fitness obtained in the potentially smaller stable groups resulting from the split. Presumably, the steeper (more negative) the slope of the fitness function for n > n* the more fissible the group (Kramer 1985). Similarly, the larger the group, the more fissable it becomes. Aggression and dominance are other means by which groups can be kept from reaching n̄(max). The effect of both of these will be the subject of the following sections.

III.   VIOLATIONS OF SOME OF THE MAJOR ASSUMPTIONS

A.   Constraints on Joining Groups

The model does not consider the group members' behavior toward recruits. It is very unlikely that group members will be indifferent to recruits. The group members' behavior will surely depend on the magnitude and direction of the recruits' effect on their w. Group members should encourage recruitment for $n < n^*$ since this would increase their w. However, the group members should discourage recruitment for $n > n^*$ because accepting a recruit for those group sizes would reduce their w. Recruits, on the other hand, should be more persistent the more group membership promises to increase their fitness. While there are numerous examples of aggression directed toward potential recruits, there are few descriptions of behaviors designed to enhance recruitment. Such behaviors may include the so-called contact calls common in foraging flocks of birds. Another example may be the drag line left by solitary, gravid emigrating females of the social spider (*Anelosimus eximius*) (Vollrath 1982). The drag line may serve to direct recruits to the site of a newly founded colony. Vollrath notes that the foundresses are surprisingly unaggressive toward even unrelated recruits.

Animals should not be expected to spend all their time either repelling recruits or attempting to join groups. It is more likely that the decisions to repel and persist will be economic ones, probably analogous in nature to the decisions of defending and intruding on territories (Brown 1964). The effect of opposing some resistance by group members to recruits, however, will always be analogous to imposing a cost to becoming a member of a group and therefore to reduce the value of joining. Since this resistance occurs at $n > n^*$ only, the value of joining supra-optimal sized groups will be reduced, making n̄ smaller (Fig. 3). Naturally, the example assumes that the cost of repelling intruders to the group members exceeds the cost to the intruder of overcoming this repulsion. This could be the case when group members stand to lose more by investing in group defense (for example, through greater depreciation in foraging efficiency) than a solitary recruit that is already doing poorly.

B.   Existence of Dominance Hierarchies

Dominance hierarchies are common in animal societies. Dominant members of a hierarchy have a higher competitive ability than fellow group members and accordingly monopolize a greater fraction of the resources. Subordinate individuals, on the other hand, obtain a smaller fraction of the resources. For individuals in a dominance hierarchy the profitability of the group size will depend on the position in the dominance hierarchy. Groups will be more profitable to dominant individuals that exploit subordinates than to the subordinates (Baker et al. 1981).

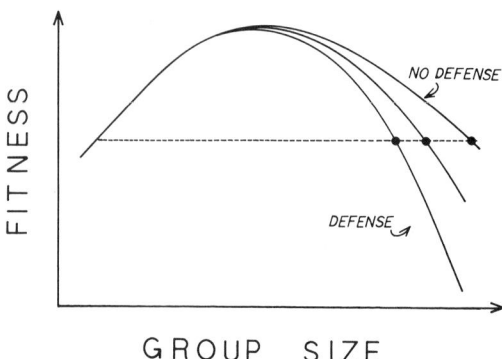

GROUP   SIZE

Figure 3. Hypothetical fitness functions for an animal facing three levels of defense of access to the groups. The greater the defense on the part of the group members the lower the value of joining a group that is larger than the optimum. Group members only gain by restricting access to a group of optimal size or larger. Thus, the defense only affects the right side of the fitness functions. The result of considering defense is increasingly smaller stable group sizes (points) with increasing quantities of defense from the group members.

Dominance therefore will affect the stable group size by modifying the value of joining a group of a given size as a function of rank.

If recruits expect to be subordinate members of a group then the break-even point, the group size at which remaining alone is just as profitable as joining, occurs at a smaller group size in a despotic population than in an egalitarian population (Fig. 4). The result still holds even if the recruit has no expectation concerning the rank it will occupy in the group. Joining a group often results in a reshuffling of dominance positions with some individuals gaining and others losing status. A recruit that joins a group that is already stable (the recruit expects to do better in the group than the most subordinate group member) will push some individual (the omega individual) below the break-even point and cause it to leave the group. This point will occur at a smaller group size in despotic populations, thus resulting in smaller group sizes with increasing amounts of despotism. Similar conclusions have been reached by Rubenstein (1978) and Vehrencamp (1985). There is some evidence that small or immature woodpigeons (*Columba palambus*) feed alone because they would otherwise occupy an extremely subordinate position in the foraging flock (Murton et al. 1971).

C.   Considering Inclusive Fitness

Selection is often wrongly pictured as operating at the level of the individual (Dawkins 1982). This frequently leads to erroneous predictions. The more appropriate level at which to model the operation of natural selection is genic (Dawkins 1982). When the players in the stable

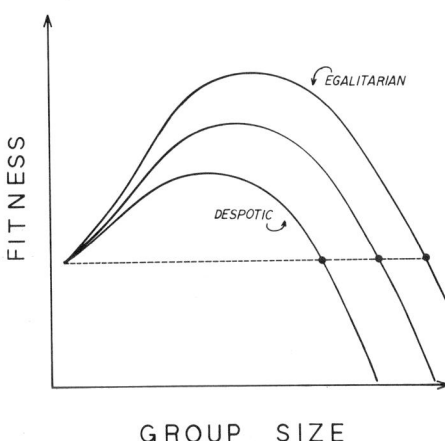

GROUP   SIZE

Figure 4.   Three hypothetical fitness functions depicting a subordinate's expected fitness in three groups with different degrees of despotism.   The greater the exploitation from dominants the lower the fitness function.   Note that the decision not to join a group for a subordinate individual (points) occurs at smaller group sizes the greater the exploitation by dominants.   Note also that the effect of dominant's exploitation is assumed to increase with group size.

group size game are related, genic and individual selection models make different predictions. When the players are related, the recruits will not be indifferent to the effect they have on the group members' fitness. Maximization of individual fitness therefore must be replaced by maximization of inclusive fitness. Joining a group of relatives of size n < n* will enhance the group members' fitness and therefore increase the recruit's inclusive fitness. Joining a group of relatives of size n ≥ n*, however, will decrease the fitness of the relatives, and therefore the player's inclusive fitness as well (Fig. 5).

When inclusive fitness is used in the model, joining is not the only behavior for which the payoff depends on the size of the group. The inclusive fitness produced by not joining will also be a function of the group size. Refraining from joining a group of relatives affects their fitness, and hence the inclusive fitness of the non-joiner, in a way that depends on the size of the group of relatives. This is different from the situation with non-related individuals where the option of remaining alone always produces the same fitness, w(1), irrespective of the size of the group that is not joined (Fig. 6). In order to estimate the stable group size of a population of related individuals, one needs to calculate the inclusive fitness of both joining and not joining for each group size. The stable group size will then be the smallest group for which the inclusive fitness of not joining is greater than the inclusive fitness of joining. As the coefficient of relatedness increases, the reduction in inclusive fitness that results from joining a group of supra-optimal size also in-

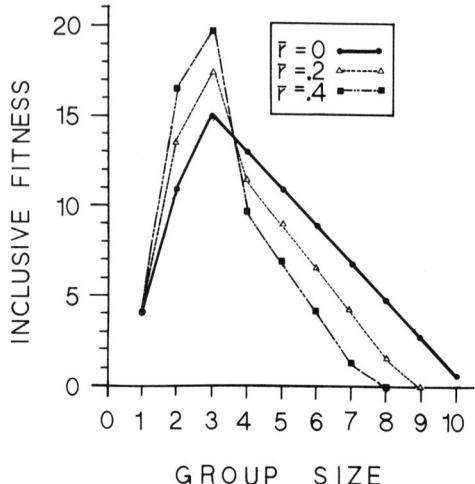

Figure 5. *Three hypothetical inclusive fitness functions for joining groups with different mean coefficients of relatedness. The X axis depicts the group size after joining. Note that the greater the coefficient of relatedness the greater the inclusive fitness for joining infra-optimal group sizes and the lower the inclusive fitness for joining supra-optimal sixed groups. Note also that individuals that can choose between groups with different coefficients of relatedness can obtain greater inclusive fitness by joining groups with lower coefficents of relatedness when groups are larger than the optimum.*

creases (Fig. 5). Similarly, as the coefficient of relatedness increases, the magnitude of the increase in inclusive fitness that results from not joining a group of supra-optimal size also increases (Fig. 6). As a result, when the coefficient of relatedness increases, the point at which not joining produces higher inclusive fitness than joining occurs at smaller group sizes (Fig. 7). Using inclusive fitness in the stable group size model therefore produces stable group sizes that are smaller the higher the coefficient of relatedness between the members of the population. This result is surprising in view of the common expectation that relatedness somehow enhances gregarious behavior.

There are at least two studies that deal with the effect of relatedness on group size (Rodman 1981; Kurkland and Beckerman 1985). Both studies argue that the effect of incorporating inclusive fitness into the model is to increase the optimal group size. Thus animals are expected to forage in increasingly large optimal groups the higher the coefficient of relatedness between group members. This conclusion, however, is based on a definition of inclusive fitness that corresponds to what Grafen (1982) called the simple weighted sum. The problem with this definition is that it counts all of the relatives' fitness into ego's inclusive fitness when it should only count the fraction of the relatives' fitness

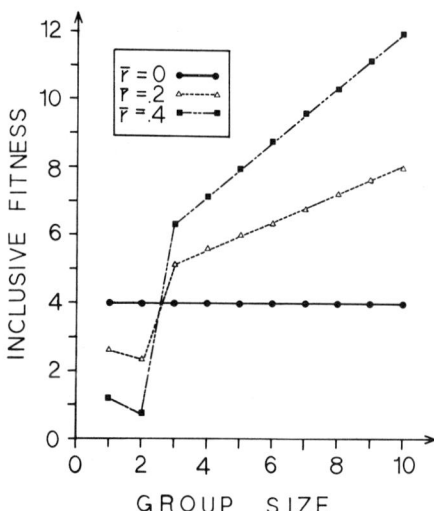

Figure 6. Three hypothetical inclusive fitness functions for not joining a group for different mean coefficients of relatedness. The X axis in this case depicts the size of the group that is not joined. Note that when the coefficient of relatedness is zero, the payoff of staying alone is independent of group size. The greater the coefficient of relatedness the greater the benefit of not joining supra-optimal sized groups. Conversely, the greater the coefficient of relatedness the lower the inclusive fitness obtained from not joining groups of infra-optimal size.

that has either been increased or decreased by ego's behavior. When the correct definition of inclusive fitness is used, the prediction of increasing optimal group size with relatedness no longer holds (Fig. 5). Unlike the optimal group size, however, the stable group size decreases to a minimum of n* with increasing relatedness (Fig. 7).

The use of inclusive fitness in the stable group size model generates a surprising prediction. The prediction concerns an individual's choice between groups differing only in the degree to which the individual and the group members are related. The individual should join the group to which it is most related when the group sizes are below the optimum. Conversely it should choose the group to which it is least related when the group sizes are above the optimum (see Fig. 5). Similarly, if optimal group sizes fluctuate as a function of season, the individuals who are expected to leave or join the groups will depend on their coefficient of relatedness with other group members as well as the direction in which the optimal group size has changed. When the optimal group size shrinks the most highly related group members should leave, reducing the average coefficient of relatedness between group members. Conversely, when the optimum increases again, those members of the solitary population who are most related to the group members should join, thus increasing

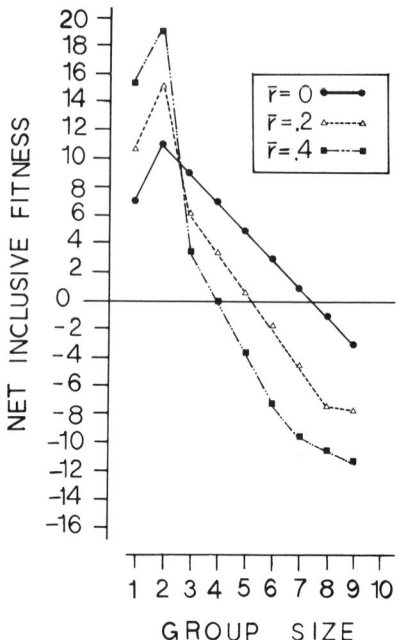

Figure 7. *The inclusive fitness of joining less the inclusive fitness of not joining. The X axis depicts the group size before the recruit joins. Individuals should join so long as the difference is positive. The stable group size is the largest group for which the difference is positive (the intersection of the net fitness functions and the horizontal axis). Note that as the average coefficient of relatedness within the group increases the stable group size decreases. The functions are calculated from the theoretical values used in Figures 5 and 6.*

the coefficient of relatedness within the group. Hence, although the optimal group size does not change with the coefficient of relatednesss, changes in the optimal group size can induce changes in group composition that will affect the group's coefficient of relatedness.

IV.   THE DETERMINANTS OF FORAGING GROUP SIZE

The stable group size model generates the largest possible group ($\bar{n}(max)$) in a population of unrelated individuals that behave in an ideal-free way. The realized stable group ($\bar{n}$), however, is likely to lie somewhere between the n^* and $\bar{n}(max)$. This is because social dominance, group defense directed against recruits and relatedness all make the realized stable group closer to n^*. Until the effects of social dominance and group defense can be quantified, however, the model will remain more qualitative than quantitative since it can only predict a range of values rather than a single value.

The goal of the following sections is to re-examine the conclusions of earlier studies of the idea that foraging success determines the size of foraging groups (Caraco and Wolf 1975; Montevecchi 1979; Smith 1981). In all these cases some measure of foraging efficiency was rejected as the determinant of group size because observed group sizes were significantly different from the optimal group size specified on the basis of foraging efficiency. These studies will be reviewed in light of the predictions of the stable group size model. For simplicity, in the present study the expected stable group sizes will be generated assuming ideal-free conditions so that the predicted stable group sizes correspond to r̄(max). The goal therefore is not so much to demonstrate that the subjects forage in stable group sizes but to illustrate the approach while determining whether the conclusions of the earlier studies need to be revised.

A.   Inuits Hunting Bearded Seals

Smith (1981) studied the foraging efficiency of 10 different types of fishing, whaling and hunting groups of Northern-Québec Inukjuamiut (Inuit) People. In 8 out of 10 foraging group types, the observed group size was larger than the optimal group size. Unfortunately, Smith only produced a benefit function for the bearded seal hunting groups. Therefore, the stable group size analysis can be done only for this hunting group type.

The ice that covers much of the ocean at northern latitudes considerably reduces the water surface from which the bearded seal (*Erignathus barbatus*) can breathe. In response to this, each seal maintains several breathing holes in the ice throughout the winter months. As a result, the seals' presence at a hole is rare and unpredictable. The chance of capturing a seal increases with the number of holes that can be monitored simultaneously (Fig. 8). The chance of success also increases with the number of hunters participating in the hunt although it is not clear from Smith's description why this would be the case, given that each hunter can only monitor one hole.

In order to calculate the optimal group size, Smith measured and/or estimated the amount of energy expended during the hunt and the amount harvested over a range of hunting group sizes for a total of 19 bearded seal hunts. The optimal group size therefore represents the group that provides the maximum net energy return per participant. According to the benefit function the optimal group size is three individuals (Fig. 8). The observed hunting group sizes were larger, ranging from 1-8 with a mean of 3.9 (mode = 4), with groups of three or more hunters accounting for 84 percent of all observations.

Jarman (1974) pointed out that the mode and the mean are observer-centered estimates of group sizes. Both measures give estimates of the most frequently encountered group size but not the group size experienced by the average individual in the population, the typical group

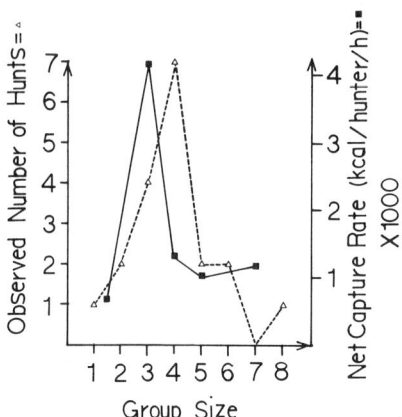

Group Size

Figure 8. The benefit function (squares) and the observed frequencies of hunting group size (triangles) of Inukjuamiut (Inuit) People hunting bearded seals (Erignathus barbatus) at their breathing holes in Northern Québec. The optimal group size is 3 hunters and the stable group size is 7 or more individuals. Modified from Smith 1980.

size $\bar{g} = (\sum_{i=1}^{N} g_i^2)/(\sum_{i=1}^{N} g_i)$ where N is the number of groups and g_i the size of each group. The typical group size is more relevant because selection affects the behavior of the individual, not the frequency of groups. The typical group size is usually larger than the mean. This reflects the fact that even if there are only a few large groups, they may contain the majority of individuals in the population. The average Inukjuamiut seal hunter was found to be in a typical group of 6.7 individuals, a value that is considerably larger than the optimal group size of three hunters.

The maximum stable group size predicted from the benefit function measured by Smith is seven hunters. Unfortunately there are no data for groups larger than seven, even though the returns at that group size are still above the returns of solitary hunting. It would be crucial to know whether the returns of foraging in groups of eight or more individuals are lower than the returns for solitary hunters. The model would certainly predict this. Smith does not provide quantitative information on the relatedness of the hunters although he does mention (p. 59) that in some cases the hunters of a group were related. Although the stable group size model may have overestimated the size of the stable group, it suggests that maximization of net energy return per individual cannot be rejected as a determinant of the seal hunting group size of the Inukjuamiut People.

B.   Ravens and Kittiwakes

Ravens often patrol the cliffs on which kittiwakes (Rissa tridactyla) are nesting. When a raven comes close to the nests the kittiwakes leave either to flee or attack the raven. When they flee, the eggs or young are exposed to predation by the ravens. The likelihood of displacing kittiwakes from their nests increases with raven patrol size. The number of kittiwakes involved in the chase, however, does not change with patrol size. The effect is such that ravens in patrols of three or more have a considerably higher individual predation success than ravens either in pairs or alone (Montevecchi 1979). Moreover, pairs are least successful of all because the added bird does not increase success sufficiently to counter-balance its cost as an extra competitor (Fig. 9). Surprisingly, Montevecchi observed that only 4 percent (26/625) of his observations consisted of three or more ravens. The vast majority of his observations consisted of solitary ravens (543/625, 87%). The mean observed patrol size was 1.2 birds (mode = 1) and the average bird was found in a typical group of 1.4 individuals.

This result is somewhat less surprising in the light of the stable group argument. Because pairs of ravens are considerably less successful than either single individuals or trios, no individual can gain fitness by joining another individual. Pairs therefore are unlikely to occur. If pairs formed, they would be very unstable, either breaking up or attracting a third individual. Assuming that groups form as a result of individuals' decisions, the low incidence of pairs would make trios extremely unlikely despite the trios' higher stability and profitability. Montevecchi's results therefore remain compatible with the hypothesis that the

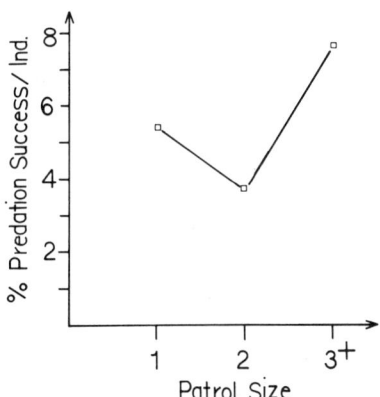

Figure 9.   The fitness function of ravens (Corvus corax) raiding the nesting colonies of kittiwakes (Rissa tridactyla). The benefit is depicted as predation success (the number of successful predation attempts divided by the total number of attempts observed) divided by the number of individuals in the patrol. Note that the optimal group size is 3. The most stable group size is also 3. Pairs are extremely unlikely to form so that groups of 3 cannot be formed easily. As a result the expected group size is 1. Modified from Montevecchi (1979).

group size of patrolling ravens is determined by the predation success of its individuals.

C. Lions Hunting Gazelle, Wildebeest and Zebra

In the Serengeti, lions are organized into prides that average 9.4 adults (Schaller 1972, p. 415). The prides are composed of several related adult females, their young and one or a few males that can be related to each other but are generally not closely related to the females (Packer 1986). Bertram (1976) reports that the average coefficient of relatedness between pride members may be of the order of 0.2. The adults of one pride are unlikely to be found all together at any one time. Rather, the pride breaks up into smaller groups where individuals either hunt alone or in small parties. The composition of these smaller groups changes on a daily basis (Schaller 1972; Packer 1986) so that within a pride individuals are not restricted in their associations. It is the size of these smaller foraging groups and not the size of the whole pride which this section will address.

Schaller reported that there may be considerable intra- and interspecific competition for food at a kill. The efficiency of both the hunting and subsequent defense of the prey from scavengers (Lamprecht 1978) is a function of foraging group size. These observations led Caraco and Wolf (1975) to hypothesize that the daily quantity of food obtained per individual may determine the lions' foraging group size. Using Schaller's data they calculated benefit functions for lions hunting different prey types. Since hunting success is not only a function of the prey species but also of the season and habitat in which the hunt occurs, Caraco and Wolf calculated different benefit functions for different prey types hunted in different habitat/season combinations (Fig. 10).

The optimal hunting group size for any prey in any season is two adult individuals. Although hunting and feeding groups may be considerably different in size (Parker 1986) Caraco and Wolf compared the predicted hunting group sizes with observed feeding group sizes and found that the feeding groups were considerably larger than the optima in all cases except the gazelles. However, since lions forage in kin groups, inclusive rather than individual benefits should be used (Rodman 1981). The individual benefit functions reported by Caraco and Wolf must be transformed to inclusive benefit. Since both joining and staying alone produce inclusive benefits that are a function of group size, inclusive benefits for both decisions must be calculated for each group size. The inclusive benefit of an individual joining a group of size n - 1 and bringing it to size n can be calculated as:

$$W(n) = w(n) + [w(n) - w(n - 1)] (n - 1)\hat{r} \qquad \text{(eq. 1)}$$

where w(n) is the individual benefit obtained in group n and the difference w(n) - w(n - 1) gives the net benefit that joining group n - 1 has had on the n - 1 group members and, as always, $\hat{r}$ is the coefficient of relatedness. The inclusive benefits of not joining a group of n - 1 individuals can be calculated as:

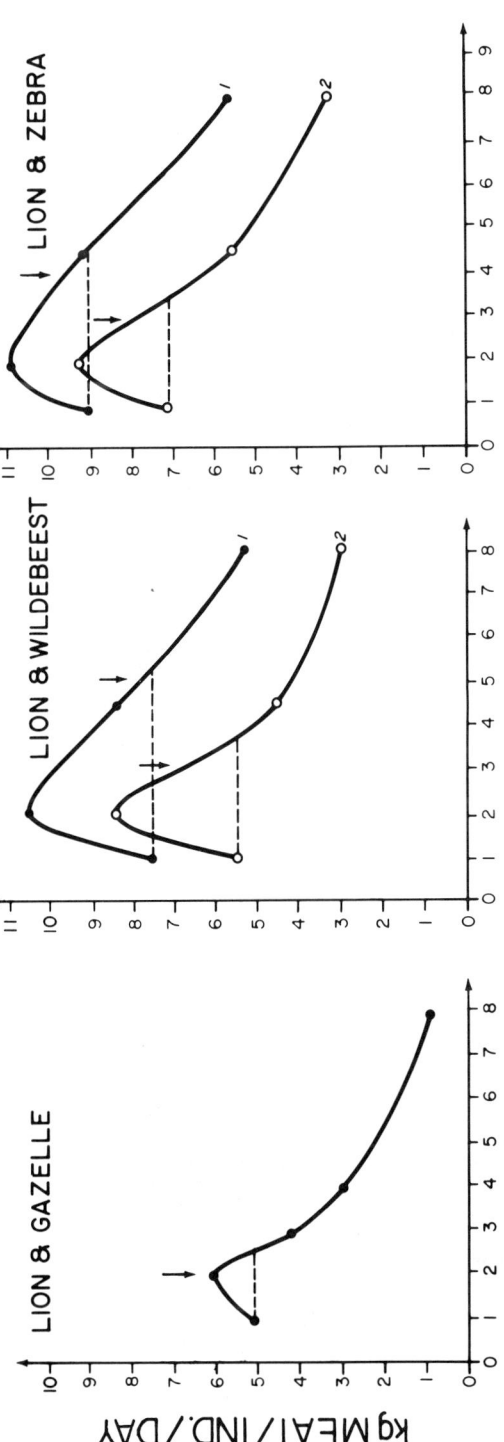

Figure 10. The individual benefit functions for lions (Panthera leo) hunting gazelle (Gazella thompsonii), wildebeest (Connochaetes taurinus) and zebra (Equus burchelli). Line 1 gives the result for the wet season while line 2 concerns eastern plains and western woodlands. The optimal group size is 2 for hunting all prey types in both habitat-season combinations. The maximum stable group sizes are given by the arrows. Modified from Caraco and Wolf (1975).

$$W(1) = w(1) + [w(n - 1) - w(n)] (n - 1)r \qquad \text{(eq. 2)}$$

where $w(1)$ is the benefit obtained by remaining alone and the difference $w(n - 1) - w(n)$ gives the net benefit caused to the $n - 1$ group member by not joining the group. The inclusive benefit functions (Fig. 11) for both decisions were calculated using a coefficient of relatedness of 0.2. The values of w were taken from figures in Caraco and Wolf. The stable group size is the size at which not joining provides more inclusive benefits than joining.

In all cases the stable group size was three adults except for gazelles where it was two (Table I). Had Schaller not lumped his observations for both large prey types the typical group sizes could have been calculated for all three prey types. The results, however, although based on fewer prey-predator types, suggest that lions hunt in typical group sizes that could be slightly larger than the group sizes predicted by the stable group model (Table I). The extent of this difference cannot be appreciated statistically because there are no appropriate statistical tests for typical group size. The extent of the variance, however, suggests that the difference may be small. Whatever the case, the results suggest that Caraco and Wolf's (1975) rejection of hunters' daily food return as a determinant of hunting group size may not be as strong as was first thought.

Further support for the idea that the size of lion hunting groups are based on food returns has been provided by Clark (1987). He showed that the sizes of lion groups hunting gazelle and zebra were consistent with predictions of a Markovian model that minimizes probability of starvation for a given fat reserve level and gut capacity. Whether lion hunting group size maximizes hunters' daily food returns or minimizes chances of starvation still needs to be determined. The point I wish to make here is that the observed hunting group sizes could be the result of maximization of hunters' daily food returns subject to the constraint of stability.

V. CONCLUSIONS AND SUMMARY

Foragers are not likely to be found in optimal group sizes. Instead they will forage in groups whose size makes them unattractive to others, the stable group size. The extent to which stable group sizes are larger than the optimum depends in part on the shape of the function relating fitness to group size. In addition, the realized stable group will approach n^* when (1) group members defend access to the group to potential recruits, (2) social dominance exists, and (3) the individuals of a population are related. Inclusive fitness does not explain the existence of supraoptimal groups. The model predicts that individuals that share the highest degree of relatedness with the rest of the group will be the first to leave when the optimum recedes but they will be the first to join when the optimal size increases.

The review of three earlier tests of the determinants of foraging group size using the stable group size model has questioned the authors'

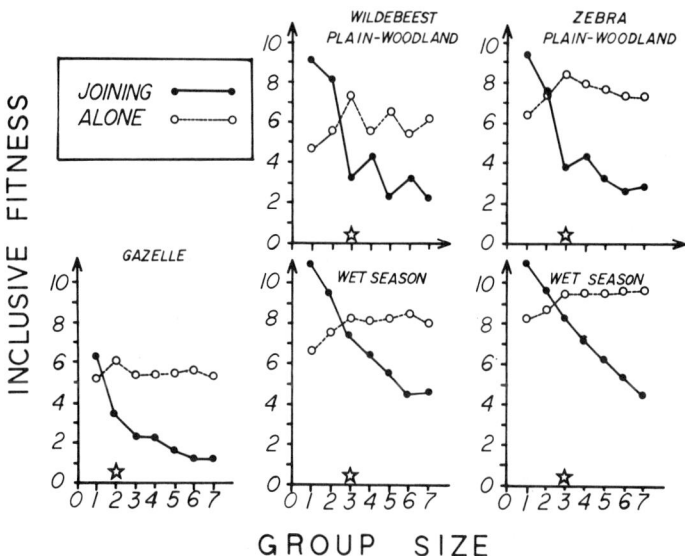

Figure 11.   The inclusive fitness functions for P. leo hunting G. thompsoni, C. taurinus and E. burchelli in different season–habitat combinations.   Note that the inclusive fitnesses for joining and not joining a group are presented on the same graph.   The stable group size is given by the star and corresponds to the group size at which not joining provides more inclusive fitness than joining.   The stable group size is 3 in all cases except for gazelles where it is 2 adults.

conclusions in all three cases.   Some measure of foraging efficiency could not be strongly rejected as a determinant of foraging group size of Inuits hunting seals, ravens raiding kittiwake colonies, and lions hunting gazelle, wildebeest and zebra.   The stable group size approach, however, is not without problems.   For instance, the predictions depend on the shape of a net benefit function.   Such a function is at best difficult to measure, a factor that may explain its rarity in the literature.   In addition, the model can often only predict a range of expected group sizes because the effects of dominance and resistance to recruits are difficult to quantify.   Nonetheless even the model's crudest predictions were more successful at predicting the size of the foraging groups than the optimal group size model.   It appears possible therefore that foraging groups of Inuits, ravens and lions all share the common attribute of being of that size which maximizes some foraging return subject to a constraint of stability.

Table I. The predicted optimal and stable group sizes and the typical hunting group sizes ($\bar{g}$) its variance (s) for *P. leo* hunting *G. thompsoni, C. taurinus,* and *E. burchelli* in the wet season (a) and in the Eastern Plains-Western Woodlands region (b).

Prey	Predicted		Observed Hunting	
	Optimal	Stable	g	s
Gazelle	2.0	2.0	3.03	1.69
Wildebeest[a]	2.0	3.0	4.23	1.78
Wildebeest[b]	2.0	3.0	--	--
Zebra[a]	2.0	3.0	4.23	1.78
Zebra[b]	2.0	3.0	--	--

Note: The stable group sizes are calculated from inclusive benefits using a coefficient of relatedness of 0.2. The hunting group sizes are taken from Schaller's (1972) Table 59, assuming that the sample sizes for each prey type reflect the frequency of hunting for these prey. Schaller lumped the observations of hunting group sizes for wildebeest and zebra. The variance of g is calculated as:

$$S^2 = \Sigma\, g_i^3 - [(\Sigma g_i)^2/\Sigma g_i]/(\Sigma g_i - 1) \text{ (Jarman 1982).}$$

ACKNOWLEDGMENTS

During this study I was financially supported by an NSERC Postdoctoral Fellowship and by an NSERC Operating Grant to Jerry Hogan. I wish to thank Martin Daly, Darren Gillis, Don Kramer, Shannon Hamm, Chris Plowright, Doug Shapiro, David Sherry, Margo Wilson as well as Ron Ydenberg and the students of the Evolutionary Ecology Seminar at Simon Fraser University for helping me develop and clarify some of the ideas contained in this study.

REFERENCES

Baker, M.C., C.S. Belcher, L.C. Deutsch, G.L. Sherman and D.B. Thompson. 1981. Foraging success of junco flocks and the effects of social hierarchy. *Animal Behaviour* 28:295-309.

Barnard, C.J. and D.B.A. Thompson. 1985. *Gulls and plovers: The ecology and behaviour of mixed-species feeding groups.* Beckenham, England:Croom Helm.

Bertram, B.C.R. 1976. "Kin selection in lions and evolution." pp. 281-301. In: P.P. Bateson and R.A. Hinde, eds. *Growing points in ethology.* Cambridge:University Press.

Brown, J.L. 1964. The evolution of diversity in avian territorial systems. *Wilson Bulletin* 76:160-169.

Caraco, T. 1979. Time budgeting and group size: A test of a theory. *Ecology* 60:618-627.

Caraco, T., S. Martindale and H.R. Pulliam. 1980. Avian flocking in the presence of a predator. *Nature* 285:400-401.

Caraco, T. and L.L. Wolf. 1975. Ecological determinants of group sizes of foraging lions. *American Naturalist* 109:343-352.

Clark, C.W. 1987. The lazy adaptable lions: A Markovian model of group foraging. *Animal Behaviour* 35:361-368.

Clark, C.W. and M. Mangel. 1984. Foraging and flocking strategies: Information in an uncertain environment. *American Naturalist* 123:626-641.

Dawkins, R. 1982. *The extended phenotype.* Oxford:Freeman.

Elgar, M.A. and C.P. Catterall. 1981. Flocking and predator surveillance in house sparrow: A test of an hypothesis. *Animal Behaviour* 29:868-872.

Feare, C.J., G.M. Dunnett and I.J. Patterson. 1974. Ecological studies of the rook (*Corvus frugilegus* L.) in North-East Scotland: Food intake and feeding behaviour. *Journal of Applied Ecology* 11:867-896.

Fretwell, S.D. 1972. *Populations in a seasonal environment.* Princeton: Princeton University Press.

Giraldeau, L.-A. and D. Gillis. 1985. Optimal group size can be stable: A reply to Sibly. *Animal Behaviour* 33:666-667.

Giraldeau, L.-A. and L. Lefebvre. 1986. Exchangeable producer and scrounger roles in a captive flock of feral pigeons: A case for the skill pool effect. *Animal Behaviour* 34:797-803.

Godin, J.-G.J. and Morgan, M.J. 1985. Predator avoidance and school size in a cypronodontid fish, the banded killifish (*Fundulus diaphanus* Lesueur). *Behavioral Ecology and Sociobiology* 16:105-110.

Grafen, A. 1982. How not to measure inclusive fitness. *Nature* 298:425-426.

Hatch, J.J. 1975. Piracy by laughing gulls *Larus atricilla*: An example of the selfish group. *Ibis* 177:357-365.

Hoogland, J.L. 1981. The evolution of coloniality in white-tailed and black-tailed prairie dogs (Sciuridae: *Cynomys leucurus* and *C. ludovicianus*). *Ecology* 62:252-272.

Horn, H.S. 1968. The adaptive significance of colonial nesting in the Brewer's blackbird (*Euphagus cyanocephalus*). *Ecology* 49:682-694.

Howell, D.J. 1979. Flock foraging in nectar-feeding bats: Advantages to the bats and to the host plants. *American Naturalist* 114:23-49.

Jarman, P.J. 1974. The social organization of antelope in relation to their ecology. *Behaviour* 48:216-267.

Jarman, P.J. 1982. "Prospect for interspecific comparison in sociobiology." In: King's College Sociobiology Group, ed. *Current problems in sociobiology.* Cambridge:Cambridge University Press.

Kenward, R.W. 1978. Hawks and doves: Factors affecting success and selection in goshawk attacks on woodpigeons. *Journal of Animal Ecology* 47:449-467.

Kramer, D. 1985. Are colonies supra-optimal groups? *Animal Behaviour* 33:1031-1032.

Krebs, J.R. 1973. Social learning and the adaptive significance of mixed-species flocks of chickadees. *Canadian Journal of Zoology* 51:1275-1288.

Krebs, J.R., M.H. MacRoberts and J.M. Cullen. 1972. Flocking and feeding in the Great Tit *Parus major*--an experimental study. *Ibis* 114:507-530.

Kurland, J.A. and S.J. Beckerman. 1985. Optimal foraging and hominid evolution: Labor and reciprocity. *American Anthropologist* 87:73-93.

Lamprecht, J. 1978. The relationship between food competition and foraging group size in some larger carnivores. *Zeitschrift für Tierpsychologie* 46:337-343.

Lazarus, J. 1979. The early warning function of flocking in birds: An experimental study with captive quelea. *Animal Behaviour* 27:855-865.

Magurran, A.E. and T.J. Pitcher. 1983. Foraging, timidity and shoal size in minnows and goldfish. *Behavioral Ecology and Sociobiology* 12:147-152.

Montevecchi, W.A. 1979. Predator-prey interactions between ravens and kittiwakes. *Zeitschrift für Tierpsychologie* 49:136-141.

Morse, D.H. 1970. Ecological aspects of some mixed-species foraging flocks of birds. *Ecological Monographs* 40:119-168.

Murton, R.K., A.J. Isaacson and N.J. Westwood. 1971. The significance of gregarious feeding behaviour and adrenal stress in a population of Wood-pigeons *Columba palambus*. *Journal of the Zoological Society of London* 165:53-84.

Nentwig, W. 1985. Social spiders catch larger prey: A study of *Anelosimus eximius* (Araneae: Theridiidae). *Behavioral Ecology and Sociobiology* 17:79-85.

Nudds, T.D. 1978. Convergence of group size strategies by mammalian social carnivores. *American Naturalist* 112:957-960.

Packer, C. 1986. "The ecology of sociality in fields." pp. 429-451. In: D.I. Rubenstein and R.W. Wrangham, eds. *Ecological aspects of social evolution*. Princeton:Princeton University Press.

Page, G. and D.F. Whitacre. 1975. Raptor predation on wintering shorebirds. *Condor* 77:73-83.

Pitcher, T.J., A.E. Magurran and I.J. Winfield. 1982. Fish in larger shoals find food faster. *Behavioral Ecology and Sociobiology* 10:149-151.

Powell, G.V.N. 1974. Experimental analysis of the social value of flocking by starlings (*Sturnus vulgaris*) in relation to predation and foraging. *Animal Behaviour* 22:501-505.

Pulliam, H.R. and T. Caraco. 1985. "Living in groups: Is there an optimal group size?" pp. 122-147. In: J.R. Krebs and N.B. Davies, eds. *Behavioural ecology: An evolutionary approach*. Second edition. Sunderland:Sinauer.

Rodman, P.S. 1981. Inclusive fitness and group size with a reconsideration of group sizes in lions and wolves. *American Naturalist* 118:275-283.

Rubenstein, D.I. 1978. On predation, competition, and the advantages of group living. pp. 205-231. In: P.P.G. Bateson and P.H. Klopfer, eds. *Perspectives in ethology: Social behavior*. New York:Plenum.

Siegfried, W.R. and L.G. Underhill. 1975. Flocking as an anti-predator strategy in doves. *Animal Behaviour* 23:504-508.

Smith, D.R.R. 1983. Ecological costs and benefits of communal behavior in a presocial spider. *Behavioral Ecology and Sociobiology* 13:107-114.

Smith, E.A. 1981. "The application of optimal foraging theory to the analysis of hunter-gatherer group size." pp. 36-65. In: B. Winterhalder and E.A. Smith, eds. *Hunter-gatherer foraging strategies.* Chicago:University of Chicago Press.

Treherne, J.E. and W.A. Foster. 1981. Group transmission of predator avoidance behaviour in a marine insect: the Trafalgar effect. *Animal Behaviour* 29:911-917.

Vehrencamp, S.L. 1985. "Exploitation in cooperative societies: Models of fitness biasing in cooperative breeders." pp. 229-266. In: C.J. Barnard, ed. *Producers and scroungers: Strategies of exploitation and parasitism.* New York:Chapman and Hall.

Vollrath, F. 1982. Colony foundation in a social spider. *Zeitschrift für Tierpsychologie* 60:313-324.

Wilson, E.O. 1975. *Sociobiology: The modern synthesis.* Cambridge: Harvard University Press.

Williamson, P. and L. Gray. 1975. Foraging behavior of the starling (*Sturnus vulgaris*) in Maryland. *Condor* 77:84-89.

Wolf, N. 1985. Odd fish abandon mixed-species groups when threatened. *Behavioral Ecology and Sociobiology* 17:47-52.

Chapter 4

MONOGAMY IN RELATION TO RESOURCES[1]

George W. Barlow

Department of Zoology and Museum of Vertebrate Zoology
University of California, Berkeley 94720

I.   INTRODUCTION

In his monumental and influential book, E.O. Wilson (1975) set out to define a new field of research, sociobiology. In doing so, he used kin selection and the allied concept of altruism as a linchpin. While ecology

[1]The writing was supported by a grant from the National Institute of Child Health and Human Development, HD 18496.

played an important part in his impressive essay, it did not receive the emphasis it deserves (Barlow 1976). The ecological basis of social systems and social behavior must be more general than kin selection because it applies to all species whereas kin selection can only operate where the actions of the actor have direct consequences for kin; many kinds of animals, however, live among conspecifics who are not closely related.

Crook (1964) recognized the fundamental influence of ecology in his dissertation comparing social systems among weaver birds. Mating systems were related to the spatio-temporal distribution of food and to defense against predators. Later, Emlen and Oring (1977) spelled this out in an influential and more general paper (see Slobodchikoff 1984 for a recent statement of this thesis). The basic theme is that the pattern of distribution of resources, in space and time, determines the pattern of spacing of animals. Mating systems are adapted to this particular pattern. Such mating systems can also be interpreted, insightfully, in terms of demography (Murray 1981), but the demography itself is a consequence of the ecology of the species.

It is not enough to know the distribution of resources. One has to take into account the fundamental differences between the sexes. Williams (1966) elaborated the situation using the concept of anisogamy, the unequal investment in gametes by males and females. Females produce few large gametes during their lifetime and often have limited mobility. The reproductive success of females is not limited by fertilizations so females need not compete among themselves for males, unless paternal care is involved. Additionally, females ought to be selective, choosing the "best" males. "Best" could mean genetic superiority (Trivers 1972, but see Boake 1986), or some trait that has a direct effect on the reproductive success of the female, such as defending a territory, behavioral complementarity, or quality of paternal care (Halliday 1983).

In contrast, males produce vast quantities of small gametes that are readily renewed, and they are usually more mobile than the females of their species. Their reproductive success turns on the number of gametes they fertilize. Consequently, males are predicted to compete aggressively for females and to monopolize them when that is possible. Males should not be choosy but should instead focus on minimizing the time between fertilizations.

Trivers (1972) has extended the anisogamy argument to a comparison of investment between the sexes. The one that invests the most in reproduction, typically the female, follows the syndrome just described. But this is a more general statement, taking into account further investment in offspring such as gestation and lactation. Usually it is relatively easy, and intuitive, to establish which sex invests the most. However, it is difficult if not impossible to use this as a general construct because there is no common quantitative currency for investment that applies to both sexes, and that extends across species, the more so when risk of predation is included in the equation.

Emlen and Oring (1977) have generalized further to the concept of the limiting sex. Whichever sex limits the reproduction of the other shows the female syndrome, although they did not develop in detail how this might apply outside the context of investment. Baylis (1978, 1981) did, however, making explicit that the limiting sex is the one that creates a bottleneck in time: ". . . the fundamental difference between the sexes that leads to exclusive maternal or paternal care is the relative rate at which the two sexes can create gametic copies (spermatogenesis requires less time than oogenesis)" (Baylis 1978:738).

Thus the sex that is most quickly able, or free, to produce another set of gametes shows the male syndrome, and that need not be the male. This is illustrated with a pleasing twist in the unpublished research of Terrance Lim on the bay pipefish, *Syngnathus leptorhynchus*, in California.

Bay pipefish have a long breeding season and reproduce several times in one season. A female inserts her ovipositor into the brood pouch of the male and deposits her eggs there. The eggs take about *three* weeks to hatch and leave the brood pouch. The female, meanwhile, needs only *two* weeks to produce another clutch of eggs. Because the male's pouch is filled by the eggs of one female, about two-thirds of the females fill all the males with eggs, assuming a sex ratio of unity (actually, there are often fewer adult males than females). The females compete for males and the males are coy. Thus the sex roles have been reversed even though the female produces relatively few, large, hence expensive, eggs.

The important point from the foregoing is that there is usually an asymmetry in the interests of males and females, whichever theoretical basis one chooses, that has to be considered when making predictions about mating systems in relation to resources. In most instances males will show the standard "masculine syndrome" (Williams 1966) and compete for females. Females will be less aggressive and tolerate more grouping unless they are competing for a limiting resource. The general situation, then, is that males are either territorial or show strong dominance relationships when competing for females.

At the one extreme resources, especially foods, are of high quality and are clumped in space or time. Then females are predicted to aggregate around the food, and to breed more or less in phase. That sets the stage for monopolization by males. If the aggregation of females is economically defensible (Brown and Orians 1970) there exists high environmental potential for polygyny (Emlen and Oring 1977).

At the other extreme, food is of low quality and is relatively uniformly distributed in space and time. Females are dispersed and basically solitary. Given other conditions (see below) a male might maximize his reproductive success by searching out a female, remaining with her, and driving other males away. The outcome is monogamy.

Other aspects of the environment will also affect the structure of the mating system. If the animals feed in an open environment with relatively uniform dispersion of food they may congregate to minimize predation (Jarman 1974); that would increase the chances of a male monopolizing a group of females and forming a harem society. Likewise, limited and concentrated breeding sites (Crook 1964) could produce polygyny if the females are breeding in phase.

As will be pointed out later, this is a basic starting point. Other features of the environment and of the biology of the species in question must be considered to understand the evolution of its mating system. Among these are the quality of the resources, migratory behavior and dispersal, phyletic "burdens of history," and general constraints such as absolute body size. Only then can one understand the diverse pathways that have led to the evolution of monogamy.

It is both surprising and understandable that monogamy has received so little attention as a mating system. There is but one recent review, an excellent essay by Wittenberger and Tilson (1980). Most of the literature pursues the issue of how males increase their reproductive success through polygyny. Monogamy is then regarded as polygyny failed--a male can get but one female (facultative monogamy of Kleiman 1977). Yet the basic sex ratio is unity in most species, and monogamy prevails in birds (Kendeigh 1952; van Rhijn 1984).

This lack of attention may be due, in part, to having defined monogamy as a system based on biparental care of the offspring (e.g., Wilson 1975). Then the interest is transferred to parental caretaking and deflected from the mating system itself. This is in spite of the fact that there are numerous species that are monogamous and do not care for their offspring (Wickler and Seibt 1981, 1983; Barlow 1984, 1986). Clearly, a more comprehensive definition of monogamy is needed, and more general ones exist in the literature (e.g., Wittenberger and Tilson 1980). Nonetheless, my own definitions of monogamy will be provided, and each of the two definitions is operationally defined.

Each definition reflects two basic features, the continuing association between a male and a female, and the genetic consequences of their reproductive behavior:

1. Exclusive: A pair is monogamous if the male and female confine most of their spawnings (copulations) to the same partner.

2. Biparental: A pair is monogamous if they remain partners after fertilization until the young no longer require their care.

These definitions are not meant to be absolutes. Monogamy lies on a continuum from polygyny to monogamy to polyandry, or even promiscuity (Wickler and Seibt 1981). However, polygyny remains a tolerated concept even when some males cannot obtain more than one female. In the same spirit, exclusive monogamy allows for occasional extra-pair fertilizations, in either direction, though it excludes polygynous situations in which each of a number of females mates repeatedly with the same male. Likewise, biparental monogamy admits to occasional secondary mates. Furthermore, it is unreasonable to demand that monogamy be for life for a given pair.

It has been common practice in the field to classify as monogamous a male and female that remain together or move about as a well-coordinated pair when nothing is known of their reproductive behavior. This sociographic definition (Wickler and Seibt 1981) is useful but tentative. Also, animals kept in captivity sometimes appear monogamous because of the circumstances but are not monogamous when observed in the field (e.g., Sussman and Kinzey 1984). Finally, it is possible for a male and female to mate monogamously but to spend almost no time together (Gronell 1984).

One of the problems in dealing with the evolution of monogamy is overcoming animal-group provincialism. It is understandable that authors take as their starting point the basic biology of the group with which they work. But if the group is reproductively conservative, then the obvious constraints become givens. Thus birds are perceived as basically monogamous and the interesting question is seen as how males become "emancipated" from parental care to permit polygyny (e.g., Beehler 1985; Mock 1985). Likewise, surveys of mammals start with the fact that females gestate and lactate and that predisposes mammals to polygyny (e.g., Kleiman 1977).

Another problem is that perceived benefits of monogamy, such as biparental care, are commonly regarded as having produced monogamy. Such benefits, and also costs, are important in maintaining and possibly reinforcing monogamy. But they can be misleading when seeking the critical selective forces that led to monogamy in the first place.

In the following, mammals and fishes will be compared, seeking evidence for the origin of monogamy in both groups. When possible, this will be related to the distribution of resources. Mammals and fishes have been chosen because they offer an intriguing reciprocal contrast. Both are mainly polygynous, but when caretaking of young is involved it is done primarily by the female among mammals (Kleiman 1977), and by the male among fishes (Baylis 1981). Further, fishes are by far the most diverse group of vertebrates and are monogamous in entirely different ways depending on the major environment: tropical freshwater species are all biparental and coral-reef species are all exclusive, save one (Barlow 1984, 1986).

II.   MONOGAMY AMONG SELECT MAMMALIAN GROUPS

This is not a review. Rather, the focus is on three groups of mammals--the antelopes, microtine rodents, and primates--with no pretension of complete coverage. Monogamy occurs among other mammals, notably canids, mustelids, mole shrews, beavers and muskrats, and other rodents (see Kleiman 1977), but there was too little time between the invitation to write this paper and the deadline to give them the attention they deserve. It is likely, however, that monogamy can be related to resources in each of those groups. For instance, canids that cooperatively prey on mammals much larger than themselves, and therefore become aggregated around a rich patch of briefly available food, are monogamous.

Relatively few species of mammals, however, display monogamy. Kleiman (1977) gives a figure of 3 percent and Dunbar (1984) a figure of

5 percent. Internal fertilization, gestation, and lactation by the female almost assure parental care by the female alone.

A. Antelopes

Ungulates challenge the notion that monogamy results from biparental care because the possibility that the male can effectively provision the young is virtually nil. Consequently it was of great interest when Jarman (1974) published his detailed analysis of the social organization of antelopes in relation to their ecology and established that one group shows frequent monogamy. Why they do so stems from the biology of their feeding.

They are browsers who are exceedingly selective, ingesting only high-quality parts of the vegetation. That calls for considerable moving from plant to plant, feeding from a variety of plants in a fine-grain environment. Since they remove all of the quality items as they move on, any group member following the leader would find little to eat, in contrast to grass feeders who remove only parts of their food items (Jarman 1974).

The consequence of such feeding behavior is that these species are either solitary or monogamous and live in small territories. When observed intensively, the male klipspringer was found to stay in close attendance of its mate, even though estrus occurs briefly only once or twice each year (Dunbar 1984). However, when a rich patch of food appears, antelopes such as the dikdik may form small groups (Jarman 1974). They rely on stealth and crypticity to evade predators; group living would be incompatible with this tactic.

These are among the smallest of antelopes. And their mouths are relatively even smaller than those of larger antelopes, an adaptation to precise selective browsing. Sexual dimorphism is slight, and females in some species are even a little larger than the males; however, only males have horns.

The food habits of these antelopes, therefore, have resulted in selection for small size, spacing and territoriality, secretive behavior, and solitary living or sequestering of a single female by a male. One can only presume that it is either economically not feasible for a male's territory to encompass two females, or that the risks of predation are too great to do so. At first glance they do not appear to fit the general scheme of monogamous species feeding on dispersed food of low quality. However, because they are so selective and have to move between small portions of preferred plant parts, the effect is to require much time to eat.

B. Microtine Rodents

There are a number of reasons one might expect rodents to engage in monogamy. They commonly live in a burrow which the female may or may not share with a male for some period of time. While the female is away, the male could assist by huddling over the pups and thereby providing warmth, by grooming the pups, by retrieving displaced pups, by interacting with them socially, and by defending against small predators (Waring and Perper 1980; Dewsbury 1985). While unlikely because of the low value of their plant food, and the male's morphology, the father could conceivably provision the pups to some degree after weaning if they stayed with their parents.

A number of species of rodents are suspected of being monogamous (Dewsbury 1980) and in some cases the evidence is strong (Foltz 1981). Nonetheless, it came as something of a surprise when it was reported that the prairie vole *Microtus ochrogaster* is monogamous (Getz 1978; Getz and Carter 1980; Getz et al. 1981; Thomas and Birney 1979). This is probably because voles are regularly regarded as classically *r* selected animals (Ostfield 1985). One needs to know, then, how the prairie vole differs from the other voles that are polygynous or promiscuous.

Microtus are among the smallest mammals to rely on green vegetation for their food (Eisenberg 1981). Vegetative parts of plants tend to be low in energy and difficult to digest because of the fibrous material they contain. Further, the small size of voles makes it impossible for them to process large amounts of ingested food at one time. Consequently, they must acquire food efficiently (Ostfield 1985).

Their reproductive biology places high demands on the metabolism of females, a further pressure to feed efficiently. The energetic requirement of pregnant and lactating females is 30 to 130 percent greater than that of nonreproducing females. They are polyestrus with postpartum estrus. The mean litter size is about six, and each female produces three to five litters within a season at intervals of about 21 days (references in Ostfield 1985).

Spacing among females appears to reflect the distribution and richness of food, although the data to test this are incomplete. Ostfield (1985) emphasized that females compete for food. He predicted that females will be territorial when food is sparse, patchy, and slowly renewable and when population density is low to moderate (population density fluctuates widely in voles).

Males, in contrast, do not bear the high costs of gestation and lactation. They increase their reproductive success by seeking out estrus females. "Copulations occur as discrete and widely scattered events in time and space, and thus require different strategies of acquisition from more or less continuously available resources, such as food and cover" (Ostfield 1985:8). And, since reproduction is less costly for males, they can invest more in territorial defense.

The social systems of voles, then, can be understood first in the way food determines the spacing of females. At one extreme, the taiga vole *M. xanthognathus* resides in an environment that is patchy both in time and space. Females occur close together, and a single male is able to sequester more than one mate. In the winter, communal burrows are formed with cached food (Wolff 1980).

The meadow vole *M. pennsylvanicus* has an intermediate social system. It eats forbs, though, unfortunately, little has been done to relate the distribution and richness of its food to its social system. The females are territorial and the males are not. At the onset of the breeding season the males increase their home range three fold as they search out and fight over estrus females (Madison 1980).

The feeding biology of the monogamous prairie vole is suggestive but not reported in sufficient detail. The original habitat was the expansive tall and mid-grass prairies. The prairie vole feeds on grass, but requires considerable succulent portions of forbs. The forbs are patchily distributed and relatively limited, and most are not succulent. Thus the forbs present a dispersed, low-quality diet. The lack of abundant forbs is thought to limit the population density. As a result, the females are territorial and sufficiently spaced that it pays a male to map his territory onto that of the female and remain with her. That he renders some degree of parental care to the pups (Getz and Carter 1980) probably evolved secondarily to monopolizing the female.

As a further indication of the poor diet of the prairie vole, the offspring may remain with the parents in the presence of another litter. Then, even though the female offspring become old enough to reproduce, they do not do so until departing the natal burrow. Getz and Carter (1980) view this as an adaptation to avoid overtaking the energetic resources of the territory. Offspring of polygynous voles do not have such an inhibition.

The prairie vole stands out among the voles through its monogamy, a relationship that persists even when not breeding (Getz et al. 1981). The dispersed food of low quality probably produced this social system in which the territories of the males and females coincide.

C. Primates

If one were to seek monogamy among mammals one might look first among the primates because male parental care presents no obstacles. The male is well equipped to assist in rearing the young because of well-developed hands and the ability to carry the young, share food, play, and protect it. Indeed, about half of the nonhuman primates in the New World have male parental care (Wright 1984). And 12 percent of hominoid primates are monogamous (Kinzey 1986). In spite of that, monogamy is relatively rare, occurring sporadically in tarsiers and lemurs, in small New World monkeys (particularly the Pitheciinae), in three species of Old World monkeys (Cercopithecidae), and among the lesser apes (MacKinnon and MacKinnon 1984; Kinzey 1986).

The tiny Callitrichinae of South America, consisting of the marmosets (seven species of *Callithrix* plus *Cebula pygmaea*) and the tamarins (11 species of *Sanguinus* plus *Leontopithecus rosalia*), are noted for paternal care of offspring. They are also regularly cited as examples of monogamous primates. That they are monogamous is apparently incorrect.

Sussman and Kinzey (1984) claimed that the conclusion of monogamy is based on captive studies of small groups of marmosets and tamarins. In nature, groups consist of 6-7 individuals and only one female can breed. She has twins, and these are carried not only by the putative father but by other males and juveniles. Individuals change groups fairly commonly, particularly younger members.   Sussman and Kinzey concluded that the social system in this subfamily finds its parallel in the avian cooperative polyandry of acorn woodpeckers (*Melanerpes formicivordes*).

It is important to establish that the callatrichins are not monogamous, and some may yet prove to be.   An examination of how they resemble and differ from the small monogamous monkeys of South America can help pinpoint the plausible ecological correlates of monogamy.   The marmosets eat predominantly insects plus exudates of trees, and the tamarins eat mainly insects plus fruit.   Some tamarins are territorial and others are not, and some are territorial in only part of their distribution.   So far, marmosets are not known to be territorial (Sussman and Kinzey 1984).

Wright (1984) studied two biparentally monogamous monkeys in South America, the night monkey *Aotus trivirgatus* and the dusky titi monkey *Callicebus moloch*.   Both are arboreal and territorial, feeding on insects, leaves, fruits and flowers.   They favor small crowned trees that are dispersed and ripen a few fruit at a time.   While no data were given, Wright suggested that the energy costs of ranging widely for dispersed crops were too great to support more than one female with young in a male's territory.   Further, she thinks the territories are too uniform to produce a local concentration of small female territories that one male could monopolize.

Wright (1984) considered that small size *per se* could necessitate biparental care and hence produce monogamy.   The smaller the female the higher the ratio of infant weight to maternal weight.   At some critical level, the energetic drain of lactating and carrying the infant might become too great.   The females in both species weigh less than one kilogram.

As Wright pointed out, however, the female of another South American primate, the squirrel monkey *Saimiri*, is even smaller and the litter is relatively even heavier, yet the female is the sole caretaker of its infant.   *Saimiri* females may be able to forgo male assistance because they feed in tree crowns where food is abundant.   This type of feeding has also produced aggregating behavior among females, facilitating polygyny.   Subadult females sometimes help in carrying the young.

The monogamous lesser apes offer a better situation for relating monogamy to resources than do the New World monkeys.   There is no direct paternal care (Brockelman and Srikosamatara 1984), so that cannot be the cause of monogamy.   And the nine species of gibbons and the siamang (Hylobatidae) may be the only consistently monogamous hominoids in the narrow sense of Kinzey (1986).

Gibbons are characteristic of moist, stable forests that have persisted for long periods of time (Chivers 1984; MacKinnon and MacKinnon 1984).   They feed on juicy pulp of fruit that is small and scattered among trees that are too small for large groups of monkeys.   They also eat leaves.   To harvest this food they travel about 1,500 km daily within their territory that is modally around 32 ha (Chivers 1984).   Without precise data, it seems that their food resource can be classified as thinly dispersed and of low quality (MacKinnon and MacKinnon 1984). The female interbirth interval is potentially two years. In the field, however, births are spaced at intervals of three to four years. Consequently, estrus is a rare event. Brockleman and Srikosamatara (1984) suggested that the long birth interval reflects the low productivity of their habitat. They also felt that this explains why gibbons are so territorial over food that the male drives out intruding females.

Gibons, then, fit the standard picture for monogamy in relation to resources. Monogamy appears to have arisen in response to females being widely distributed because of a poor but stable source of food in a stable habitat. A male can monopolize but one female in an exclusive feeding territory they share.

III.   MONOGAMY AMONG FISHES

It is almost a handicap to use fishes to reflect on how resources affect mating systems, particularly with regard to monogamy. There are few data to call upon, though there are some exemplary analyses relating feeding tactics and risk of predation to polygynous mating systems, e.g., Hoffman's (1983) field studies on wrasses of the genus *Bodianus*.

The attempt here is to make comparisons at a more global level. Thus pervasive effects of the environment are sought that relate ultimately, nonetheless, to resources. Therefore, instead of treating fishes by taxa, they are considered by two major environments: the warm fresh waters and coral reefs. When dealing with the marine fishes, some examples of monogamy will be described that relate directly to the distribution of resources.

A.   Freshwater Fishes

Freshwater environments are mainly riverine (Barlow 1981). Flowing waters are inherently unstable. Even in the tropics there are commonly pronounced wet and dry seasons. The rivers and streams fluctuate enormously in response to the pattern of rainfall: the levels rise and fall, paths of flow change, and the input of nutrients fluctuates (Lowe-McConnell 1975). The consequences to fishes, in terms of spatial patterning of food resources, is poorly understood.

An additional and important feature of freshwater habitats is that they tend to be continuous. Thus a fish of any size can often disperse throughout the water system because of uninterrupted association with an appropriate habitat situation.

When monogamy among fishes was reviewed (Barlow 1984, 1986), however, one unexpected generalization emerged. Feeding territories in the field are virtually unreported, especially when one excludes lake fishes. It is not clear why this should be. Fishes taken from those regions commonly set up feeding territories when confined to aquaria, so they have the behavioral mechanisms necessary for feeding territories, i.e., site tenacity and aggressive exclusion of conspecifics and others. The lack of reports could reflect nothing more profound than the absence of observers in those regions, and that is bound to be at least part of the answer. However, my own limited experiences led me to suspect that it has to do as well with the instability of the riverine habitat. Food could fluctuate enormously with the seasons, and within the rainy season from rainfall to rainfall. Thus fishes could aggregate around rich patches of food, deplete them, and then move on. If that is so, why then are any riverine fishes monogamous? Several are.

Six families are known to display monogamy. These are the bony-tongues (Osteoglossidae), the North American catfishes (Ictaluridae), the Asian catfishes (Heteropneustidae), the African catfishes (Bagridae), the snakeheads (Channidae), and the cichlids (Cichlidae). The Ictaluridae of the temperate zones is included here because the species known to show monogamy is a warm-water fish.

These families share a number of adaptations. The most remarkable of these is that all are biparental and that caretaking extends to protecting free-swimming young, not just to eggs. In addition, all possess breeding territories, and the eggs are deposited on the substrate. They tend to be seasonal breeders. Generally they are large fish. In most of these families, there are but a few species that are monogamous, except for the cichlidae.

Cichlids are truly exceptional in that monogamy is ancestral. The Cichlidae is a large family with around 1,000 species. A number of lines have radiated off into mouth brooding of eggs, and sometimes young, in both the Old and New World tropics. Mouth brooding usually, but not always, results in polygyny, typically lekking (Loiselle and Barlow 1978). But the predominant theme among cichlids is biparental care and monogamy.

Biparental care needs as a precursor male care of eggs, which is relatively common among fishes. Baylis (1978, 1981) has argued that paternal care prevails because of the scarcity of suitable breeding sites. Most fishes fertilize their eggs externally. Then, "males are favoured as site-constant individuals because . . . males are able to produce new batches of fertile gametes and remate at shorter intervals than females. A male monopolizing a site will leave more descendants than a female occupying that site for the same duration" (Baylis 1978:738). A female free of parental care is able to continue feeding and produce more gametes. Thus she increases her reproductive success and that of the male as well (Loiselle and Barlow 1978). At the level of male care, therefore, limiting resources may be an important factor. Extending this to the evolution of monogamy is not as straightforward.

The key feature seems to be time-constrained fecundity of females (Barlow 1984, 1986). Given that the male cares for the eggs, the important evolutionary decision for the female is whether she can increase her reproductive success by leaving the male and producing another clutch of eggs, or by remaining to help defend the young. If the season is too short to permit producing another clutch, she should assist.

Large size might affect the decision because larger fish have lower metabolism per unit weight (e.g., Barlow 1961); thus, large females might take longer to produce a new clutch and, if so, ought to be more inclined to stay and protect their offspring.

Another important ingredient is the ability to disperse at any age because of the continuity of the environment. Freshwater fishes can disperse at any stage in their life history because dispersal does not require a specialized propagule restricted to the initial phase of life, i.e., planktonic eggs and larvae. We are left with a less-than-satisfactory effort to relate monogamy among freshwater fishes to the spatio-temporal distribution of resources, except in an indirect fashion. If Baylis is correct, breeding sites are a limiting resource that promotes paternal care of eggs. Given that, the temporal pattern of resources may play an important role in restricting the female to breeding but once per season--that facilitates biparental care and thus monogamy.

B.   Coral-reef Fishes

Monogamy among coral-reef fishes presents a completely different picture. Only one species is known to show biparental care, the plankton-feeding damsel *Acanthochromis polyacanthus* (Robertson 1973). All the other examples are of exclusive monogamy without biparental care.

Parental care of young is selected against because coral-reef fishes live in a hierarchically structured patchy environment (Barlow 1981). They are so closely adapted to their reef habitats that they are ill-equipped to move from one patch to another. For a small species, a patch can be a coral head and for a large species an atoll or major segment of a barrier reef. Care of young requires young capable of maneuvering about and staying within the parents' territory. That calls for well-developed young, and that would be maladaptive for planktonic dispersal. The result is that most coral-reef fishes spawn abundant planktonic eggs at short intervals. In many of the smaller species the eggs are demersal; they develop into planktonic larvae.

Thirteen families of reef fishes are counted in which there is reasonable evidence of monogamy (Barlow 1984): ghost pipefishes (Solenostomidae), pipefishes (Syngnathidae), sea basses (Serranidae), tilefishes (Branchiostegidae), butterfly fishes (Chaetodontidae), angelfishes (Pomacanthidae), damsels (Pomacentridae), wrasses (Labridae), gobies (Gobiidae), surgeonfishes (Acanthuridae), triggerfishes (Balistidae), filefishes (Monacanthidae) and sharpnose puffers (Canthigasteridae). The evidence is weak for some of these families (Serranidae, Branchiostegidae, and Labridae). But it is likely that other families will be added as more is known (possibly the Blenniidae, Eleotridae, Caracanthidae, Cirrhitidae). Pseudochromids are sometimes mentioned as monogamous species, but the evidence (Lubbock 1975) is not persuasive.

Monogamous coral-reef fishes differ from their freshwater counterparts in several fundamental ways (Barlow 1984, 1986). Many of them are strongly territorial over food or over a place of refuge. They spawn repetitively over long periods of time, often daily. Spawning may occur within a territory, as in the damsels and gobies, or the fish may ascend, out of their territory, as in the surgeonfishes and butterfly fishes. They tend to be relatively small species, though the triggerfishes are large. With the one exception mentioned, there is no biparental care of free-swimming young, though male care of eggs occurs in some families.

It is possible, in some instances, to relate this type of exclusive monogamy to the distribution of resources. The most obvious examples involve a critical resource other than food (Barlow 1984, 1986). Several species in different families feed on plankton passing over their place of refuge, which may be a burrow, a crevice, or an anemone. These refuges may be used by an individual; such species (e.g., damsels of the genus *Chromis* and *Dascyllus*) are usually polygynous. Often, however, the refuge is shared by a pair, as among gobies, and they are monogamous. Why this difference exists is not known.

The case of the anemone fishes (*Amphiprion*) is easier to understand in that one anemone supports but a pair and a few associated juveniles, and the anemones are usually spaced. The male is smaller than the dominant female in this protandrous genus. He provides most of the care of the eggs while the female produces another clutch of eggs. When a non-breeding season exists, as at the northernmost limit of their distribution, the larger female of *A. clarkii* tries to evict her mate when they are not breeding (Moyer 1980).

At the other extreme are species that are thinly distributed across the reef and roam widely. Monogamous pairs of such species are found most notably among the butterfly fishes but also among the sharpnose puffers, e.g., *Canthigster benneti* in Tahiti (Barlow unpublished). Differences in food habits producing this thin distribution of pairs have not been studied.

At least one case of monogamy has been related to the spatial distribution of food. Robertson et al. (1979) compared the social systems of three species of surgeonfishes (*Acanthurus lineatus*, *A. leucosternon*, *Zebrasoma scopas*) in relation to feeding territoriality. All graze on microalgae though *lineatus* also takes larger filamentous and leafy algae. *A. lineatus* is the largest species, is solitary, and has the smallest territory (7 m^2) in the most productive habitat, the shallowest areas on the reef front; the algae grows as a continuous dense carpet. *A. leucosternon* is intermediate in body size, defends a territory of intermediate size (17 m^2), and forms a monogamous pair; the algae occurs in discontinuous patches of richer, red algae. *Z. scopas* is the smallest species and defends the largest territory (37 m^2) as, typically, trios of one male and two females; the algae in its territory is similar to that of *leucosternon* except it is sparser and not in distinct patches.

There were 14 percent more females than males of the monogamous species, *leucosternon*, so some females held solitary territories. These females had no perceptible fat bodies in the mesenteries of their guts. In contrast, the paired females had visible, often large, fat deposits suggesting higher fecundity as a direct benefit of monogamy.

Paired males of *leucosternon* sometimes consorted with neighboring solitary females. And one male defended the territories of two females even though their territories were separated by that of another pair. When Robertson et al. (1979) removed the females of some pairs, the unmated females fought over the bachelor males. But when males were removed from pairs, the females remained unmated for one month.

Given just the information about the territoriality and feeding of these surgeonfishes, one would have predicted that *leucostemon* would be polygynous, especially in light of the excess females. Its food is apparently patchily distributed and the population density is high, but not too high to make male defense of territory too costly.

On the other hand, *scopas* would have been expected to be monogamous. The algal resource appears to be thinly but uniformly distributed within its defensible territory, resulting in a large territory, especially relative to its smaller size. Clearly, one needs to understand their ecology in more detail.

Hourigan (ms.) has explored monogamy experimentally in a Hawaiian butterfly fish, *Chaetodon multicinctus*. In keeping with the results of Robertson et al. (1979) he hypothesized that a female benefits from the territorial defense provided by a male. That allows her more time for feeding and she can thereby increase her fecundity. The male benefits from an assured mate who is also more fecund (see also Barlow 1984).

Females of pairs of *C. multicinctus* did in fact feed more than did the males, and the males spent more time and energy defending the territory. When the male of a pair was removed, the territory shrank, and she defended much more and fed half as much; some females lost their territories. Removing females had no such effect; the males successfully maintained their territories at little additional cost.

The last example is the longnose filefish *Oxymonacanthus longirostris*. It is exclusively monogamous and territorial, spawning as frequently as each day (Barlow in press). The longnose filefish feeds almost entirely on polyps of corals of the genus *Acropora*. In the backreef area of Guam, ramose *Acropora aspera* occur in extensive beds that are broken up by small areas of sand and dead coral. Coral is low in calories and also deficient in nutrients. Occasionally the filefish break off from feeding on coral and glean from algae and dead coral.

It is an ideal situation to examine the role of food as a resource for a number of reasons. Only one species of coral is involved, and the abundance of the coral can be manipulated. The fish are diurnal, small, slow moving, restricted to a local area, and not disturbed by the close approach of an observer. They are also easily captured and individually marked. We (Linda Barlow and I) have recently initiated an analysis of their monogamous behavior.

The fish move about as closely coordinated pairs. A large portion of their time is spent feeding, and the female feeds only slightly more often than the male. As the day progresses the female becomes obviously swollen with eggs, and the rate of feeding slows. Both sexes defend against conspecific intruders, and each attacks especially like-sex intruders. Large bachelor males sometimes challenge the male of a pair. The bachelors drift between territories and shadow pairs. Small females sometimes approach the pairs and are usually driven away by the female.

When the male of several pairs was removed one at a time, a new large male appeared within 24 hours and became the female's mate. When the original male was then restored a fight, often prolonged, ensued; the original male usually won. When the female of several pairs was removed none of their males obtained a new mate within 24 hours. In one instance a small female persistently followed the 'widow' male but he showed no interest in her.

We concluded from this that there is a shortage of large females. Large males sequester females, even driving them into hiding while defending them against the advances of other large, unmated males. This is despite our preliminary data indicating a sex ratio of unity in the population when fish of all sizes were counted. We hypothesize that the growth of females is slowed by the energetic drain of producing a hefty clutch of eggs daily on a calorically poor diet. That results in a shortage of big females. We found no indication of differential mortality between the sexes, but that cannot be ruled out. Fecundity is a function of size, so the reproductive success of males is a direct function of the size of the one female he can monopolize.

For her part, the female benefits from the territorial defense provided by the male. To keep him in attendance, she prolongs prespawning bheavior and does not actually spawn until shortly before sunset. Then it would be difficult for the male to locate another female who has not herself spawned, and persuade her to spawn with him (see also Fischer 1980).

If we are correct, monogamy in this filefish can be viewed as a consequence of their poor diet. Large females become a critical limiting resource. Despite the skew in the operational sex ratio toward an excess of males, polyandry does not occur because of the intense aggression between males.

Although the examples among coral-reef fishes are few, they are consistent with the precept that the nature of limiting resources molds the mating systems of the fishes. In many respects they present more favorable material for testing hypotheses about the causation of mating systems than do terrestrial vertebrates.

IV. DISCUSSION

A. Falsifying Hypotheses

The hypothesis, broadly stated, that mating systems are an outcome of the distribution of resources has a seductive charm about it, so much so that we tend to fit findings into its framework rather than attempt to falsify it. To a large degree, I have been guilty of this sin in this essay. But we cannot be comfortable believing that the explanation of mating systems requires knowing no more than the distribution of females and the competitive interactions of males. However, it is all too tempting to assume that when a species does not fit easily into this construct the problem is that we need more data. Or, conversely, when a species appears to fit, we become tolerant of inadequate data about the distribution of resources.

Different workers will be able to provide examples of poor fits from the kinds of animals they know; two are mentioned here. One is the surgeonfishes studied by Robertson et al. (1979), and described in the foregoing. I suspect that a more detailed analysis of their feeding biology would resolve the issue.

Another example from coral-reef fishes presents the opposite situation: two closely related species that feed on the same food but have different mating systems. The coralivorous butterfly fishes *Chaetodon triangulum* and *Megaprotodon* (formerly *Chaetodon*) *strigangulus* hold feeding territories over plate-like heads of corals of the genus *Acropora*. *C. triangulum* does so as monogamous pairs, but *M. strigangulus* does so solitarily (Reese 1977). There has to be a reasonable explanation for the difference. The equation for the input and output of energy must balance, whatever the complexities of the situation.

B. Constraints

Many other factors overlie the energetic substrate of mating systems. In some instances they can produce counter intuitive evolutionary outcomes. One class of constraints might be called historical burdens, characters that pervade large taxa (Eibl-Eibesfeldt and Wickler 1962). Mode of reproduction is an obvious example. All mammals gestate and lactate, and all birds must leave their eggs, and usually their young, at a fixed site. This difference explains why a bird and a monkey may have essentially the same diet but manifest different mating systems, as the following illustrates.

Beehler (1984) compared the mating systems of three species of birds of paradise (Paradisaeidae) in Papua, New Guinea. Two species are polygynous and one, the trumpet manucode (*Manucodia keraudrenii*) is monogamous. The nesting behavior is essentially the same in all three species. The polygynous species feed on the fruit of trapline plants that produce a crop comparatively rich in proteins and lipids, concentrated in a season; they also take an appreciable number of insects. Because food is relatively compressed in space and time, the females breed in phase and each can rear her young alone.

The manucode feeds on plants termed opportunists, especially figs. These plants produce bursts of fruit that are rich in carbohydrates and water, but deficient in proteins and lipids. Insects are typically a source of proteins for birds that are predominantly frugivores, but they make up no more than 1 percent of the manucode's diet. Fig trees are relatively rare; about four trees bearing fruit can be found in the home range of the nonterritorial manucode at a given time. Hence the manucode must fly considerable distances to feed on figs, and to provision its nestlings with the fruit of the fig trees. Some figs can be found bearing fruit throughout the year. And each tree produces a large crop. Consequently, a fruiting fig tree attracts numerous unspecialized foragers who arrive mostly in large flocks, devour the crop, and depart.

One would predict a polygamous mating system based on the demographics of the fig trees. And that is the case in the squirrel monkey *Saimiri* that feeds on figs in South America. However, more may be involved because the females breed in synchrony (Wright 1984), suggesting a seasonal burst of food. Of more importance, however, is the surprising monogamy of the manucode on such a diet. Beehler explains the monogamy with the low nutritive value of the figs. Because the figs are so widely dispersed, it takes two parents to provision the nestlings on such a poor diet. The necessity of laying eggs and rearing helpless young in a fixed nest imposes monogamy on the manucode when it feeds on dispersed low-quality food even though the food produces aggregating behavior. The squirrel monkey, on the other hand, takes its fetus or infant with it and thus can be continuously aggregated. In this comparison, the bird is constrained by its mode of reproduction.

Another critical constraint may prove to be body size (e.g., Wright 1984). With decreasing size, weight-specific metabolism increases exponentially and reproductive effort rises; life span decreases (May and Rubenstein 1985). Large female fishes may not be able to produce a second clutch in a breeding season, and small female fishes have difficulty producing enough eggs for planktonic propagules because egg size stays relatively constant. In the New World cichlid fishes there is a tendency for small species to become polygynous though the reasons for doing so remain obscure. Small mammals are driven to divide up their litter into a number of pups; however, small primates continue to give birth to but one or two infants, which may relate to the difficulty of managing so many offspring in an arboreal habitat. But size itself is an adaptation, even if it brings with it constraints.

One can imagine a group of closely related species, say surgeonfishes, feeding on microalgae. Selection through competition could favor diversification into species of different sizes, all still feeding on the same diet, but now getting it from slightly different places on the same part of the reef. Size of the animals alone could result in different mating systems if care of eggs became essential to the small species for planktonic dispersal (see Barlow 1981). Similarly, the largest and hence most mobile species would be better able to move long distances; that could result in shifting the diet to rich patches of leafy algae and the mating system might then shift to a haremic system (e.g., Barlow 1974).

Many aspects of the environment will also shape the social system. These include the physical environment, most obvious in species that live in extreme environments such as the intertidal, torrential streams, or the desert. Other biological factors, particularly predation, have a strong effect as well, limiting the scope of behavior of their prey.

C.   Comparative Approach

The interaction of these factors will be clarified through concerted efforts to analyze the behavior and ecology of closely related groups. Such truly comparative studies will put us in a better position to evaluate the effect of the environment in harmony with phyletic constraints when comparing distantly related organisms. On the other hand, convergences between unrelated species can also be insightful when evaluated with due appreciation for the constraints on each species.

The work to date on voles, marmosets and tamarins is a step in the right direction. But more needs to be done. It would also be instructive to examine a group with a wider range of mating systems. The dwarf cichlids (*Apistogramma*) are a case in point. There are about 50 species in South America and all known so far care for free-swimming young. However, a few species are monogamous and biparental, some are facultatively biparental or polygynous with solely female care, and some are haremic with only female care (Barlow pers. observ.). In addition, due to their tiny size--females breed when they are around 25-30 mm long-- they are readily managed and bred in captivity. Thus their social systems are open to experimentation under semi-natural conditions.

D. Experimental Approach

There are still too few field studies providing us with convincing data about the relationship between mating systems and ecology. Nonetheless, more progress could be made through experimentation both in the field and in the laboratory. For example, Slobodchikoff (1984) manipulated food within the territory of Gunnisons' prairie dog (*Cynomys gunnisoni*) to test hypotheses about the effect of richness and distribution of food on the mating system. As predicted, low-quality, evenly distributed food pushed the system toward monogamy.

Other appropriate manipulations entail altering the composition and density of the population, litter or clutch size, and the pattern of spacing. Spacing can sometimes be conveniently managed when the animals use shelters such as nest boxes or beer bottles, happily accepted by burrow-dwelling fishes such as blennies.

This is an open and promising area for research. It has the prospect of obtaining surprising findings that force a reevaluation of theory.

E. Evolved Monogamy and Experimentation

Kleiman (1977) made the useful distinction between Type I or facultative monogamy and Type II or obligate monogamy (see also Wickler and Seibt 1981). Facultative monogamy refers to mating systems in which the male's strategy is polygyny, but he has only one mate. The term obligate monogamy was coined to indicate that, without aid from the male, the female cannot rear the offspring by herself. I prefer to use the term "evolved monogamy" in the sense that Williams (1966) wrote of evolved adaptations, instead of obligate monogamy. The usage is more general and encompasses monogamy even when no caretaking is involved.

It would be a mistake to view monogamy as falling into one type or another. Rather, exclusive and biparental monogamy are fuzzy sets. Monogamy can grade into polygamy, and to varying degrees in different species. This is nicely demonstrated in a cichlid fish in Panama in which monogamy prevails when the threat of predation on the young is high.

Many males desert when the threat of predation is lower (Townshend and Wootton 1985).

As Rasmussen (1981) pointed out, behavioral mechanisms supporting pair bonding will evolve when they increase the reproductive success of the animals involved; he included polygynous stable groups as well as monogamous animals.   There should be secondary reinforcement of monogamy when it is advantageous.   This will result in a number of adaptations ranging from the sexual act (Dewsbury 1980), frequency of sexual behavior (Kleiman 1977), mate choice (Burley 1981) and fidelity, and acquisition of specific paternal duties in the male (e.g., Wright 1984).

It is here that the experimental approach can be revealing in a way descriptive studies seldom can.   And experiments can be done in the field as well as in the laboratory.   For instance, the operational sex ratio can be altered by removing mates.   And fidelity can be tested by offering new mates.   Assessment of quality of mate ought to be reciprocal in monogamous species, so experiments on mate choice are in order.   Burley (1981) has articulated some provocative hypotheses about the assessment of self in relation to mate choice in monogamous species that need testing.

Understanding the behavioral mechanisms underlying evolved monogamy will clarify arguments about whether species are monogamous or not because the issue of degree of plasticity will arise in its place.   And that, in turn, will offer insight into the type of variation that evolution works on to produce mating systems.

F.   Conclusions

When I started enquiring into the causation of monogamy I was motivated by the belief that there might be something unique about it.   However, the body of theory that has developed to account for mating systems in general has proved remarkably powerful.   Monogamy, in the broad sense, fits well.   Nonetheless, a knowledge of the spatio-temporal pattern of resources, so often limited to food, is often insufficient alone to account for the form of the mating system.   As indicated, many other factors must be considered to provide an adequate explanation.   When that is done, the distillate is that monogamy is often driven by resources, but monogamy in different species has evolved in different ways and to meet different selective pressures (Barlow 1984; Kinzey 1986).

Finally, to those who for so long have held to the belief that monogamy evolved out of the necessity for biparental care, the following: In most cases the main driving force for monogamy has been the monopolizing of a female; that made possible the evolution of biparental care (see also Wickler and Seibt 1981, 1983; Barlow 1984).

V.   SUMMARY

The starting assumption was that the spatio-temporal pattern of animals, particularly females, is set by the spatio-temporal pattern of limiting resources. Because of the asymmetry of reproductive biology between males and females, males will try to monopolize females. When food resources are dispersed and of low quality, and territoriality is still economically feasible, one male will monopolize one female, resulting in monogamy.

Monogamy was operationally defined in two different ways. In *exclusive* monogamy the male and female spawn or copulate repeatedly, but with the same partner. In *biparental* monogamy the pair remains together until the young are independent.

The basis of monogamy was contrasted in fishes and select mammalian groups because of the interesting reciprocal contrast: the starting condition in mammals is female care of young whereas in fishes the male is the caretaker when caretaking occurs.

Antelopes have exclusive monogamy. These small, territorial animals feed selectively on dispersed but high-quality browse. Among microtine rodents, several are thought to be monogamous. The territorial prairie voles appear to be in transition between exclusive and biparental monogamy; they feed on low-quality dispersed forbs.

Primates show both types of monogamy. The night monkey and the dusky titi monkey have biparental care. They are territorial and feed on low-quality dispersed food as do the gibbons and siamangs. Gibbons and siamangs have exclusive territoriality, however.

Fishes were examined with regard to two major environments, warm fresh waters and coral reefs. Monogamous freshwater fishes all have biparental care of free-swimming young in a breeding territory. Feeding territories are unknown. Caretaking evolved from male care of eggs; that the male is the caretaker probably results from limited breeding sites plus the need to free the female to feed to produce more eggs. Female participation in caretaking evolved when the breeding season was too short for her to produce a second clutch. Because of continuity of the riverine habitat, dispersal can occur during any stage of development.

With but one exception, coral-reef fishes have exclusive monogamy. They are relatively aseasonal, spawning repeatedly with the same partner, often on a daily basis. In their discontinuous environment, dispersal is restricted to eggs and larvae, precluding parental care of young. Limiting resources, such as refuges, play a role in monogamy.

The distribution of algal food correlates with monogamy in a surgeonfish. But the example is counter to theory in that the territory of the monogamous species is intermediate in size and has patches of richer algae. The surgeonfish with the largest territory, having thinly and evenly distributed algae, is haremic.

The coralivorous longnose filefish appears to fit predictions because it is territorial over low-quality food. However, the prime agent seems to be a shortage of large fecund females over which an excess of large males compete.

The basic theory provides a reasonable starting point. But to understand monogamy, or any other mating system, one must consider other factors. Among these are constraints such as mode of reproduction, body size, predation, and features of the physical environment.   An understanding of these would be promoted by truly comparative studies with quantification of limiting resources, and a willingness to apply the experimental method.   Such studies would elucidate monogamy as an evolved adaptation.

There are several paths to monogamy, but it turns around monopolization of a female by a male, often facilitated by female tactics. That provides a situation from which biparental care of offspring can evolve.

ACKNOWLEDGMENTS

This article owes a great deal to Frank Pitelka who turned my attention to the relationship between resources and social behavior.   I want to thank him, additionally, for his comments on this article.

REFERENCES

Barlow, G.W. 1961. Intra- and interspecific differences in rate of oxygen consumption in gobiid fishes of the genus *Gillichthys*. *Biological Bulletin* 121:209-229.

Barlow, G.W. 1974. Contrasts in social behavior between Central American cichlid fishes and coral-reef surgeon fishes. *American Zoologist* 14:9-34.

Barlow, G.W. 1976. Review of *Sociobiology: The new synthesis*, by E.O. Wilson. Belknap Press of Harvard University, Cambridge, Massachusetts. *Animal Behaviour* 24:700-701.

Barlow, G.W. 1981. Patterns of parental investment, dispersal and size among coral-reef fishes. *Environmental Biology of Fishes* 6:65-85.

Barlow, G.W. 1984. Patterns of monogamy among teleost fishes. *Archiv für FischWissenschaft* 35(Beiheft 1):75-123.

Barlow, G.W. 1986. A comparison of monogamy among freshwater and coral-reef fishes. *Proceedings of the Second Indo-Pacifid Fish Conference*.

Barlow, G.W. (in press). Spawning, eggs and larvae of the longnose filefish *Oxymonacanthus longirostris*, a monogamous coralivore. *Environmental Biology of Fishes*.

Baylis, J.R. 1978. Paternal behaviour in fishes: A question of investment, timing or rate? *Nature* 276:738.

Baylis, J.R. 1981. The evolution of parental care in fishes, with reference to Darwin's rule of male sexual selection. *Environmental Biology of Fishes* 6:223-251.

Beehler, B. 1985. "Adaptive significance of monogamy in the trumpet manucode *Manucodia keraudrenii* (Aves:Paradisaeidae)." pp. 83-99. In: P.A. Gowaty and D.W. Mock, eds. *Avian monogamy*. American Ornithological Union, Ornithological Monograph 37. Lawrence, Kansas:Allen Press.

Boake, C.R.B. 1986. A method for testing adaptive hypotheses of mate choice. *American Naturalist* 127:654-666.

Brockleman, W.Y. and W. Srikosamatara. 1984. "Maintenance and evolution of social structure in gibbons." pp. 298-323. In: H. Preuschoft, D.J. Chivers, W.Y. Brockelman, and N. Creel, eds. *The lesser apes. Evolutionary and behavioural biology.* Edinburgh:Edinburgh University Press.

Brown, J.L. and G.H. Orians. 1970. Spacing patterns in mobile animals. *Annual Review of Ecology and Systematics* 1:239-262.

Burley, N. 1981. Mate choice by multiple criteria in a monogamous species. *American Naturalist* 117:515-528.

Chivers, D.J. 1984. "Feeding and ranging in gibbons: A summary." pp. 267-269. In: H. Preuschoft, D.J. Chivers, W.Y. Brockelman, and N. Creel, eds. *The lesser apes. Evolutionary and behavioural biology.* Edinburgh:Edinburgh University Press.

Crook, J.H. 1964. The evolution of social organisation and visual communication in the weaver birds (Ploceinae). *Behaviour Supplement* 10:1-178.

Dewsbury, D.A. 1980. An exercise in the prediction of monogamy in the field from laboratory data on 42 species of muroid rodents. *The Biologist* 63:138-162.

Dewsbury, D.A. 1985. Paternal behavior in rodents. *American Zoologist* 25:841-852.

Dunbar, R. 1984. The ecology of monogamy. *New Scientist* 103:12-15.

Eibl-Eibesfeldt, I. and W. Wickler. 1962. Ontogenese und Organisation von Verhaltensweisen. *Fortschritte der Zoologie* 15:354-377.

Eisenberg, J.F. 1981. *The mammalian radiation.* Chicago:University of Chicago Press.

Emlen, S.T. and L.W. Oring. 1977. Ecology, sexual selection, and the evolution of mating systems. *Science* 197:215-223.

Fischer, E.A. 1980. The relationship between mating system and simultaneous hermaphroditism in the coral reef fish *Hypoplectrus nigricans* (Serranidae). *Animal Behaviour* 28:620-623.

Foltz, D.W. 1981. Genetic evidence for long-term monogamy in a small rodent, *Peromyscus polionotus. American Naturalist* 117:665-675.

Getz, L.L. 1978. Speculation on social structure and population cycles of microtine rodents. *The Biologist* 60:134-147.

Getz, L.L. and C.S. Carter. 1980. Social organization in *Microtus ochrogaster. The Biologist* 62:56-69.

Getz, L.L., C.S. Carter and L. Gavish. 1981. The mating system of the prairie vole, *Microtus ochrogaster:* Field and laboratory evidence for pair-bonding. *Behavioral Ecology and Sociobiology* 8:189-194.

Gronell, A.M. 1984. Courtship, spawning and social organization of the pipefish, *Corythoichthys intestinalis* (Pisces: Syngnathidae) with notes on two congeneric species. *Zeitschrift für Tierpsychologie* 65:1-24.

Halliday, T.R. 1983. "The study of mate choice." pp. 3-32. In: P. Bateson, ed. *Mate choice.* New York:Cambridge University Press.

Hoffman, S.G. 1983. Sex-related foraging behavior in sequentially her-maphroditic hogfishes (*Bodianus* spp.). *Ecology* 64:798-808.

Hourigan, T.F. (ms.) Pair bonding and monogamy in the banded butterfly-fish (*Chaetodon multicinctus*).

Jarman, P.J. 1974. The social organisation of antelope in relation to their ecology. *Behaviour* 48:215-267.

Kleiman, D.G. 1977. Monogamy in mammals. *Quarterly Review of Biology* 52:39-69.

Kendeigh, S.S. 1952. Parental care and its evolution in birds. *Illinois Biological Monographs* 22:1-356.

Kinzey, W.G. 1986. "A primate model for human mating systems." In: W.G. Kinzey, ed. *Evolution of human behavior*. Ithaca:State University of New York Press.

Loiselle, P.V. and G.W. Barlow. 1978. "Do fishes lek like birds?" pp. 31-75. In: E.S. Reese and F.J. Lighter, eds. *Contrasts in behavior*. New York:Wiley.

Lowe-McConnell, R.H. 1975. *Fish communities in tropical freshwaters*. New York:Longman.

Lubbock, R. 1975. Fishes of the family Pseudochromidae (Perciformes) in the northwest Indian Ocean and Red Sea. *Journal of Zoology, London* 176:115-157.

MacKinnon, J.R. and R.S. MacKinnon. 1984. "Territoriality, monogamy and song in gibbons and tarsiers." pp. 291-197. In: H. Preuschoft, D.J. Chivers, W.Y. Brockelman and N. Creel, eds. *The lesser apes. Evolutionary and behavioural biology*. Edinburgh:Edinburgh University Press.

Madison, D.M. 1980. An integrated view of the social biology of *Microtus pennsylvanicus*. *The Biologist* 62:20-33.

May, R.M. and D.I. Rubenstein. 1985. "Reproductive strategies." pp. 1-23. In: C.R. Austin and R.V. Short, eds. *Reproduction in mammals. Book 4: Reproductive fitness*. 2nd ed. New York:Cambridge University Press.

Mock, D.W. 1985. "An introduction to the neglected mating system." pp. 1-10. In: P.A. Gowaty and D.W. Mock, eds. *Avian monogamy*. American Ornithological Union, Ornithological Monograph 37. Lawrence, Kansas:Allen Press.

Moyer, J.T. 1980. Influence of temperate waters on the behavior of the tropical anemonefish *Amphiprion clarkii* at Miyake-Jima, Japan. *Bulletin of Marine Science* 30:261-272.

Murray, B.G. 1981. "A demographic theory on the evolution of mating systems as exemplified by birds." pp. 71-140. In: M.K. Hecht, B. Wallace and G.T. Prance, eds. *Evolutionary Biology*. Vol. 18. New York:Plenum Press.

Ostfield, R.S. 1985. Limiting resources and territoriality in microtine rodents. *American Naturalist* 126:1-15.

Rasmussen, D.R. 1981. Pair-bond strength and stability and reproductive success. *Psychological Reviews* 88:274-290.

Reese, E.S. 1977. Coevolution of corals and coral feeding fishes of the family Chaetodontidae. *Proceedings of the Third International Coral Reef Symposium*, pp. 267-274.

Robertson, D.R. 1973. Field observations on the reproductive behaviour of a pomacentrid fish, *Acanthochromis polyancanthus*. Zeitschrift für Tierpsychologie 32:319-324.

Robertson, D.R., N.V.C. Polunin and K. Leighton. 1979. The behavioral ecology of three Indian Ocean surgeonfishes (Acanthurus lineatus, A. leucosternon and Zebrasoma scopas): Their feeding strategies, and social and mating systems. *Environmental Biology of Fishes* 4:125-170.

Slobodchikoff, C.N. 1984. "Resources and the evolution of social behavior." pp. 227-251. In: P.W. Price, C.N. Slobodchikoff and W.S. Gaud, eds. *A new ecology: Novel approaches to interactive systems*. New York:Wiley.

Sussman, R.W. and W.G. Kinzey. 1984. The ecological role of the Callitrichidae: A review. *American Journal of Physical Anthropology* 64:419-449.

Thomas, J.A. and E.C. Birney. 1979. Parental care and mating system of the prairie vole, *Microtus ochrogaster*. *Behavioral Ecology and Sociobiology* 5:171-186.

Townshend, T.J. and R.J. Wootton. 1985. Variation in the mating system of a biparental cichlid fish, *Cichlasoma panamense*. *Behaviour* 95:181-197.

Trivers, R.L. 1972. "Parental investment and sexual selection." pp. 136-179. In: B. Campbell, ed. *Sexual selection and the descent of man*. Chicago:Aldine Press.

van Rhijn, J. 1984. Phylogenetical constraints in the evolution of parental care strategies of birds. *Netherlands Journal of Zoology* 34:103-122.

Waring, A. and T. Perper. 1980. Parental behaviour in Mongolian gerbils (*Merioines unguiculatus*). II. Parental interactions. *Animal Behaviour* 28:331-340.

Wickler, W. and U. Seibt. 1981. Monogamy in crustacea and man. *Zeitschrift für Tierpsychologie* 57:215-234.

Wickler, W. and U. Seibt. 1983. "Monogamy: An ambiguous concept." pp. 33-50. In: P.P.G. Bateson, ed. *Mate choice*. New York:Cambridge University Press.

Williams, G.C. 1966. *Adaptation and natural selection*. Princeton: Princeton University Press.

Wilson, E.O. 1975. *Sociobiology: The new synthesis*. Cambridge:Belknap Press of Harvard University.

Wittenberger, J.F. and R.L. Tilson. 1980. The evolution of monogamy: Hypotheses and evidence. *Annual Review of Ecology and Systematics* 11:197-232.

Wolff, J.O. 1980. Social organization of the Taiga vole (*Microtus xanthognathus*). *The Biologist* 62:34-45.

Wright, P.C. 1984. "Biparental care in *Aotus trivirgatus* and *Callicebus moloch*." pp. 60-75. In: M.F. Small, ed. *Female primates: Studies by women primatologists*. New York:Liss.

III. ECOLOGY OF SOCIAL MAMMALS

Chapter 5

RESOURCES AND GROUP SIZES OF PRIMATES[1]

Peter S. Rodman

Department of Anthropology, and
California Primate Research Center
University of California, Davis 95616

[1]*The research in 1977-80 was supported by NSF grant DEB76-21413; support is acknowledged by USPHS grant RR00169; and the Committee on Research of the University of California, Davis.*

I.   INTRODUCTION

Animals are dispersed over the earth in minimal persistent units that vary in size from single individuals to large groups.   In species whose members form groups, the larger units vary in constancy of membership and in number of members from time to time for the same group, from group to group in the same population, from population to population within a species, and from generation to generation in evolutionary time.   The size and composition of each group at any time result from interaction of a set of rates of birth, immigration, death and emigration (Cohen 1968).   These rates can produce highly labile groups as in the formation of feeding flocks of birds (transient membership = high immigration and emigration), or they may produce very stable groups such as those of the primates.

Diurnal primates, with the important exceptions of the orangutans of Southeast Asia (Rodman and Mitani 1987), chimpanzees of Africa (Wrangham 1986), and spider monkeys of Central and South America (Klein and Klein 1975; Milton 1984) live in heterosexual groups of quite constant size.   Each group contains a stable core of members of both sexes, although some annual change in membership may be associated with breeding seasons and the dispersal of young of one or both sexes. Group size varies between groups in a population.   The range of variation is limited in each population, however, and there is usually a clear central tendency within that range of group sizes.   The fact that there are limits to variation in group sizes is not surprising, but it raises interesting questions about how the central tendency and range of variation have evolved and how they are maintained in the present.

My purposes in this paper are to (1) demonstrate variation in group sizes of forest primates with reference to examples from the literature; (2) to examine current theory about why group sizes differ among species; (3) to present a new model of the interaction of group size with subgrouping dynamics that addresses how and why group size may change within a population through evolutionary time; and (4) to describe briefly some interesting differences between coexisting gibbons and macaques of Borneo that shed light on possible utility of some hypotheses about the relationship of grouping to resource dispersion. Two recent, excellent papers, one by Andelman (1986) and one by Terborgh and Janson (1986), also address the problem of variation in group size among primates and provide additional breadth of examples of variation in grouping patterns of primates.   Terborgh and Janson (1986), and particularly Terborgh (1983), explicitly address the interesting question of why coexisting species of primates differ in group size.   Here this issue is the focal point for discussion in relation to the diverse theory presented below.

II.   VARIATION IN GROUP SIZE WITHIN POPULATIONS OF FOREST LIVING PRIMATES

A.   Allopatric Populations

Some primates, such as the white-handed gibbons (*Hylobates lar*) of Southeast Asia, show highly canalized distributions of group sizes with low variance and strong modal values (Fig. 1). Carpenter (1940) counted group membership in a population in Thailand and found a clear modal size of four with a range of size from two to six. In a population of the same species in southern Malaya, Ellefson (1974) found a lower mode of three members, with the same range of sizes from two to six. This comparison illustrates interesting variation within a species that should encourage further investigation. Since monogamy and territorial defense are highly stereotyped among all species of gibbons described so far (Mitani 1984), interpopulational variation must result from differences in interbirth intervals or differences in the age of emigration of offspring. The simplicity of the group structure makes very clear how variation in group sizes results from simple changes in demographic parameters. Titi monkeys (*Callicebus moloch*) of Columbia (Mason 1964) and sifakas (*Propithecus verrauxi*) of Madagascar (Richard 1977) have similar canalized ranges of group sizes with clear modal sizes and clearly limited ranges of variation in group size.

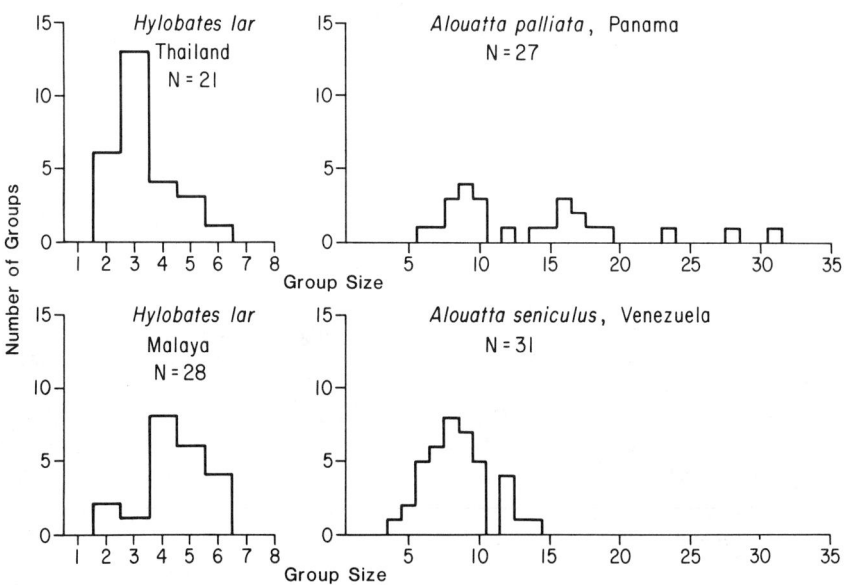

Figure 1.   Group sizes of gibbons and howler monkeys. Sources of data are: Carpenter 1940 (Hylobates lar, Thailand); Ellefson 1974 (H. lar, Malaya); Smith 1977 (Alouatta palliata, Panama); Neville 1972 (Alouatta seniculus, Venezuela).

Group sizes in populations of other forest primates vary more widely with less definition of the shape of the distribution of group sizes.
Howler monkeys of Barro Colorado (*Alouatta palliata*) live in groups of
highly variable size (Fig. 1) (Smith 1977). The group sizes appear to be
bimodal in frequency with one peak at nine members, another peak at 16
members, and a long tail at the upper end of the range. By comparison,
groups of red howlers (*Alouatta seniculus*) of Venezuela counted by
Neville (1972) vary less in size and show a clear mode at a group size of
eight.

B. Coexisting Populations

Variations in modes and ranges of group sizes between populations
in different locations such as those illustrated so far are not surprising.
Observations that might contribute to understanding why the group sizes
differ are hard to come by, however, because the objectives of the studies producing the observations of group sizes have differed. Variation
between sympatric species may offer better sources of explanation of
the patterns because studies of these communities have produced more
uniform information on each species. Several communities of primates
have been described that show the variation in group sizes among sets of
coexisting species.

Lemur catta, *L. fulvus*, and *Propithecus verrauxi* coexist in parts of
their ranges, and variation in their grouping patterns is obvious (Fig. 2A).
Although the ranges of variation in group size overlap, each species has
a distinct range that is different from the others. Mean size of groups
of *L. catta* is about twice that of *L. fulvus*, which is again twice that of
P. verrauxi. In Uganda, five cercopithecoid species coexist and have
overlapping ranges of group sizes, but distinct mean group sizes (Fig. 2B)
(Struhsaker and Leland 1979). The congeners *Colobus guereza* and *C.
badius* differ remarkably in group size, and the range of variation in
group size of *C. badius* is unusually high for a forest primate (Struhsaker
and Oates 1975). Sympatric primates of Columbia also vary in group
sizes (Fig. 2C) (Klein and Klein 1975), and Terborgh (1983) has described
the similar community of primates at Cocha Cashu, Peru, in a book that
concludes with cogent analysis of the ecological sources of differences
in grouping among coexisting species there. I discuss his work and his
model explaining interspecies differences in group size below (see IIID).
Finally, in East Kalimantan (Borneo), Indonesia, four species differ quite
clearly in sizes of groups within a small section of forest (Fig. 2C)
(Rodman 1978). This community includes the large orangutan (*Pongo
pygmaeus*; females 35 kg, males 80 kg), as well as grey gibbons (*Hylobates moloch*; 6 kg), Sunda Island leaf monkeys (*Presbytis aygula*; 6 kg),
and long-tailed macaques (*Macaca fascicularis*; males 5.5 kg, females
3.5 kg). The latter three species are similar in body size but quite
different in group size.

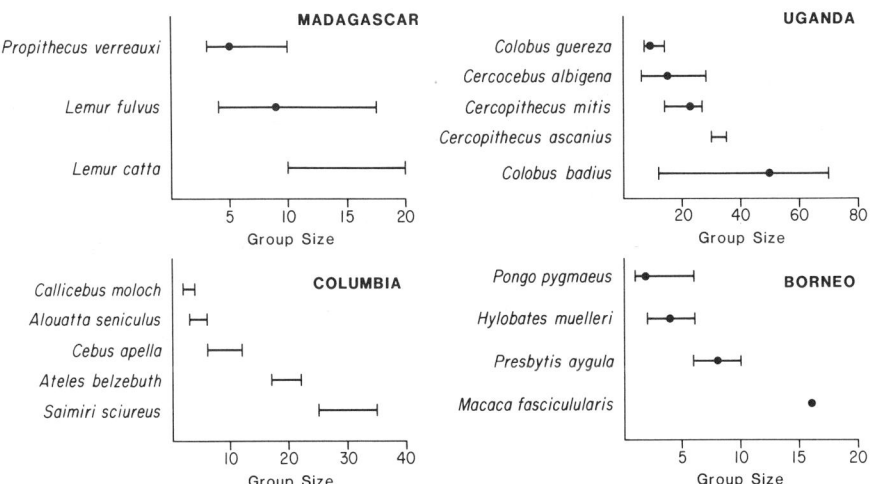

Figure 2.   Group sizes in four communities of primates. Bars show the range of variation in group size; solid circles show mean group size for each species if reported. Sources of data are: Jolly 1966, Sussman 1974 and Richard 1978 (Madagascar); Klein and Klein 1975 (Columbia); Struhsaker and Leland 1978 (Uganda); and Rodman 1978 (Borneo).

III.   HYPOTHESES ABOUT THE ADAPTIVE BASIS OF VARIATION IN GROUP SIZES

Why do primate populations in forests throughout the world take on particular ranges of group sizes? Variation in grouping is normally attributed by theoreticians to efficiency of obtaining resources or to variable influences of predation (Alexander 1974) and to the trade-offs among costs and benefits of each. Alexander (1974) asserted that risk of predation is probably the only reason primates live in groups since feeding competition always increases in groups. Van Schaik and van Hoof (1983) concur. (I would add another, possibly independent factor leading to grouping, which is the advantage through inclusive fitness effects of adding descendants or relatives, even at the expense of individual fitness, that can find a place in the population only "at home" [Rodman 1981, but see Giraldeau, this volume, for counter argument]). The "predation hypothesis" is relatively simple to describe. Hypotheses accommodated under the heading of foraging efficiency and group size are more complex and diverse.

A.   The Predation Hypothesis

1.   Predation causes variation in group size

It has been proposed that variation in grouping among primates is a response to variation in risks of predation and further that group living necessarily is costly to individual feeding success (Alexander 1974; van Schaik and van Hooff 1983; Terborgh 1983; Terborgh and Janson 1986). Since groups provide individuals with defense from predation in several ways, we expect groups to be larger where risks of predation are higher and smaller where predation risk is relaxed because individuals feed more effectively in smaller groups. Some circumstantial evidence supports this proposal. Van Schaik and van Noordwijk (1986) have shown that group sizes of long-tailed macaques (*Macaca fascicularis*) are larger at Ketambe, North Sumatra, where there are tigers (*Panthera tigris*), golden cats (*Felis temminckii*), and clouded leopards (*Neofelis nebulosa*) that may prey on primates, than on the island of Semeulue (off the coast of Sumtra in the Indian Ocean), where there are no felids present. Other evidence from Ketambe and Semeulue indicates that behavior of the macaques is sensitive to differences in group size in ways that can be interpreted to depend on risks of predation (van Schaik et al. 1983a, 1983b). Smaller parties were found higher in trees at both sites, for example, and females and young animals constituted a larger proportion of members of small parties on Semeulue than at Ketambe, which suggests that in the cat-free area the macaques behaved as though they were less vulnerable.

2.   Critique of effects of predation on grouping

These observations and the logic of inference are compelling, but it is important to note there is no direct evidence from Sumatra that any of the felids actually preys on macaques. Attacks by raptors have been observed in Borneo (L. Berenstain, pers. comm.), and similar avian predators occur in Sumatra. I do not know whether a full complement of avian predators is present on Semeulue, and I am not arguing that variation in predation is irrelevant to variation in group size of long-tailed macaques. But the "natural experiment" offered by comparison of Sumatran macaques with those of Semeulue hinges on presence and absence of felids only. It is very important, therefore, that group sizes of long-tailed macaques of Borneo (Fittinghoff and Lindburg 1980) are similar to those in Sumatra, but that neither tigers nor golden cats are present in Borneo. Despite van Schaik and van Noordwijk's compelling argument and interesting observations, the relationship of group size to predation is not supported as clearly as the comparison of Ketambe and Semeulue might suggest at first glance.

Note also the difference in group sizes between black howler monkeys of Barro Colorado where larger felids are not present and red howlers of Venezuela, where such felids are present (Fig. 1). Neville (1972) described in detail his reasons for suspecting that red howlers in Venezuela were subject to attacks by cats. Barro Colorado, on the other hand, is free of predators on howler monkeys. Since red and black howlers are quite similar in other ecological regards, the comparison suggests that the difference in group size is not directly related to effects of predation, which intuitively should lead to larger groups where predation is more intense (Alexander 1974; Terborgh 1983; Terborgh and Janson 1986).

Finally note the differences in group sizes among coexisting species of similar body size in forests of Uganda, Kalimantan, Madagascar, and Peru (Fig. 2). Members of these communities all face the same set of predators so that, all else being equal, we cannot attribute differences in their group sizes to differences in risks of predation. All other things probably are not equal, it is true. For example, risk may be mediated by differences among species in perceptual abilities so that the same predators pose less risk to a gibbon than to a long-tailed macaque, or different locomotor capacities may offer different potential escape patterns (Denham 1971). Despite such possibilities, we have no evidence that such variation between these two species, or among other sets of primates illustrated in Figure 2, mediates relations to predators. Thus, although there is reasonably good evidence that risks of predation are correlated with group size among forest primates (van Schaik and van Noordwijk 1986), this evidence is unconvincing that predation *causes* variation in group sizes in all cases. Interspecific differences in group size *within communities* remain difficult to explain by a simple difference in risk of predation.

B.   Other Hypothesis

1.   The phylogenetic hypothesis
Different phylogenetic histories may be called on to explain interspecific differences in social patterns (Struhsaker 1969). This very unsatisfying hypothesis begs the question by asserting in effect that different species are different because they are different species, and it moves the causes of differentiation out of the present and out of reach of observation and functional analysis.

The phylogenetic hypothesis may be refined, however. It has been argued in one case that different grouping patterns evolved allopatrically and that the differences have been transported into the current environment without modification (Struhsaker and Oates 1975). This hypothesis implies that if two coexisting species differ in grouping patterns, at least one of them *is not adapted* to the current environment. Logically this suggests that one of the two species is living inadaptively with respect to group size, which does not make sense because in evolutionary time we expect more efficient species to replace less efficient species in the community (Darwin 1859; Hutchinson 1957).

Another refinement of the phylogenetic hypothesis may be that past selection has produced invariant behavioral mechanisms that generate a canalized grouping pattern. Pair bonding and intrasexual intolerance among gibbons could be examples of such mechanisms. Together the two patterns set a very small group size by insuring that only two adults are present in a group and that offspring emigrate when they mature. Larger group size might be advantageous to gibbons in some forests, but under this hypothesis the fixed behavioral mechanisms prevent change. I am inclined to argue in this case that the species with a fixed pattern living under suboptimal conditions would be replaced by one or more other species that are more efficient under local conditions, or that the "invariant" behavioral mechanisms would be selected out and replaced with other patterns. The concept of invariant behavioral mechanisms-- "fixed action patterns"--has become somewhat outmoded in modern behavioral study. Instead we are likely to find individual variation that given appropriate opportunity could spread in a population, even at the expense of the most common current forms of behavior. I readily admit this is a bald statement of an adaptationist view, but I believe it is an appropriate one since the more closely any species is studied, the more "individualistic" its members appear. Demonstration that this individual variation can, let alone does, lead to evolutionary change may be elusive, but Darwin's logic and the modern theoretical and empirical demonstration of effectiveness of very small heritabilities in directional evolution give me confidence that the concept of true fixed action patterns should not be the basis for expectation that social evolution cannot occur.

2.   The stochastic hypothesis
     Under the stochastic hypothesis, group size is an *emergent phenomenon* and not directly relevant to individual fitness. Instead, rates of birth, immigration, death and emigration have evolved in response to independent factors and together produce a stable frequency distribution of group sizes. Of course, demographic parameters may be adjusted *because* they adjust group size and ultimately affect individual fitness, but under this hypothesis they do not. So, for example, group size may be small because rates of death are high or because rates of birth are low when risk of disease is high (e.g., yellow fever among black howlers of Barro Colorado Island in the 1950s; Collias and Southwick 1952). Van Schaik and van Noordwijk (1986) have shown that numbers of infants per female vary in relation to group size of a large number of primate species, and they conclude that reproductive success decreases with increasing group size. Such a relationship is opposite to what we would expect if group size is generated by birth rate, which suggests that group size is the independent variable relative to this demographic parameter rather than vice versa.

C.   Foraging Hypothesis

     The general foraging hypothesis is that group size is adjusted to dispersion and abundance of resources. This hypothesis is attractive for two reasons. First, we know from numerous studies that dietary patterns differ among coexisting species within a community, and there is thus reason to expect that resource dispersion and abundance will differ among the species. A priori, then, the foraging hypothesis has potential explanatory power. Second, the foraging hypothesis should be testable with sufficient effort in the forest, while other hypotheses will be less amenable to plain hard work. Several more precise hypotheses about the relationship of grouping to foraging patterns and dispersion of resources have been proposed, and I summarize them below.

1.   Harvest efficiency 1: the Cody/Altmann model

Altmann (1974), in an argument reminiscent of one by Cody (1971) regarding mixed species flocks of finches, suggested that for primates, feeding in a group rather than independently may maximize individual feeding efficiency by minimizing returns to exhausted patches.   By spreading out and feeding in parallel, the members of the feeding group regulate their return times to patches of food previously fed upon by themselves or others. Such a mechanism would cause variation in group size depending on the dispersion of food resources. We would expect that when many food patches are small, but not far apart, individuals would feed most efficiently in large groups with individual foragers moving in parallel along a wide swath. Terborgh (1983) invoked the Cody/ Altmann model to explain heterospecific associations of species of Callitrichidae at Cocha Cashu, Peru.

2.   Harvest efficiency 2: the Horn model

Horn (1968), in a model dealing with blackbirds that are central place foragers, used simple geometry to examine distance traveled to harvest resources. When resources are clumped in space and unpredictable in time, he found that a set of individuals will minimize travel to harvest resources by living in a large colony at the center of a large range, while if resources are uniformly dispersed in space and time, the same number of individuals will minimize travel costs by living in several small groups at the centers of small ranges.

No diurnal primates are true central place foragers, so the geometry of life for primates is quite different than for the Brewer's blackbirds of interest to Horn. The model has intuitively broader application, however. Wilson (1975) called it "the Horn principle" and suggested that it will ultimately explain all variation in primate gregariousness. So far there have been few studies that have measured resource dispersion in a manner that allows tests of this hypothesis for primates. Careful measures of resource dispersion independent of the observations of the animals are necessary to test this and most other foraging hypotheses, and so far appropriate measures have rarely been taken. Work by M. Leighton (1982, 1986), D.R. Leighton (1987), and Leighton and Leighton (1982, 1983) is exceptional in this regard, and unpublished results of their studies on primates of Borneo remain the most valuable data set for examination of the foraging hypotheses presented here.

3.   Information exchange

Shared knowledge about resources may benefit individuals living in groups. Once again, dispersion of resources should determine the benefits to be gained. If resources occur in widely separated clumps so that travel to one is costly, particularly if it is "empty," and if clumps are large enough to be shared, then daily fission of subgroups and return to a common roosting spot where information about foraging success can be shared may be beneficial to individuals living in large groups (Ward and Zahavi 1973). Fittinghoff and Lindburg (1980) suggested that long-tailed macaques of Kalimantan may fit this model because they sleep in a group near a river bank at night and frequently break into smaller foraging groups during the day. Sophisticated use of loud calls to share information about food resources may promote close cooperation of male chimpanzees (*Pan troglodytes*) (Wrangham 1977, 1986; Ghiglieri 1984), and so the communal relations of these apes may relate partially to information exchange.

4.   Feeding competition and resource defense

Groups offer individuals power in numbers, and a quick review of literature shows that in most studies of primates, larger groups displace smaller groups when there is conflict. For example, among red colobus monkeys (Struhsaker 1975), black howlers (Carpenter 1934), vervet monkeys (Cheney 1981), and rhesus macaques (Southwick et al. 1965), larger groups displace smaller grups. Thus if the resource patches are large enough to accommodate multiple feeders, individuals in larger cooperative groups may hold resources effectively against outsiders in smaller groups. This effect of grouping is the basis of Wrangham's (1980) analysis of grouping in "female-bonded" species of primates in which groups of relatives cooperate to defend resources against other groups.

Power in numbers is effective in interspecific competition as well, and Buss (1981), in a short paper on sessile marine organisms, suggested that interspecific competition is a major cause of gregariousness in a most general sense. Ghigilieri (1984) observed that groups of red colobus monkeys of Uganda displace chimpanzees from feeding trees, and in Kalimantan, a large group of long-tailed macaques occasionally displaces single orangutans and families of gibbons from fruit trees (pers. obs., unpub.).

Once again, the effect of intergroup competition on grouping depends on the dispersion of food. We expect groups to be *as large as possible* given the normal size of clumps of food in order to "defend" those clumps effectively. For a given density of food in a large area of habitat, groups should be larger the more clumped the food is in large patches. Despite the possible negative effect of local feeding competition on individual feeding success, the net effect on individual feeding in larger groups can be positive. Demonstration that individual feeding rates decrease with increasing group size (van Schaik et al. 1983a) only supports the hypothesis that individual feeding rates decrease with increasing group size. Unfortunately measuring the net effect of group size on individual feeding is a much less tractable task.

D.   Costs of Larger Groups

The functional hypotheses (i.e., excluding the stochastic and phylogenetic hypotheses) above suggest that groups should be as large as possible, and other more theoretical analyses that have little to do with specific relations between ecological factors and group size suggest a similar direction (Rodman 1981; Slobodchikoff 1984; Clark and Mangel 1986). Thus regardless of what the function of group size may be--i.e., related to defense or to resources--the weight of theory suggests that groups should be as large as possible. Because of this, we should find that group sizes depend on limits set by the costs of living in larger groups. Terborgh's (1983) simple graphical analysis, repeated and refined by Terborgh and Janson (1986), shows how the constant benefit of larger groups interacts with escalating costs of grouping to produce groups of different sizes in species with different slopes of cost (Fig. 3). According to this model, species in which costs increase more rapidly with group size should live in smaller groups. The question is what are the costs associated with grouping that impose the limit? Here it seems likely that we must deal with feeding economics alone, at least among diurnal species (Terborgh 1983).

Waser (1977) showed that for mangabeys (*Cercocebus albigena*) of the Kibale forest, Uganda, daily travel increases with group size. The mean day range for a group of 25 to 30 members was approximately

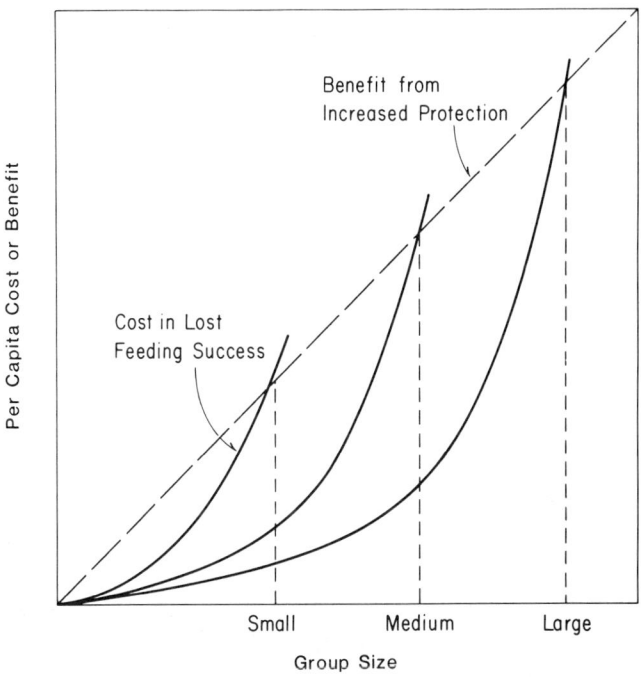

Figure 3. J. Terborgh's model for regulation of group size. Redrawn after Figure 10.1 of Terborgh (1983). Per capita benefit due to increased protection from predators (dashed line) increases monotonically with group size. The figure shows per capita cost of increased foraging costs in three hypothetical species (solid lines) that differ in rate of increase in cost with group size. When costs increase rapidly, group size must be small. When costs increase more slowly, group size may be larger. Terborgh and Janson (1986) present a modified model that shows group size as an optimization determined by the cost curve of foraging costs as drawn here and the benefit curve of defense from predators.

twice the mean day range for a group of five members, and for six groups there was monotonic increase of day range with group size. In addition, Waser found that there was ". . . a clear tendency for increasing aggression in small food patches with increasing group size . . ." (1977, p. 218). Van Schaik et al. (1983a) found that in a population of long-tailed macaques (*Macaca fascicularis*) in groups ranging in size from 11 to 34, the day range and time spent traveling as well as individual time spent searching for dispersed food items increased with group size. "Social tension" also increased with increasing group size. These observations show that there are true economic costs to individuals living in larger groups and that these costs result from greater travel and increased competitiveness among group members. They do not show, however, that the *net foraging benefit* to individuals in a larger group is always smaller than the net foraging benefit in a smaller group.

IV. SUBGROUPING AND RESOURCE DISPERSION

Waser (1977), referring to Kummer's (1971) earlier point, briefly described the possible response of groups to different sizes of food patches. As long as the normal patch size is large enough for the group, foraging in groups will not be costly. If the group size exceeds the capacity of patches, members will do better to split into subgroups to forage or to form permanent smaller groups. The groups of mangabeys Waser studies do split into smaller feeding parties occasionally, as do groups of many other species.

Leighton and Leighton (1982) reported a close positive relationship between the size of "feeding aggregates" of howler monkeys of Barro Colorado Island and the size of *Trichilia cipo* trees in which howlers fed on ripe fruit. They concluded that, ". . . feeding aggregate size was adjusted to, and not merely limited by patch size" (Leighton and Leighton 1982, p. 81). They attributed this result to the fact that feeding aggregates were subgroups of larger foraging groups and that as patches were encountered by the larger unit, subgroups formed that suited the size of the patch. In their analysis, Leighton and Leighton pointed out that the size of the larger unit might be related to other selective influences.

Waser's (1977) discussion and Leighton and Leighton's (1982) observations and analysis suggest a possibly useful diachronic perspective on group size and its evolution in relation to density and dispersion of food. My aim in the following discussion is to consider why evolution from one group size to another would take place and what changes in resource dispersion would affect the nature of group size by adding economic costs *due to splitting and rejoining of subgroups.* A critical point in this argument is that, for the variety of reasons identified by diverse theoretical considerations, there is selective pressure for groups to be as large as possible.

I define "permanent groups" loosely as sets of individuals that get together approximately daily. Such permanent groups may be small or large and may vary in cohesiveness during the day. I propose three levels of permanent groups that differ in degree of cohesion of members:

A. Members seldom separate
B. Group spreads out into several patches, but members travel in parallel and regroup frequently
    1. Members assort ad lib into parties or subgroups
    2. Subgroups have constant membership
C. Members form subgroups that often are not in contact and forage independently for long parts of the day
    1. Members assort ad lib into parties or subgroups
    2. Subgroups have constant membership

No primate population with which I am familiar is purely a population of A's, although gibbons approach this condition. Numerous species fall into level B, including howler monkeys (Milton 1980; Leighton and Leighton 1982) and long-tailed macaques (Berenstain in press). Mangabeys (Waser 1977), banded leaf monkeys (*Presbytis obscura*) (Curtin 1977), and squirrel monkeys (*Saimiri sciureus*) (Terborgh 1983) seem to fit level C. Few descriptions of primate groups allow differentiation between ad lib and consistent membership in subgroups, although Curtin's (1977) description of banded leaf monkeys suggests that their subgroups have consistent membership.

Evolution from a population that assorts into permanent groups that fragment frequently (level C) to a population of smaller permanent groups that are more cohesive should depend on the *cost of subgrouping* to individuals. At some cost ceiling it will be relatively advantageous to individuals to remain permanently in smaller groups *rather than regrouping*. On the other hand, *keeping in mind the premise that groups should be as large as possible*, evolution from smaller groups to larger, somewhat less cohesive groups should occur as the costs of subgrouping decrease over evolutionary time. Figure 4 illustrates the relationships.

How are the dynamics of subgrouping related to resources? The answer is suggested by Leighton and Leighton's (1982) analysis of howler feeding aggregations in relation to larger foraging units. Imagine that the core of the permanent group travels along some minimal path from resource patch to resource patch to satisfy daily needs of its members. Subgroups will be forced to split off from this core when the patch is too small to accommodate the whole group. Per capita costs of subgrouping will depend on (1) the rate at which the individual is part of a subgroup; and (2) what the subgroup does or must do when it forms.

The rate at which individuals fall into subgroups depends on two variables: (1) the frequency of subgrouping; and (2) the likelihood that the individual will be in the subgroup. Frequency of subgrouping depends on the relationship between the group size N and the patch size P (measured in the same units as N). When a patch is encountered that will hold less than all the members of the group (P < N), the patch can either be ignored or it can be filled by some members while the "extra" group members either wait or travel off to another patch. Frequency of subgrouping depends on the number of patches encountered of size P < N.

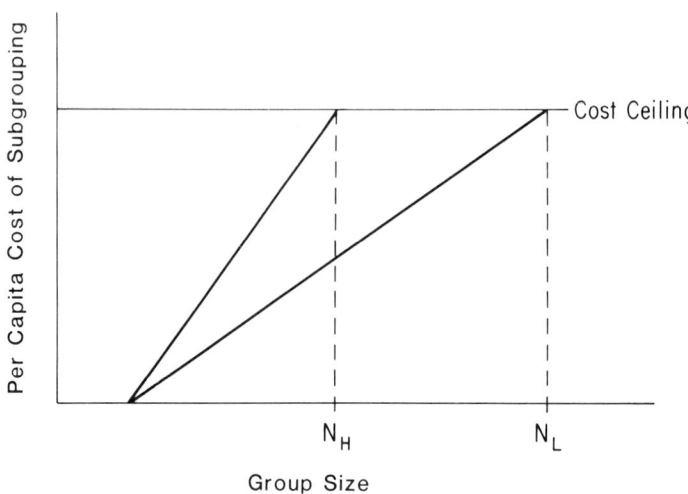

Figure 4. Regulation of group size by net costs of subgrouping. A cost ceiling is imposed by the net flux of energy for each individual, which must be null over some period of time. For similar animals (same size, similar diets), this cost ceiling is the same. Costs of subgrouping will increase with increasing group size, but at different rates depending on dispersion of resources (and "danger" associated with traveling longer distances in smaller groups). When subgroups form frequently (small resource patches) and individuals in subgroups must move long distances (low patch density), costs of subgrouping increase rapidly with group size and the average permanent group should be small (N_H). Under the opposite conditions, per capita costs of subgrouping increase more slowly, so that permanent groups may be larger (N_L).

As this proportion increases, rate of subgrouping must increase. If there is a cost to subgrouping (such as not feeding, feeding on less desirable food, or traveling farther to feed than members of the group who remain in the patch) then costs of subgrouping and regrouping increase with increasing proportion of small patches encountered. As has been pointed out much more succinctly by Kummer (1971) and Waser (1977), the consequence is that when resources are distributed in small patches, we expect group size to be small.

Frequency of membership in a subgroup--thus average cost of subgrouping to individuals--depends on the chance that an individual will be part of a subgroup when it forms. Assuming that all individuals are equal (even though they are not), the chance of falling into a subgroup is equal to mean subgroup size (S) divided by the size of the permanent group ([N-P]/N). Mean subgroup size depends directly on the difference between permanent group size (N) and mean patch size (P), so that frequency of being in a subgroup increases as N increases or as P decreases.

We should expect that individual competitive ability and cooperative coalitions will allow some individuals to minimize their costs of sub-grouping at the expense of others.

One outcome of this analysis is a less obvious prediction about the relationship between group size and resource dispersion: part of the variation in frequency of membership in a subgroup depends on *the shape of the frequency distribution of patch sizes around the mean* (Fig. 5). As the distribution becomes flatter, the proportion of patches closer to the mean decreases and the proportion of patches toward the ends of the distribution increases. The set of patches of interest is the set for which $P < N$. The mean of this set will be smaller *the more uniform the distribution of patches* across size classes. Subgroups will therefore be larger, and the chance of falling into a subgroup will increase, when the frequency distribution of patch sizes is platykurtic. Thus I expect that when the dispersion of patch sizes is flatter, the costs of subgrouping are higher on average and permanent group size should be smaller.

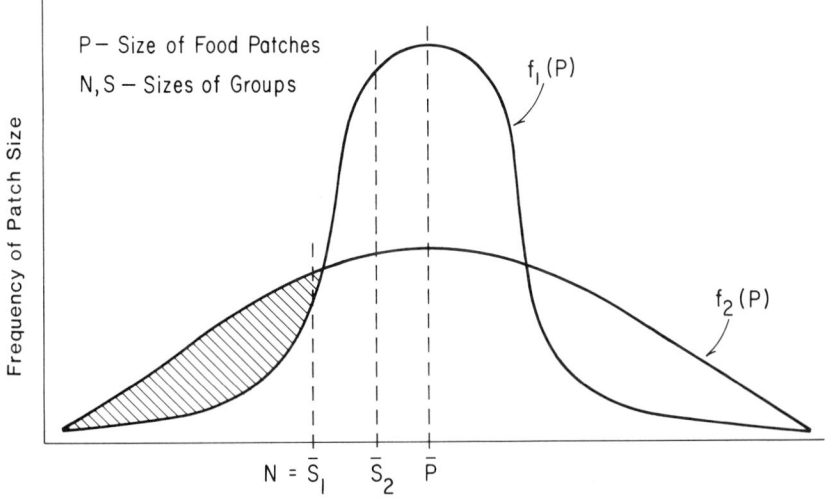

Figure 5.   Relationship of mean subgroup size to the distribution of patch sizes.   Two frequency distributions of patch sizes are shown, $f_1(P)$ and $f_2(P)$, of which the second is a more uniform distribution.   Assuming mean permanent group size $N = P_1 = P_2$, then the mean size of subgroups is the difference between P and S in which S is the mean patch size for all $P < P$.   Under $f_1(P)$, S is larger than under $f_2(P)$.   The result is that individuals foraging on resource patches dispersed as $f_1(P)$ would fall in subgroups more frequently than individuals foraging on resource patches dispersed as $f_2(P)$ and the per capita cost of subgrouping would be lower in the latter case for the same permanent group size.   As explained under Figure 5, larger permanent group size is expected in the latter case.   In other words, groups should be larger for similar species foraging on resources with a uniform frequency of patch sizes.

Finally, the costs of subgrouping for a given set of patch sizes will vary depending on what the subgroup must do when it forms. If the subgroup must seek out other food sources, costs will depend on the distance to the next food source both because of the travel costs incurred and because of increased risk if members of the subgroup are more vulnerable to predation while they are separated from the core. As density of patches decreases, regardless of the mean or other parameters of the distribution of patch sizes, permanent groups should be smaller. For similar-sized animals I expect those with a lower density of patches to have smaller groups because the costs of subgrouping due to increased travel by subgroups will be higher (see Fig. 4).

To summarize, then, we expect that through evolutionary time, if mean patch size changes due to habitat change or to adaptation by a species to use new food sources, individual behavior should be selected that adjusts group size up or down depending on changes in patterns of subgrouping. Under the model discussed, if two species differ in group size and pattern of subgrouping and if group size has evolved under regulation by per capita costs of subgrouping, I expect the distribution and abundance of patches to differ in specific ways. A species characterized by larger groups should have either a set of larger patches available, or a higher density of patches, or both, than a species similar in other respects but living in smaller groups. If there is no other difference between dispersions of resources, we should expect the species living in smaller groups to have a more uniform distribution of patch sizes available.

V. COMPARISON OF GREY GIBBONS AND BORNEAN MACAQUES

Two coexisting primate species found in East Kalimantan, Indonesia, make an interesting case for consideration in relation to the hypotheses enumerated and discussed above. These are the grey gibbon (*Hylobates muelleri*) studied by D.R. Leighton (1987 and in prep.) from June 1978 through August 1979, and the long-tailed macaque (*Macaca fascicularis*) studied by L. Berenstain (in press and unpub.) from June 1982 through August 1983 at the Mentoko research site (Fig. 6). The brief comparison made here is subjective. Leighton and Leighton (unpub.) have collected valuable quantitative data on dispersion of resources of these and other animals that may provide appropriate critical tests of hypotheses about group size and resource dispersion.

A. Bornean Gibbons

Grey gibbons are a species limited to Borneo. They are monogamous and territorial like all gibbon species described so far (Leighton and Whitten 1984; Mitani 1984 D.R. Leighton 1987). Groups thus consist of a male, a female and their young. Apparently up to four successive offspring may be with the parental pair on a territory. In 1978-79, there were either three or four members of each of nine families in the study

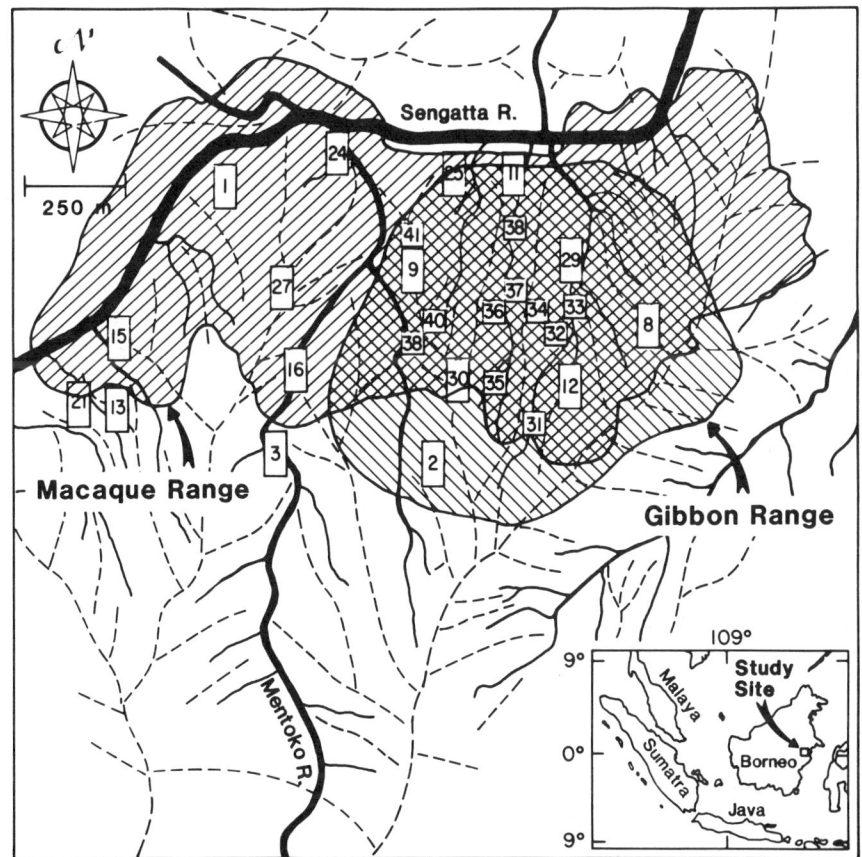

Figure 6. The study area with home ranges of gibbon and macaque groups. Dashed lines show ridge tops, and solid lines show streamlets and larger streams. Topography redrawn from Leighton 1982; gibbon and macaque ranges provided by D.R. Leighton and L. Berenstain.

area. The focal group described here consisted of the adult pair with one infant. This group used a range of 44 ha, and traveled 598 to 1232 m (range is for monthly means, 11 months) per day (Table I) (D.R. Leighton, unpub.). The body weight of adult grey gibbons is approximately 6 kg (Napier and Napier 1968).

B.   Long-tailed Macaques

Long-tailed macaques live in multi-male breeding groups along rivers and streams of Southeast Asia (Fittinghoff and Lindburg 1980). Typically the groups roost in trees near the edge of a river or stream and forage within 500 m of the river during the day until they return to

TABLE I. Characteristics of gibbons and macaques

	Gibbons[1] (*Hylobates* *muelleri*)	Macaques[2] (*Macaca* *fascicularis*)
Home range area	44 ha	112.5 ha
Day range length (monthly means)	598-1232 m (11 mo)	1203-2142 m (15 mo)

Body weight
    Adult male
    Adult female
	
	6 kg
	6 kg
		
		5.5 kg
		3.5 kg

Group size	1 male 1 female 1 infant	3 adult males 5 adult females 1 subadult male 8 juveniles
Group weight	12 kg	51 kg
Group weight/Home range area	0.27 kg/ha	0.45 kg/ha

[1]Study by D.R. Leighton, 1978-1979
[2]Study by L. Berenstain, 1982-1983

sleeping trees at the river (Fittinghoff and Lindburg 1980). There has been a single group occupying a constant home range in the alluvial area of the Mentoko study site since 1970 (Rodman 1978; Berenstain in press; Leighton and Leighton unpub.), and the group has varied in size from 11 to 20 members. In 1982-83, the modal size of the group was 17 adults, subadults and juveniles (Table I) (Berenstain unpub.). This group used a home range of 112.5 ha and traveled 1203 to 2142 m per day (movement of the center of the group; range is for monthly means for 15 months) (Berenstain unpub.). The macaque group's range overlaps approximately 90 percent of the focal gibbon group's range (Fig. 6). Body weights of adults are 5.5 kg for males and 3.5 kg for females in north Sumatra (van Schaik and van Noordwijk 1986), and those of Borneo appear to be very similar (pers. obs.).

C.   Comparative Grouping and Ranging

The gibbon pair is highly cohesive. The two adults were within 20 m of each other in approximately 80 percent of records of the distance (D.R. Leighton unpub.). In contrast, the macaques spread out over considerable distance and break into subgroups that occasionally separate widely. In order to characterize the macaques' pattern for this paper, Berenstain (unpub.) picked a random time in the morning (between 0600 and 1200 hr) of each of 104 days the group was followed, and provided me with the dimensions of the area over which the group was spread and the distances between the group followed and any parties or subgroups that might have split off (Table II). Parties or subgroups were identified if the boundaries of two subgroups were separated by 100 m or more. No subgroups were identified in 78 percent of samples, and in these samples, the median long dimension of the group spread was 60 m (range 15 to 200 m). The median group count when there were no subgroups identified was 16 (range 6 to 20). Two or more subgroups (i.e., the group followed and one or two other groups) were identified in 22 percent of samples. When there were subgroups, the median distance between the edges of the group followed and the nearest subgroup was 200 m (range 100 to 580 m). The median count of the group followed when subgroups had formed was 8 (range 3 to 16), and the median group spread of the group followed was again 60 m (range 5 to 100 m).

D.   Resource Utilization and Foraging Patterns

Both gibbons and macaques are predominantly frugivorous; for both species, feeding on fruits constitutes more than 60 percent of observations (Rodman 1978). Gibbons make more use than macaques of other parts of vegetation, and macaques spend considerable time apparently searching carefully for insects. Feeding habits of gibbons and macaques thus differ in three important ways that probably contribute to differences in grouping. First, gibbons feed heavily on leaves. My data on diets (Rodman 1978) probably underestimate the contribution of leaves to the diet of the macaques, but there is nonetheless a strong difference between the two primate species in this regard. Presence of useful leaf patches among the fruit patches would increase the density of food sources for gibbons. On the other hand, use of leaf patches should increase spatial and temporal uniformity of food dispersion for the gibbons over that of the macaques. Using leaves may therefore be an important cause (or effect) of small group size for gibbons under "the Horn principle."

Second, the macaques feed on very small patches quite frequently. As they move through the forest, members of the group stop at single

TABLE II. The state of one group of long-tailed macaques (*Macaca fascicularis*) at a randomly chosen time on a sample of 104 mornings.[1]

		N	Modal group size	Median group spread	Median distance between subgroups
Number of subgroups	0	81	16	60 m	--
or parties:	2	21	8		
				60 m	200 m
	3+	2	-		

[1]Unpublished data of L. Berenstain

fruits, or stop and pick through the bases of single epiphytic plants. Gibbons, on the other hand, move from larger patch to patch of fruit or leaves. The important effect on resource dispersion is that if all food sources were taken into account, the macaques should have *many very small patches* dispersed at relatively high density through the range.

A final important difference in feeding habits is in the patterns of feeding on insects. Although ingestion of insects takes up a very small part of feeding time of either primate species, the macaques spend considerable time searching for insects as they travel. Typically each group member follows a slightly different path through the tangle of lianas and branches in the middle canopy, carefully opening dry leaves, examining bark, and breaking up dry twigs. Occasionally the searcher discovers and eats an insect. This pattern of insect searching would add to the frequency of very small patches dispersed over the habitat. Gibbons also take insects, but their foraging does not appear to be organized around such a search as does foraging by the macaques (pers. obs.).

E.   Group Size of Macaques and Gibbons in Relation to Foraging Patterns

Gibbons live in small families, and macaques live in relatively large groups. They coexist throughout forests of Southeast Asia, and since they are similar in size, they should be appropriate prey for the same predators in all communities. Differential threat of predation alone should not account for the difference in their modal group sizes. Conceivably their contrasting modes of locomotion (Fleagle 1980) may allow different patterns of escape, and a careful compendium of anecdotes about their relations to predators may be revealing in this regard. But perhaps the difference in grouping is understandable purely with respect to different use of resources. Approaching this question here is very preliminary without using specific quantitative data on resource dispersion, but some tentative comments are possible.

First, given the difference in use of resources, it seems quite unlikely that the size of macaque groups is larger than the size of gibbon groups because resource patches are larger for the macaques as suggested by Horn's model and by my subgrouping model or by the general model proposed by Kummer (1971) and Waser (1977: small patches--small groups). This eliminates most of the foraging hypotheses discussed above as explanations for the difference in group size. The foraging patterns described--very subjectively--suggest instead that the "Cody/Altmann" model (Harvest efficiency I, III above) may provide the best explanation because it is the one foraging hypothesis that suggests that groups should be larger when resources are dispersed in small patches. The macaques' pattern of use of isolated small fruit patches and other pieces of vegetation as well as their pattern of insect search suggest they may well be foraging together to minimize individual returns to "exhausted" small patches. Their use of large fruit trees might best be viewed as opportunistic relative to their other foraging, rather than vice versa. This subjective discussion awaits assault by appropriate data.

VI. A DIDACTIC DISCUSSION AND CONCLUSION

What has been gained by this review? No doubt in many readers'
minds there persists the conviction that grouping is related to the risks
of predation. Others remain convinced that primates live in groups because they are primates: phylogeny cannot be denied! Others favor female bonding and damn the predators! For me the result is the conviction that we need much more careful evaluation of the meaning of our
general hypotheses. So far, foraging hypotheses are considerably more
sophisticated than other hypotheses in considering alternative mechanisms. This does not make them "correct" or even more likely to be correct in some sense than "predation" hypotheses, but it does accentuate
the fact that supposed "tests" of the predation hypothesis are bound to
be overly simplistic until the complexity of relations between organismal
characteristics and the behavior of predators is as thoroughly considered
as the relations of organisms to resources. The phylogenetic hypothesis
is remarkably simplistic until appropriate measures are taken on all species compared to "test" it. When we knew very little about each species,
phylogenetic schemes accounted for much variation. Phylogenetic
schemes may still reduce variance, but the variables considered need
careful definition and scrupulous sorting to be sure they are the same for
each species entered in the scheme.

There is an interesting difference among the positions of advocates
of the influence of predation on variation in group size. Van Schaik and
van Noordwijk (1986) apparently argue for risk of predation as a true
independent *variable* in the complex equation predicting group size. Interestingly, they have been concerned primarily with intraspecific comparison of different populations. Terborgh, on the other hand, has been
concerned with interspecific comparison of coexisting species, and his
model (Terborgh 1983, modified appropriately by Terborgh and Janson
1986) (Fig. 3) explicitly suggests that risk of predation is a constant
function of group size--always decreasing with increasing group size,
regardless of species. In the latter case, for a given community of similar-sized primates, risk of predation is a "constant" function of group
size and interspecific variation must depend on other extrinsic variables.
Others (Wrangham 1980; and I in this chapter) choose to ignore predation
as a factor in the evolution of diferences in group size.

Despite several stimulating papers on the evolution of group size in
primates (Alexander 1974; Wrangham 1980; van Schaik 1983; Terborgh
1983; Andelman 1986; Terborgh and Janson 1986), two epistemological
issues remain for thoughtful analysis. All of us concerned with the issue
of variation in group size are caught in the sticky problem of sorting out
the evolutionary consequences of natural selection from the potential
for individual facultative response to current environmental variation,
and so far none of us has addressed this issue adequately (of course, such
an issue is not limited to our interests in group size and has only occasionally been properly considered in any functional analysis). In addition,
we are often puzzled by differentiation of *first causes* of grouping from
independent variables affecting variation in group size. Continuing

evaluation of hypotheses about grouping will require dispassionate evaluation of alternatives, which means even-handed application of standards for acceptance of evidence and/or interpretation to avoid subtle "advocacy" of one or another view. So, for example, Terborgh and Janson (1986) criticize Wrangham's (1980) model of female-bonded groups because it ignores the role of predation in maintaining grouping and because, they assert, ". . . a growing body of observational evidence indicates that coalitions of females are only occasionally involved in the defense of food resources and in aggressive intergroup encounters" (p. 117). The latter assertion is at least highly debatable, and might be weakened by the same logic used by Terborgh and Janson in partial support of predation as a pervasive influence on grouping: The fact that it is seldom seen *may* be interpreted as manifesting effectiveness of the strategy.

Despite its unsatisfying nature, the phylogenetic hypothesis may be correct. We have no good way of knowing whether we are observing a species in decline resulting from subtle selective disadvantages relative to other species in a community. But the phylogenetic hypothesis should be one of last resort *because it leaves us with nothing to do,* and because tests of phylogenetic effects should be based on the most thorough understanding of natural behavior possible. We should first be concerned with functional explanations of how current differences in group sizes are allowed or maintained. Pursuit of functional explanations will at the very least stimulate good, thorough field work that may provide a wealth of data for comparative studies testing the phylogenetic component of variation in group size. If all functional explanations fail, perhaps we will be left with the phylogenetic hypothesis, but until all other explanations fail, this hypothesis should be put aside. Comparative phylogenetic analysis may ultimately be significant, but let it be conducted by our unfortunate descendants in armchairs long after we functionalists and the primates and their forests have passed into memory.

ACKNOWLEDGMENTS

I am deeply indebted to D.R. Leighton for providing data on gibbons. I thank L. Berenstain for his generous contribution of data on subgrouping and range use by macaques gathered during a difficult two-year study. The analysis presented here is my own and the interpretation is my sole responsibility; thus, all errors are mine. The research in 1977-1980 was supported by NSF grant DEB76-21413. L. Berenstain was supported by NSF grant BNS81-19444. I have been supported by USPHS grant RR00169 and by the Committee on Research of the University of California, Davis, during preparation of this chapter. The Leightons, L. Berenstain and I have been sponsored in Indonesia by the Indonesian Ornithological Society and the Indonesian Institute of Sciences. We have consistently received cooperation and support from the Department of Nature Conservation and Wildlife Management and the Department of Forestry of Indonesia. It is always a pleasure to acknowledge the gener-

ous hospitality and logistic assistance of Mr. C.L. Darsono who has been
a constant friend and supporter in Jakarta.

REFERENCES

Alexander, R.D. 1974. The evolution of social behavior. *Annual Review
of Ecology and Systematics* 5:324-383.

Altmann, S.A. 1974. Baboons, space, time and energy. *American Zoologist* 14:221-248.

Andelman, S.J. 1986. "Ecological and social determinants of cercopithe-
cine mating patterns." pp. 201-206. In: D.I. Rubenstein and R.W.
Wrangham, eds. *Ecological aspects of social evolution*. Princeton:
Princeton University Press.

Berenstain, L. In press. Responses of long-tailed macaques to drought
and fire in eastern Borneo: A preliminary report. *Biotropica*.

Buss, L.W. 1981. Group living, competition and the evolution of cooperation in a sessile invertebrate. *Science* 213:1012-1014.

Carpenter, C.R. 1934. A field study of the behavior and social relations
of howling monkeys. *Comparative Psychological Monographs* 10:1-
168.

Carpenter, C.R. 1940. A field study in Siam of the behavior and social
relations of the gibbon, *Hylobates lar*. *Comparative Psychological
Monographs* 16:1-212.

Cheney, D.L. 1981. Intergroup encounters among free-ranging vervet
monkeys. *Folia Primatologica* 35:124-146.

Clark, C.W. and M. Mangel. 1986. The evolutionary advantages of group
foraging. *Theoretical Population Biology* 30:45-75.

Cody, M.L. 1971. Finch flocks in the Mohave desert. *Theoretical Population Biology* 2:142-158.

Collias, N. and C. Southwick. 1952. A field study of population density
and social organization in howling monkeys. *Proceedings of the
American Philosophical Society* 96:143-156.

Cohen, J.E. 1971. *Casual Groups of Monkeys and Men*. Cambridge, MA:
Harvard University Press.

Curtin, S.H. 1977. Niche separation in sympatric Malaysian leaf monkeys
(*Presbytis obscura* and *Presbytis melalophos*). *Yearbook of Physical
Anthropology 1976, Yearbook Series* 20:421-439.

Darwin, C.R. 1859. *On the Origin of Species*. London:MacMillan.

Denham, W.D. 1971. Energy relations and some basic properties of primate social organization. *American Anthropologist* 73:77-95.

Ellefson, J.O. 1974. "A natural history of gibbons in the Malay peninsula." pp. 1-136. In: D. Rumbaugh, ed. *Gibbon and siamang*, vol. 3.
Basel:S. Karger.

Fittinghoff, N.A., Jr. and D.G. Lindburg. 1980. "Riverine refuging in
East Bornean *Macaca fascicularis*." pp. 182-214. In: D.G. Lindburg,
ed. *The macaques: Studies in ecology, behavior and evolution*. New
York:Van Nostrand and Reinhold.

Fleagle, J.G. 1980. "Locomotion and posture." pp. 191-207. In: D.G. Chivers, ed. *Malvan Forest primates: Ten years' study in tropical rainforest.* New York and London:Plenum Press.

Ghiglieri, M.P. 1984. *The chimpanzees of Kibale Forest.* New York: Columbia University Press.

Hamilton, W.J. III and K.E.F. Watt. 1970. Refuging. *Annual Review of Ecology and Systematics* 1:263-286.

Horn, H.S. 1968. The adaptive significance of colonial nesting in the Brewer's blackbird (*Euphagus cyanocephalus*) *Ecology* 49:682-694.

Hutchinson, G.E. 1957. Concluding remarks. *Cold Spring Harbor Symposia on Quantitative Biology* 22:415-427.

Jolly, A. 1966. *Lemur behavior.* Chicago:University of Chicago Press.

Klein, L.L. and D.B. Klein. 1975. "Social and ecological contrasts between four taxa of neotropical primates." pp. 59-86. In: R.H. Tuttle, ed. *Socioecology and psychology of primates.* The Hague:Mouton.

Kummer, H. *Primate societies.* Chicago:University of Chicago Press.

Leighton, D.R. 1987. "Gibbons: territoriality and monogamy." In B.B. Smuts, D.L. Cheney, R.A. Seyfarth, T.T. Struhsaker and R.W. Wrangham, eds. *Primate societies.* Chicago:University of Chicago Press.

Leighton, D.R. and A.J. Whitten. 1984. "Management of free-ranging gibbons." pp. 32-43. In: H. Preuschoft, D.J. Chivers, W.Y. Brockelman and N. Creel, eds. *The lesser apes: Evolutionary and behavioral biology.* Edinburgh:Edinburgh University Press.

Leighton, M. 1982. Fruit resources and patterns of feeding, spacing and grouping among sympatric Bornean hornbills (Bucerotidae). Doctoral dissertation. Davis:University of California.

Leighton, M. 1986. "Hornbill social dispersion: variations on a monogamous theme." pp. 108-130. In: D.I. Rubenstein and R.W. Wrangham, eds. *Ecological aspects of social evolution.* Princeton:Princeton University Press.

Leighton, M. and D.R. Leighton. 1982. The relationship of size of feeding aggregate to size of food patch: howler monkeys (*Alouatta palliata*) feeding in *Trichilia cipo* fruit trees on Barro Colorado Island. *Biotropica* 14:81-90.

Leighton, M. and D.R. Leighton. 1983. "Vertebrate responses to fruiting seasonality within a Bornean rain forest." pp. 181-196. In: S.L. Sutton, T.C. Whitmore and A.C. Chadwick, eds. *Tropical rainforest: Ecology and management.* Oxford:Blackwell.

Maxon, W.A. 1966. Social organization of the South American monkey *Callicebus moloch. Tulane Studies in Zoology* 13:23-28.

Milton, K. 1980. *The foraging strategy of howler monkeys: a study in primate economics.* New York:Columbia University Press.

Milton, K. 1984. Habitat, diet and activity patterns of free-ranging woolly spider monkeys (*Brachyteles arachnoides* E. Geoffroy 1908). *International Journal of Primatology* 5:419-514.

Mitani, J.C. 1984. The behavioral regulation of monogamy in gibbons (*Hylobates muelleri.*) *Behavioral Ecology and Sociobiology* 15:225-229.

Napier, J.R. and P.H. Napier. 1967. *A handbook of the living primates.* London:Academic Press.

Neville, M.K. 1972. The population structure of red howler monkeys (*Alouatta seniculus*) in Trinidad and Venezuela. *Folia Primatologica* 17:56-86.

Richard, A.F. 1978. *Behavioral variation: Case study of a Malagasy lemur.* Lewisburg:Bucknell University Press and London:Associated University Press.

Rodman, P.S. 1978. "Diets, densities and distributions of Bornean primates." pp. 465-478. In: G.G. Montgomery, ed. *The ecology of arboreal folivores. Symposia of the National Zoological Park.* Washington:Smithsonian Institution Press.

Rodman, P.S. 1981. Inclusive fitness and group size with a reconsideration of group sizes in lions and wolves. *American Naturalist* 118:275-283.

Rodman, P.S. and J.C. Mitani. 1987. "Orangutans: Sexual dimorphism in a solitary species." In: B.B. Smuts, D.L. Cheney, R.A. Seyfarth, T.T. Struhsaker and R.W. Wrangham, eds. *Primate societies.* Chicago: University of Chicago Press.

Slobodchikoff, C.N. 1984. "Resources and the evolution of social behavior." pp. 227-251. In: P.W. Price, C.N. Slobodchikoff, and W.S. Gaud, eds. *A new ecology.* New York:John Wiley & Sons.

Smith, C.C. 1977. "Feeding behaviour and social organization in howling monkeys." pp. 97-126. In: T.H. Clutton-Brock, ed. *Primate ecology: Studies of feeding and ranging in lemurs, monkeys and apes.* London:Academic Press.

Southwick, C.H., M.A. Beg and M.R. Siddiqi. 1965. "Rhesus monkeys in North India." pp. 111-159. In: I. DeVore, ed. *Primate behavior: Field studies of monkeys and apes.* New York:Holt, Rinehart and Winston.

Struhsaker, T.T. 1969. Correlates of ecology and social organization among African cercopithecines. *Folia Primatologica* 11:80-118.

Struhsaker, T.T. 1975. *The red colobus monkey.* Chicago:University of Chicago Press.

Struhsaker, T.T. and L. Leland. 1979. Socioecology of five sympatric monkey species in the Kibale Forest, Uganda. *Advances in the Study of Behavior* 9:159-228.

Struhsaker, T.T. and J.F. Oates. 1975. "Comparison of the behavior and ecology of red colobus and black and white colobus monkeys in Uganda: A summary." pp. 103-124. In: R.H. Tuttle, ed. *Socioecology and psychology of primates.* The Hague:Mouton.

Sussman, R.W. 1974. "Ecological distinctions in sympatric species of *Lemur.*" pp. 75-108. In: R.D. Martin, G.A. Doyle and A.C. Walker, eds. *Prosimian biology.* London:Duckworth.

Terborgh, J. 1983. *Five New World primates: A study in comparative ecology.* Princeton:Princeton University Press.

Terborgh, J. and C.H. Janson. 1986. The socioecology of primate groups. *Annual Review of Ecology and Systematics* 17:111-135.

van Schaik, C.P. 1983. Why are diurnal primates living in groups? *Behaviour* 87:120-144.

van Schaik, C.P. and J.A.R.A.M. van Hoof. 1983. On the ultimate causes of primate social systems. *Behaviour* 85:91-117.
van Schaik, C.P. and M.A. van Noordwijk. 1986. The evolutionary effect of the absence of felids on the social organization of the Simeulue monkey (*Macaca fascicularis fusca*, Miller 1903). *Folia Primatologica* 44:138-147.
van Schaik, C.P., M.A. van Noordwijk, R.J. de Boer and I. den Tonkelaar. 1983a. The effect of group size on time budgets and social behaviour in wild long-tailed macaques (*Macaca fascicularis*). *Behavioral Ecology and Sociobiology* 13:173-181.
van Schaik, C.P., M.A. van Noordwijk, R.J. de Boer and I. den Tonkelaar. 1983b. Party size and early detection of predators in Sumatran forest primates. *Primates* 24:211-221.
Ward, P. and A. Zahavi. 1973. The importance of certain assemblages of birds as "information centres" for food finding. *Ibis* 119:517-534.
Waser, P. 1977. "Feeding, ranging and group size in the mangabey *Cercocebus albigena*." pp. 183-222. In: T.H. Clutton-Brock, ed. *Primate ecology: Studies of feeding and ranging in lemurs, monkeys and apes*. London:Academic Press.
Wilson, E.O. 1975. *Sociobiology: The new synthesis*. Cambridge, MA:Harvard University Press.
Wrangham, R.W. 1977. "Feeding behaviour of chimpanzees in Gombe National Park, Tanzania. pp. 504-538. In: T.H. Clutton-Brock, ed. *Primate ecology: Studies of feeding and ranging in lemurs, monkeys and apes*. London:Academic Press.
Wrangham, R.W. 1980. An ecological model of female-bonded primate groups. *Behaviour* 75:262-300.
Wrangham, R.W. 1986. "Ecology and social relationships in two species of chimpanzee." pp. 352-378. In: D.I. Rubenstein and R.W. Wrangham, eds. *Ecological aspects of social evolution*. Princeton:Princeton University Press.

Chapter 6

RESOURCES, PHILOPATRY, AND SOCIAL INTERACTIONS AMONG MAMMALS

Peter M. Waser

Department of Biological Sciences
Purdue University, West Lafayette, IN 47907

I. INTRODUCTION

Only a small proportion of mammalian species are gregarious, and even fewer live in stable, cooperative groups. The thesis of this chapter will be that in understanding why the few gregarious species got that way, much is to be learned by studying the majority that did not.

Ecological factors can influence the evolution of social structure on at least three levels. Most obviously, ecological factors can create the opportunity for individuals to gain by association or cooperation.

At this level, many causal relationships have been postulated to link ecology and behavior. For instance, spatiotemporal dispersion of resources might influence the costs and benefits of group foraging (Rodman this volume; Slobodchikoff and Schulz this volume), or of sharing information about food locations (Heinrich this volume). Habitat characteristics might influence how far predators can detect prey, and vice versa, thereby determining whether gregariousness increases or decreases predatory risk. The nature of food and of predators determines whether it is feasible for nonreproductives to bring food back to the den, or drive predators off, or defend a feeding territory, and thereby determines whether cooperative rearing increases juvenile survival rates (Ligon this volume).

However, it has frequently been pointed out that whatever benefits gregariousness has, it also always carries costs, most prominently that individuals in a group must share a common resource set and therefore are likely to compete (Alexander 1974). Therefore, there is a second, perhaps more basic level at which ecological factors can influence the evolution of social behavior: by decreasing the costs of home range sharing.

Finally, it is now well established that thresholds for the evolution of cooperative behaviors will be lower if cooperating individuals gain indirect, as well as direct increments of fitness, i.e., if they are related (e.g., Brown 1986). Moreover, a precondition for the initial spread of cooperation by reciprocation among nonrelatives--the tit-for-tat strategy in an iterated prisoner's dilemma game--is that individuals that share the strategy are clustered, in effect that neighbors are kin (Axelrod 1984). This raises still a third level at which ecological factors can influence social evolution: by increasing the likelihood that individuals sharing home ranges are related.

This chapter will concentrate on the last two levels, that is, how ecology might influence the cost of sharing home ranges and the probability that neighbors are kin. These conditions are likely to be closely intertwined. Surveys of mammalian species (Greenwood 1980; Waser and Jones 1983) indicate that home range sharing is most often associated with "natal philopatry," the prolonged attachment of individuals to their natal home range, and that natal philopatry is nearly always the mechanism by which kin end up as neighbors. Range sharing and natal philopatry are both far more widespread among mammals than are either gregariousness or cooperation. Therefore, I will catalog the mechanisms by which ecological conditions might tip the selective balance in favor of natal philopatry, concentrating on that subset of mechanisms which could lead to philopatry in combination with home range sharing. I will then assess the relative importance of these mechanisms in a species that shows both natal philopatry and home range sharing, but not gregariousness (the bannertailed kangaroo rat *Dipodomys spectabilis*).

II.   WHAT ECOLOGICAL CONDITIONS FAVOR PHILOPATRY?

A.   Opportunistic Philopatry

The most basic, but probably least-explored mechanism by which ecology might influence the probability that individuals have related neighbors is through its effects on age-specific birth and death rates.

This follows because demography determines how rapidly home range ownership can turn over: death rates determine how often new openings arise, and birth rates determine the number of contestants for those openings.   In many, perhaps most mammals, juveniles practice what might be termed "opportunistic philopatry," remaining on their natal range after reproductive maturity if competing adults (generally the same-sexed parent) have died.   Sharing a home range with other adults, including parents, will in most cases reduce reproductive success, so that in these same species juveniles are not philopatric when they reach reproductive maturity unless their natal range is uncontested.   In 1967, Bertram Murray pointed out that animals practising opportunistic philopatry and dispersing in a straight line to the nearest available site should produce a characteristic negative exponential distribution of dispersal distances.   The exact form of the expected distribution depends on the rate with which home ranges turn over, in other words on demography.

Figure 1 illustrates the consequences of opportunistic philopatry if home range turnover rate is intermediate: the probability of a given site's being uncontested when a juvenile reaches maturity on it is 0.4.   If juveniles disperse in a straight line, the dashed line shows the expected proportion of juveniles that will settle 1, 2, ..., n home ranges away from the natal site.   Of course an animal dispersing in a straight line might miss some empty sites if it started off in the wrong direction; the solid line shows what happens if it finds the absolutely nearest empty site, say by searching in a spiral out from its natal home range.   Most mammals do something intermediate, for instance moving to the nearest empty site if it is adjacent to the natal range but searching in a relatively straight line if not.   With an intermediate turnover rate, most juveniles will settle on natal or adjacent home ranges.   Under these conditions, sibs are likely to end up as neighbors.

The likelihood that neighbors are kin will be lower in a population experiencing high rates of home range turnover, e.g., one living in unstable habitats and characterized by high mortality rates (Fig. 1C).   Under these conditions most juveniles will be philopatric if they survive to adulthood, but their parents will be dead and surviving siblings will be rare.

On the other hand, if ecological conditions are highly stable so that mortality rates are low, home range turnover will be slow and opportunistic philopatry alone will not lead relatives to be neighbors (Fig. 1B). Neither the natal nor an adjacent home range is likely to be available, and most juveniles will have to disperse far from relatives before finding an opening.

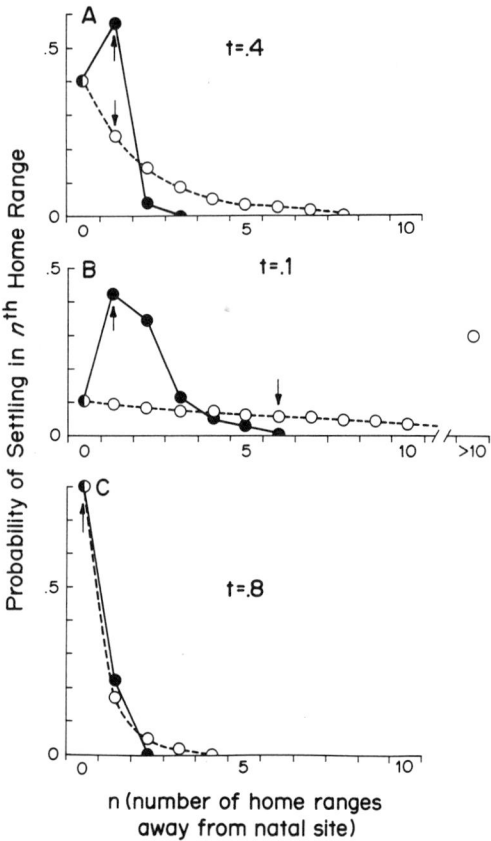

Figure 1: Probabilities of philopatry and dispersal to various distances if philopatry is opportunistic. Demographic characteristics determine t, the probability that a home range becomes available between birth and reproductive maturity. The probabilities of philopatry and of dispersal to an immediately adjacent home range are particularly relevant to the likelihood that an individual has related neighbors. (a) Mortality rates are moderate, most juveniles find openings either on or adjacent to the natal range, and the likelihood that relatives are neighbors is maximal. (b) Mortality rates are high, young that survive to maturity are likely to be philopatric but without surviving relatives. (c) Mortality rates are low, juveniles can rarely settle either on or adjacent to the natal site so that relatives are again unlikely to be neighbors. See Waser (1986) for further explanation.

B.   Philopatry by Parental Consent

Opportunistically philopatric animals can live adjacent to kin, but by definition they do not share home ranges with them. More interesting from the point of view of social evolution is the phenomenon of philopatry by parental consent--philopatry that arises through tolerance by the parent(s) of grown offspring on the natal range.

What ecological conditions could, in principle, favor selection of parental tolerance for philopatric young? One answer is that parental tolerance should evolve when the presence of philopatric young does not decrease parental survival or fecundity, i.e., tolerating them requires no parental investment (Trivers 1972). Under this set of conditions selection should favor not only philopatry by parental consent, but any form of home range sharing.

Alternatively, parental tolerance of philopatric young could evolve even if it decreases parental survival or fecundity, as long as that parental investment pays off in increased survival or reproductive success of the offspring.

1.   Nondepreciable resources

Stuart Altmann (1974, p. 235) early suggested a link between the spatial dispersion of resources and the potential for partial sharing of home ranges: "home range overlap . . . will be low in relatively uniform habitats and will be extensive if several essential resources have very restricted distributions." Altmann specifically discussed sleeping trees or cliffs and water holes as examples of such essential, but potentially shared, resources.

Other lines of argument have suggested conditions under which not only particular resource patches, but complete home ranges, might be shared. In one of the earliest of these, Horn (1968, p. 688) demonstrated that it was energetically most efficient for central-place foragers to adopt overlapping home range if "enough food to support four nests is distributed among 16 points at a . . . food is assumed to be present at only one of the points at any particular time, but . . . at different points at different times." There have been surprisingly few attempts to generalize Horn's model to more complex or realistic resource distributions, but numerous variants of Horn's approach have led to the same general conclusion, that home range sharing is likely when resource distributions are both patchy and "moving" (cf. MacDonald 1983; Waser 1976; Waser and Wiley 1980; Wittenberger and Dollinger 1984).

In contrast, Von Schantz (1984) has argued that social groups (and thus implicitly, home range sharing) can occur even if resources are spatially homogeneous, given the right pattern of temporal variation in resources: when overall resource levels fluctuate interannually and bottlenecks occur at least once during an individual's life span, home range overlap can arise during the "better" years. In addition, spatial homogeneity of resources need not prevent home range sharing if resources renew (Waser 1984).

These apparently disparate lines of argument share a common theme: home ranges (or parts of home ranges) can overlap if the use of a limiting resource by one individual does not reduce its availability to a second. The special case of overlap between parent and mature offspring would then require no parental investment. Under these conditions, the resource can be termed "nondepreciable" (Altmann et al. 1977; cf. Charnov et al. 1976). The less depreciable limiting resources are, the less strongly selection should oppose home range sharing--including sharing by parent and philopatric young.

a. Nonconsumable resources. Unlike most resources, dens, sleeping sites, or hibernaculae are not consumed by their users. Unless the site is physically too small to share, or unless sharing it increases its conspicuousness to predators, one individual may tolerate another there without cost. Yet such sites often vary widely in quality, and occupying a good one can markedly increase an individual's chances of escaping predators or surviving adverse environmental conditions (e.g., Armitage this volume).

It is important to note that sharing a den or similar resource will be cost-free only if shared use does not lead to competition for other critical resources. For instance, the dispersion and density of food sources will determine whether sharing dens lead to sharing food sources, and whether sharing food sources reduces survival or fecundity.

b. Time-limited resources. While food and most other resources are consumed by their users, it does not follow that when one individual allows another access to the resource it necessarily cuts its own intake in half. This is true for two reasons: (1) processes other than consumption by the individuals concerned may deplete the resources, and (2) resources may renew themselves.

Arguments that link home range sharing to patchy, "moving" resource distributions assume that resources in a patch disappear whether or not foragers consume them; availability of the resource is time-limited. This assumption, sometimes implicit, is in fact a critical one, as the following examples suggest. Even for homogeneously distributed resources, costs of sharing are decreased if resource availability is time-limited.

A raven finding a newly opened carcass (Heinrich this volume) will never be able to consume the entire carcass itself, because other species, from microorganisms to mammalian carnivores, will exploit it too. The first raven loses nothing by sharing with a second *as long* as the second consumes only food that would otherwise be consumed by coyotes, or rot. Paradoxically, competition from other species decreases the cost to a forager of sharing the resource with a conspecific.

Competition from other species is not the only process that will deplete food resources: seeds germinate, edible buds mature into inedible leaves, prey escape into inaccessible refuges. Thus a single gibbon encountering a newly fruiting tree (Rodman this volume) would not have time to consume its entire crop before at least some fruit ripened and fell to the ground. The fact that fruits become unavailable through processes other than consumption decreases the cost to a gibbon of sharing the fruit crop.

The cost of sharing a resource can even be reduced by side effects of a predator's own foraging. A bat-eared fox encountering harvester termite workers cannot consume the entire termite population, because its foraging provokes a pheromone-mediated escape response (Lamprecht 1979). Again, a fox loses nothing by foraging with another fox as long as it removes only termites that would otherwise escape underground, where they are effectively inaccessible.

c. Renewing resources. Where resources renew themselves, availability to a forager is dictated by equilibrium, rather than initial resource levels. Where renewal rates are not influenced by the rate of harvesting--water flowing out of a spring, for instance, or detritus washing up on a shoreline--exploitation by a single forager will reduce the equilibrium resource density but the addition of a second forager has little additional effect. Sharing is particularly "cheap" if resources renew relatively rapidly or if resource depression caused by any single forager is slight (Waser 1984). The details depend on the dynamics of renewal, and where renewal rate is influenced by the intensity of harvesting, a second individual's foraging could even *increase* the equilibrium prey level. Such effects are well known in studies of predator-prey population dynamics (e.g., optimum yield) and have been documented in some plant-herbivore systems (McNaughton 1979). The effects of renewal and time limitation can reinforce each other.

d. "Permanently" superabundant patches. Altmann's (1974) prediction of home range overlap at local concentrations of essential resources, unlike most "patchy resource" arguments, does not assume that resource availability in a patch is time-limited. Instead, it depends on the presence of multiple resource constraints. For instance, the availability of food might keep baboon populations low enough that water in the local water hole, though essential, would always be superabundant. The effects of multiple resource constraints on the costs of sharing particular resources remain largely unexplored.

2.   Philopatry involving parental investment

An alternative to the "nondepreciable resource" hypothesis is that parents tolerating philopatric offspring *do* pay a cost in future survival and reproduction. In this view philopatry by parental consent requires postweaning parental investment, but parental investment has a payoff in more grandchildren.

Two variations of this theme can be imagined. In one, home ranges vary widely in quality, but a good home range one season is also good the next. Under these circumstances, a parent tolerating its own offspring on a better-than-average home range thereby increases its offspring's reproductive success even if dispersal is cost-free. Alternatively, if dispersal risks are extremely high, a parent may be able to increase its offspring's survival by tolerating philopatry, at least until an adjacent home range has opened up or the offspring is large or experienced enough to make it through dispersal successfully.

a. Heterogeneous habitats. Imagine that female home ranges can be ranked by the reproductive success of their occupants (Fig. 2), that the

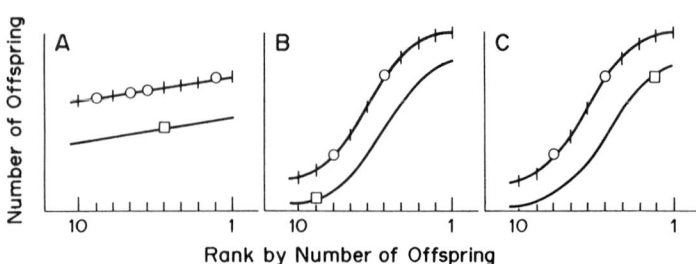

Figure 2: How heterogeneous habitats might allow parents to increase offspring reproductive success by tolerating philopatry: the "philopatry threshold" model. Female home ranges are ranked by the number of offspring produced in them and this ranking is assumed to be related to home range quality and stable between breeding seasons. In the simplest case, sharing a home range is assumed to reduce both parent and offspring reproductive success and dispersing offspring are assumed always to find an empty home range. Circles denote home ranges that become available while the juvenile in the home range indicated by a square is maturing. (a) Habitat homegeneous, home ranges differing only slightly in "quality." (b) Habitat heterogenous, home ranges differing markedly in quality, offspring on low-quality home range. (c) Habitat heterogeneous, home ranges differing markedly in quality, offspring on high-quality home range. Under condition (c), but not (a) and (b), philopatric young would expect higher reproductive success than dispersers.

rank of a particular home range is a reflection of the resources it contains, and that home range rankings are stable from one reproductive season to the next. For simplicity, assume also that sharing a home range decreases reproductive success substantially for both residents (i.e., it requires parental investment) and that dispersal, when it occurs, is always successful.

Whether a maturing juvenile would maximize its own reproductive success by dispersal or philopatry will depend on (among other things) habitat heterogeneity and the quality of its natal range. The following examples illustrate the logic of the model:

(1) Little habitat heterogeneity, home ranges similar in "quality" (Fig. 2A): If the average unoccupied home range provides enough resources to ensure higher reproductive success than can be achieved by sharing the best home range, offspring on any home range would do best to emigrate.

(2) Considerable habitat heterogeneity, high variance in home range quality, offspring on a low-quality home range (Fig. 2B): The resources on an average unoccupied range would again ensure higher reproductive success than attainable on the shared natal range, and offspring should disperse.

(3) Considerable habitat heterogeneity, offspring on a high-quality home range (Fig. 2C): The average unoccupied range would have a lower payoff than the shared natal range, and offspring should be philopatric.

This model might be termed the "philopatry threshold" model, by analogy to Orians' (1969) polygyny threshold model, particularly the variant outlined by Brown (1975; cf. Koenig and Pitelka 1981). If the cost of sharing home ranges is decreased, or if successful dispersal is not guaranteed, philopatry becomes a more attractive option. Logic would be qualitatively similar from the parent's perspective: philopatry by parental consent should be favored in superior home ranges where habitat heterogeneity is high.

b. Risky dispersal. A final route by which ecological circumstances can tip the balance towards philopatry by parental consent is by decreasing the likelihood of successful dispersal. For instance Andrew Smith (1974) has shown that pikas living on isolated talus slopes face thermoregulatory collapse, as well as predation, starvation, and aggression from established residents if they attempt to disperse. Not surprisingly, pikas often stay home with their relatives (Smith and Ivins 1983).

For pikas, it is primarily the patchy distribution of suitable habitat that makes dispersal risky. The costs of dispersal are compounded by small size and consequent vulnerability. Patchy habitat distribution is not, however, the only ecological condition that might drive animals toward philopatry (Waser and Jones 1983). Species that require extensive home range "improvements" such as burrow systems or food caches suffer a parallel problem: dispersers that must undertake such improvements themselves (rather than finding empty, "improved," home ranges) are less likely to survive and reproduce. Dispersers are under a similar handicap where predators or resource types are such that familiarity with the home range markedly increases safety or foraging success. Finally, low mortality rates and habitat saturation, by increasing the distances dispersers have to move to find or successfully compete for an opening, stack the cards in favor of philopatry (cf. Brown 1974).

III.   ECOLOGICAL PRECONDITIONS FOR PHILOPATRY: A CASE STUDY

Since 1979, W. Thomas Jones and I have been following a population of several hundred bannertailed kangaroo rats (*Dipodomys spectabilis*) on a 64 ha site in Southeastern Arizona (Jones 1982, 1984, 1986, 1987). Ecological conditions that allow the first steps towards the evolution of cooperative, stable groups--those that influence the cost of sharing home ranges and the probability that neighbors are kin--are best investigated in species at the edge of gregariousness, but not quite there. *D. spectabilis* is such a species.

A. Background and Methodology

Dipodomys spectabilis are among the largest of the heteromyid rodents, and like all their congeners, they are solitary and granivorous. After the summer rains, annuals and herbaceous perennials set a large seed crop which the kangaroo rats, foraging singly, clip and store for the winter. Each animal dens and amasses its fall seed cache in a large mound, again to a first approximation singly. Males and females live in separate mounds, interdigitated in the population (Vorheis and Taylor 1922; Holdenreid 1957; Monson 1943; Schroder and Geluso 1975).

Inside each mound is a labyrinth of tunnels and seed caches, storing up to 6 kg of seed. Mounds and their seed caches take many months to build (Best 1972; Jones 1982). New mounds are rarely constructed--on our study site, only about once in 90 mound-years. Most adults have only one mound, but if a neighbor dies the mound is quickly occupied and used as a secondary mound by an adjacent adult. Individual rats may live as long as five years, but mounds last for decades.

Mating begins in midwinter, and the modal female produces two litters of one young each year (Jones 1982, 1984). The young are weaned at about four weeks but remain in the natal mound for months. Two successive litters can occupy their mother's mound simultaneously and in good years, a few young females breed in the natal mound concurrently with their mother. When young animals do disperse it is often to one of their mother's secondary mounds. Dispersal distances are correspondingly short for both sexes, the great majority of individuals breeding less than 50 m from their natal mounds (Fig. 3). During most of the year, mound- and particularly home-range sharing between mothers and

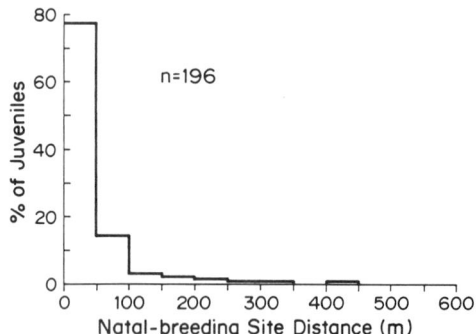

Figure 3:   Observed probabilities of philopatry and dispersal for Dipodomys spectabilis. Histograms plot the proportion of 196 juveniles born in the central study area between 1979 and 1983 that reached sexual maturity and age one in mounds at different distances from the natal site. Since the mean distance between occupied mounds during this period was approximately 50 m, the histogram illustrates the probabilities of philopatry and of successful dispersal to adjacent home ranges. During this period, the central study area is 600 m on a side.

weaned offspring is common, natal philopatry is the rule, and neighbors are often related.

B.   Is Philopatry Opportunistic?

Whether philopatry is solely opportunistic, its high incidence dictated by kangaroo rat demography is readily determined from survivorship and fecundity data.   Quarterly censuses of all mound occupants (e.g., Jones 1984) allow us to estimate fecundity, survival and dispersal distances.   For adult females in our main study population, the annual mortality rate is 0.55.   During the first five years of the study, 196 juveniles in 115 families survived to at least one year of age; given the observed adult mortality rate, 63 should have been opportunistically philopatric.   In reality, 156 of these 196 juveniles bred within 50 m, approximately one home range diameter, of their natal mound (Fig. 3). Clearly, demography alone does not "explain" the observed high frequency of philopatry.   The extra philopatry is a consequence of three phenomena: mothers sharing mounds with mature offspring, mothers donating secondary mounds within their home ranges to offspring, and mothers abdicating their mounds and moving elsewhere.   All three phenomena clearly involve philopatry by parental consent (Jones 1987; Waser 1987).

C.   Does Philopatry by Parental Consent Occur Because Habitats are Heterogeneous?

D. spectabilis fulfill two of the preconditions suggested by the habitat heterogeneity hypothesis to lead to philopatry (and parental tolerance of philopatry) in high-quality home ranges.   First, a small number of females' mounds produce 4-5 young in a year, while the majority produce only one (Fig. 4).   Second, as will be discussed below, sharing a mound appears to cost a female relatively little.   Therefore, females owning mounds in which they produce large numbers of young would do better to tolerate philopatry, while females owning less productive mounds should not.

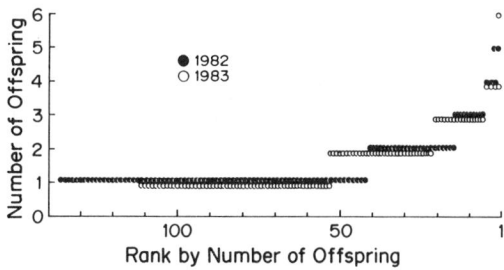

Figure 4:   Variation in home range "quality."   Number of D. spectabilis juveniles born in each mound in the central study area, 1982 and 1983.

However, we have found no tendency for juveniles in "high-quality" home ranges, as measured by their mothers' success in reproducing there, to be more inclined to stay home. In families that produce at least one young surviving to breed, the proportion with at least one *philopatric* young is independent of the number of young produced (Table I). This is true whether one defines philopatry narrowly, as remaining at age one in the natal mound (and presumably breeding there), or whether it is defined more broadly, as remaining in a mound (50 m from the natal site; thus, using at least part of the natal home range).

TABLE I. Is philopatry more likely in "high-quality" home ranges?

(a) "Philopatry" = use of natal den at age one.

Home range quality (no. of offspring produced at den during 1 year)	Incidence of philopatry (no. of dens that produced >1 surviving young, including n that were philopatric)		Percent of dens producing $\geq$ philopatric young
	$n = 0$	$n \geq 1$	
1	68	55	45
>1	40	23	37

$G = 0.58$; $p \approx .5$, N.S.

(b) "Philopatry" = use of den; <50 m from natal den at age one.

Home range quality (no. of offspring produced at den during 1 year)	Incidence of philopatry (no. of dens that produced >1 surviving young, including n that were philopatric)		Percent of dens producing $\geq$ philopatric young
	$n = 0$	$n \geq 1$	
1	32	91	74
>1	14	49	77

$G = 0.34$; $p \approx .5$, N.S.

Note: data from all dens in central study area, 1979-1983.

In retrospect it seems likely that the philopatry threshold model fails for *D. spectabilis* because one of its assumptions is not met: good home ranges one year are not necessarily good the next. There was no correlation across mounds between the number of offspring produced in 1982 and 1983 (Table II). Either the number of offspring produced in a particular mound is determined by female, rather than habitat "quality," or else critical aspects of the mound or of food resources vary unpredictably fron one year to the next. *Eriogonum aberti* plant and seed densities were not correlated across sample quadrats between years (Table II).

D. Does Philopatry by Parental Consent Occur Because Resources are Nondepreciable?

We have approached this question in two ways: indirectly, by attempting to measure how patchy and rapidly renewing resources are; and directly, by experimentally manipulating the duration of philopatry.

To assess the distribution of critical food resources, we counted plants and/or numbers of flower buds, flowers, and seeds for 35 plant species known to be harvested by *D. spectabilis*. We sampled 1 m^2 quadrats evenly distributed 12.5 m apart. In 1980-81, we sampled the entire 36 ha central study area in early fall. In 1982-83, we used permanently marked quadrats and sampled 6.75 ha with high *Dipodomys* density in

TABLE II. Does home range "quality" remain stable across years?

	Number of Sampling Units (n)	Correlation across Years (r)	p
Measure of "Quality"	n	r	p
a) number of offspring produced in 1982 vs. 1983, for all dens on central study area producing $\geq$ 1 young in each year	62	.005	>.9
b) number of *Eriogonum* plants on 1 m^2 quadrats spaced at 12.5 m intervals in central study area, fall 1980 vs. fall 1982	2401	.06	.4
c) number of *Eriogonum* seeds on 1 m^2 quadrats spaced at 12.5 m intervals in high *Dipodomys*-density areas, fall 1982 vs. fall 1983	885	-.02	.3

spring, summer, and fall. 1982 data were grouped into 50 x 50 m samples--the size approximating that of a *Dipodomys* home range--and compared with the number of offspring produced in those areas in 1983. Regressions of spring and summer seed production on offspring production were not significant, but fall seed production by *Eriogonum aberti* was (r^2 = 0.19, p = 0.02). Together, seed production by the seven most common species in fall explained 0.37 of the variance in offspring production the following spring.

We also estimated rates of seed renewal, both into seed-free soil and into artificial depressions. In the first case, we removed seeds from the top 2 cm of 100 cm^2 soil patches by flotation, replacing the soil to its initial level; in the second, we implanted arrays of 2 cm-deep plastic cups flush with the soil surface. In both cases, we later removed samples (either 10 cc of soil after 2 months or randomly chosen seed cups at 10-day intervals), extracted seeds by flotation in potassium carbonate (Nelson and Chew 1977), and weighed and counted them.

Since 1985, we have taken a more direct approach to asking whether resources are depreciable, manipulating the timing of juvenile dispersal and following the mother's subsequent survival and reproduction. As soon as juveniles are reliably trappable--at weights of 60-90 g--we encircle their natal mound with a temporary enclosure. Juveniles of this weight are at least two weeks post-weaning and can disperse naturally, particularly when population density is low. We leave mothers in the enclosure, but trap juveniles, radiotag a subset of them, and then either release them outside the enclosure--thereby forcing them to disperse--or return them to the enclosure, where they serve as philopatric controls. The enclosure is removed after approximately one week, at which point "control" juveniles remain in the natal mound, while "forced-dispersal" juveniles are apparently unable to reenter it.

So far, manipulations of dispersal timing have not exposed any significant cost to tolerating philopatric young, either in reproduction or in survival. Mothers whose offspring were forced to disperse were not more likely to reproduce again, either during the same or the next breeding season, than control mothers whose offspring were returned to the enclosure and were thus philopatric (Table III). Mothers whose young were forced to disperse did have a slightly higher probability of surviving to the next year than control mothers, and also than unmanipulated mothers (whose young were therefore philopatric), but neither difference was statistically significant (Table III).

Experimental manipulations therefore suggest that, to a first approximation at least, *D. spectabilis* resources are nonpredictable and tolerance of philopatry requires no parental investment. However, these results should be viewed as preliminary for several reasons.

First, the experiments have been conducted during a 7-year population low; females may have found either food, or mounds, or both to be superabundant. Presumably because unoccupied mounds were common, even the control juveniles dispersed earlier than usual, thus reducing any costs to mothers from sharing mounds.

TABLE III. Is tolerance of philopatry parental investment?

(a) Effects of manipulation on reproduction: number of mothers that produce a subsequent litter the year they are manipulated.

	Litter	No Litter	Percent Producing Litter
17 mothers of offspring forced to disperse	3	14	18
13 mothers of offspring returned enclosure	2	11	15

p = 0.37, Fisher test

(b) Effects of manipulation on reproduction: number of mothers producing a litter the year *after* they are manipulated.

	Litter	No Litter	Percent Producing Litter
8 surviving mothers of offspring forced to disperse	6	2	75
5 surviving mothers of offspring returned to enclosure	2	3	40

p = 0.21, Fisher test

(c) Effects of manipulation on survival: number of mothers surviving to the next breeding season.

	Survive	Die	Percent Surviving
16 mothers of offspring forced to disperse	9	7	56
11 mothers of offspring returned to enclosure	5	6	45

p = 0.26, Fisher test

Table III (continued)

	Survive	Die	Percent Surviving
16 mothers of offspring forced to disperse	9	7	56
24 unmanipulated mothers	9	15	38

p = 0.13, Fisher test

Second, since unoccupied mounds were so common, most of the off-spring that were forced to disperse were able to settle within 50 m of their mother. Although they ceased sleeping in the natal den, it is quite possible that they continued to use part or all of the natal home range and even to raid the mother's seed cache. Thus, the results are perhaps most clearly an indication that parental investment is not required for sharing the mound *per se*. Mounds, as a source of protection from predators and from thermal stress (Kay and Whitford 1978), would seem to be "nonconsumable" and thus nondepreciable.

Third, unlike the mounds themselves, the food caches that mounds contain are preeminently depreciable, being neither time-limited nor renewing. Resources elsewhere in the home range also seem likely to be depreciable. For instance, while *Eriogonum* availability was clearly patchy on a small scale, as indicated by the overrepresentation of both empty and seed-rich 1 m^2 quadrats, on the 50 x 50 m scale of kangaroo rat home ranges seeds were evenly dispersed relative to Poisson expectations. Other plant species have patchier distributions, but the relative synchrony of seed production by most plant species suggests that the home range large enough to support one animal throughout the year is not likely to support others on occasion.

D. spectabilis not only clip seed heads, but also harvest seeds in soil, particularly when they occur in local concentrations such as those formed in soil depressions (Reichman 1984). *A priori* it seems possible that wind transport might "renew" soil seed levels rapidly enough to make them effectively nondepreciable, but our renewal measurements have not confirmed this. We found that (1) no detectable seed renewal occurred into bare soil; (2) artificial depressions, which are the type of site most likely to trap windblown seeds, do so at rates of less than one seed/day even during the fall peak of seed production; and (3) outside the fall peak, seed renewal into artificial depressions occurs at one-fourth this rate or less (Fig. 5).

In summary, experimental manipulations do not detect a cost of mound-sharing to mothers, but the cost of sharing other resources within the home range, or of sharing mounds in a saturated population, remain to be determined.

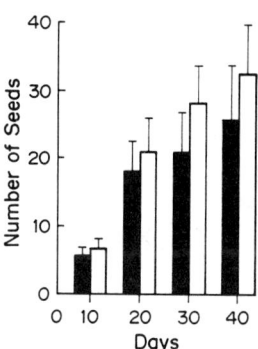

Figure 5:   Seed "renewal" in artificial depressions, illustrating the low rate of seed influx even into effective seed traps during periods of peak seed production.   Solid bars are mean (S.E.) number of Eriogonum aberti seeds and open bars are mean (S.E.) numbers of seeds from all species, recovered from 3 cm in diameter, 2 cm in depth, plastic cups at 36 locations after 10, 20, 30 and 40 days.   Seeds arrived more slowly (or were removed more rapidly) as the seed cups filled; mean rates of arrival were 0.6 seeds/day (Eriogonum aberti) and 0.8 seeds/day (all species). These data are from fall 1982, during the period of peak seed production in a productive year; seed arrival rates from spring and summer experiments at the same locations were 0.1 and 0.2 seeds/day (all species).

E.   Does Philopatry by Parental Consent Occur
Because Dispersal is Risky?

Since it remains a possibility that tolerating philopatry requires parental investment, for philopatry to spread requires that tolerant parents produce offspring with either higher reproductive success or a higher probability of survival. As noted earlier, changes in home range quality between years deny tolerant parents a payoff in increased offspring reproductive success even if they inhabit superior home ranges.   Does tolerating philopatry have a payoff in increased offspring survival?

Two sources of information bear on this question. First, the risks and causes of mortality can be compared between offspring experimentally forced to disperse and those allowed to remain in the natal mound as long as they choose. Second, survival rates can be compared between offspring that disperse naturally and those that do not. These comparisons can be used to estimate the advantages associated with philopatry at two stages: first, the advantage of not having to disperse and establish a new residence; and second, the advantage of living in the natal home range rather than elsewhere.

Both types of advantage appear to be substantial for *D. spectabilis*. Recall that we manipulated juveniles at an age at which some disperse naturally.   Nevertheless, forcing juveniles to disperse reduced the median duration of post-manipulation survival from 90 days (n = 6, excluding trap deaths, for philopatric, control juveniles) to 32 days (n = 12, U = 60,

one-tailed p).   All control juveniles survived the first month post-
manipulation, but less than half of the forced dispersers did (Table IV).
Mortality during dispersal and establishment in a new home range
was predominantly caused by predation.   Of 21 juveniles who carried
radios when they died or disappeared, the proximate cause of death was
predation by snakes, owls, or canids in seven cases and probably predation (overnight disappearance) in the rest.   Overnight disappearance is
interpreted as predation because the longest recorded nightly movement
by a radio-tagged juvenile was 200 m, and because no juveniles whose
signals were lost for more than one night were ever retrapped on the
600 x 600 m study site. Predation may be elevated in dispersers because
they have less access to or familiarity with predator-safe sites; but in
addition, dispersers establishing a new home range may suffer elevated
risks because they have difficulty finding adequate food: two of the 12

TABLE IV. How risky is dispersal?

(a) Risks associated with dispersal and establishment in a new home
range: survival through the first month post-manipulation.

	Survive	Die	Percent Surviving
12 juveniles forced to disperse	5	7	42
9 juveniles returned to enclosure	9	0	100

p = 0.01, Fisher test

(b) Risks associated with living in a non-natal mound: survival through
the first winter

	Survive	Die	Percent Surviving
110 unmanipulated juveniles that successfully established them-selves in non-natal mounds	42	68	38
37 unmanipulated juveniles that successfully established them-selves in their natal mounds	24	13	65

p = 0.01, log likelihood (G) test

forced dispersers showed weight losses before disappearing, and adding seed to mounds colonized by an additional set of forced-dispersal juveniles increased median postmanipulation survival from 32 days (n =12) to 67 (n = 6, U = 55, one-tailed p = 0.05).

Mortality rates in juveniles forced to disperse may be somewhat higher than in juveniles that disperse of their own volition (or when the mother drives them out). On the other hand, when population densities are higher than they were during the experimental period--which is most of the time--empty mounds would be in shorter supply and dispersal and establishment could be even more risky than suggested by these results.

Even after successfully establishing themselves in a new mound, dispersing *D. spectabilis* juveniles are at a disadvantage compared to those that are philopatric. Jones (1986) found that juveniles spending their first winter in their natal den are almost twice as likely to survive to their first breeding season than are juveniles that establish themselves in a non-natal mound their first fall (Table IV). Even juveniles that disperse (50 m) have significantly lower overwinter survival rates than philopatric young, so that the advantages of philopatry are associated with possessing the natal mound or its cache rather than with living or foraging in a familiar area (Jones 1986).

IV.   CONCLUSIONS: IMPLICATIONS OF PHILOPATRY FOR THE EVOLUTION OF MAMMALIAN SOCIALITY

For *Dipodomys spectabilis,* the ecological factors that promote philopatry are clearly those that decrease the odds of successful dispersal. Philopatry is also encouraged by the fact that at least certain resources at certain times appear to be nondepreciable.

For these kangaroo rats, it seems likely that the same factors are involved in increasing philopatry's advantages and in decreasing its costs. The link is that survival requires large, elaborate mounds, both because of pressure from predators and because of daily and seasonal climate extremes. Rats with mounds have much lower mortality rates, but mounds are nonconsumable so they can be shared at relatively slight cost.

This may be a common pattern. Like kangaroo rats, marmots (Armitage this volume), wood hoopoes (Ligon this volume), and termites (Myles this volume) require elaborate refuges; since refuges are in short supply, this requirement ensures both that chances of successful dispersal will be low and that cost of sharing the refuge will be relatively small. The details differ. The predominant advantages to possessing a refuge are often thermoregulatory or antipredator, but they may also lie in the suitability of such sites for storage of food caches or even (as for termites) as food sources themselves. Some species, like termites and wood hoopoes, cannot influence the availability of suitable refuges--they are treeholes created by other animals, or dead branches of suitable tree species. If refuges are critical for offspring survival and are uncommon relative to other resource constraints, selection should favor parents that tolerate philopatry. Other species, such as *D. spectabilis,* can build their own refuges so that refuge densities should slowly increase until they are no longer limiting. For species that construct their own refuges, selection should favor tolerance of philopatry only when demography makes refuges infrequently available.

Despite these differences in detail, a requirement for refuges that are expensive to build or limited in number is a widely identifiable ecological prerequisite for the evolution of philopatry. When this condition is present, neighbors are expected to be relatives and to share home ranges, setting the stage for subsequent steps in social evolution.

Dependence on and sharing of dens or tree-holes is conspicuous in taxa that have produced gregarious species; among mammals, these include sciurids, canids, viverrids, and prosimians. It is not clear, however, whether a requirement for refuges is the most common ecological prerequisite for the evolution of philopatry or only the most conspicuous one. The answer will come with difficulty, because it will require measurement or manipulation of the resource attributes postulated to influence dispersal costs and resource shareability.

The ecological reasons for philopatry in *D. spectabilis* may or may not be representative of its causes in other mammals. But philopatry's central involvement in mammalian social evolution can be illustrated by its implications for a conspicuous pattern: by and large, mammalian social groups are kin groups. Mammals are quite unexceptional in this regard, but the correlation invites comment.

The evidence suggests that the reasons for this pattern are at base ecological. Among mammals, opportunistic philopatry is more widespread than philopatry by parental consent, which is in turn more widespread than gregariousness (Waser and Jones 1983). Mammals need not stay near neighbors because of any inherent advantage of interacting with, cooperating with, or being near kin. The widespread taxonomic distribution of philopatry, as well as data on its benefits from species like *D. spectabilis,* indicate that they do so because of the risks of leaving the natal area. If ecological conditions allow home-range sharing, the individuals predisposed to share will be relatives and the mechanism will be philopatry. If ecological conditions subsequently allow cooperation or favor gregariousness, the cooperating group members will be relatives. Cooperation or altruism will arise more easily as a result. But for kin selection to act, natal philopatry must first create the opportunity.

REFERENCES

Alexander, R.D. 1974. The evolution of social behavior. *Annual Review of Ecology and Systematics* 5:325-383.
Altmann, S.A. 1974. Baboons, space, time and energy. *American Zoologist* 14:221-248.

Altmann, S.A., S.S. Wagner and S. Lenington. 1977. Two models for the evolution of polygyny. *Behavioral Ecology and Sociobiology* 2:297-310.

Axelrod, R. 1984. *The evolution of cooperation.* New York:Basic Books.

Best, T.R. 1972. Mound development by a pioneer population of the banner-tailed kangaroo rat, *Dipodomys spectabilis baileyi* Goldman, in eastern New Mexico. *American Midland Naturalist* 87:202-206.

Brown, J.L. 1974. Alternate routes to sociality in jays with a a theory for the evolution of altruism and communal breeding. *American Zoologist* 14:63-80.

Brown J.L. 1975. *The evolution of behavior.* New York:W.W. Norton and Co.

Brown, J.L. 1986. The Evolution of Cooperative Breeding. Princeton: Princeton University Press.

Charnov, E.L., G.H. Orions and K. Hyatt. 1976. Ecological implications of resource depression. *American Naturalist* 110:247-259.

Greenwood, P.J. 1980. Mating systems, philopatry and dispersal in birds and mammals. *Animal Behaviour* 28:1140-1162.

Holdenried, R. 1957. Natural history of the banner-tailed kangaroo rat in New Mexico. *Journal of Mammalogy* 38:330-350.

Horn, H.S. 1968. The adaptive significance of colonial nesting in the Brewer's blackbird *Euphagus cyanocephalus. Ecology* 49:682-694.

Jones, W.T. 1982. Natal nondispersal in kangaroo rats. Unpub. Ph.D. thesis, Purdue University.

Jones, W.T. 1984. Natal philopatry in banner-tailed kangaroo rats. *Behavioral Ecology and Sociobiology* 15:151-155.

Jones, W.T. 1986. Survivorship in philopatric and dispersing kangaroo rats (*Dipodomys spectabilis*). *Ecology* 67:202-207.

Jones, W.T. 1987. "Dispersal patterns in kangaroo rats (*Dipodomys spectabilis*)." In: B.D. Chepko-Sade and Z. Halpin, eds. *Mammalian dispersal patterns: the effects of social structure on population genetics.* Chicago:University of Chicago Press, in press.

Kay, F.R. and W.G. Whitford. 1978. The burrow environment of the banner-tailed kangaroo rat, *Dipodomys spectabilis,* in South-central New Mexico. *American Midland Naturalist* 99:270-279.

Koenig, W.D. and F.A. Pitelka. 1981. "Ecological factors and kin selection in the evolution of cooperative breeding in birds." pp. 261-282. In: R.D. Alexander and D.W. Tinkle, eds. 1982. *Natural selection and social behavior.* New York:Chiron Press.

Lambprecht, J. 1979. Field observations on the behavior and social system of the bat-eared fox *Otocyon megalotis* Desmerest. *Zeitscrift für Tierpsychologie* 49:260-284.

MacDonald, D.W. 1983. The ecology of carnivore social behavior. *Nature* 301:379-384.

McNaughton, S.J. 1979. Grazing as an optimization process: grass-ungulate relationships in the Serengeti. *American Naturalist* 113:691-703.

Monson, G. 1943. Food habits of the banner-tailed kangaroo rat in Arizona. *Journal of Wildlife Management* 7:9-102.

Murray, G.B., Jr. 1967. Dispersal in vertebrates. *Ecology* 48:975-978.

Nelson, J.F. and R.M. Chew. 1977. Factors affecting the seed reserves on the soil of a Mojave Desert ecosystem, Rock Valley, Nye County, Nevada. *American Midland Naturalist* 97:300-320.

Orians, G.H. 1969. On the evolution of mating systems in birds and mammals. *American Naturalist* 103:589-603.

Reichman, O.J. 1984. Spatial and temporal variation of seed distributions in Sonoran desert soils. *Journal of Biogeography* 11:1-11.

Schantz, T. von. 1984. Carnivore social behaviour--does it need patches? *Nature* 307:389-390.

Schroder, G.D. and K.N. Geluso. 1975. Spatial distribution of *Dipodomys spectabilis* mounds. *Journal of Mammalogy* 56:363-368.

Smith, A.T. 1974. The distribution and dispersal of pikas: influences of behavior and climate. *Ecology* 55:1368-1376.

Smith, A.T. and B.I. Ivins. 1984. Colonization in a pika population: dispersal vs. philopatry. *Behavioral Ecology and Sociobiology* 13:37-47.

Trivers, R.L. 1972. "Parental investment and sexual selection." pp. 136-179. In: B. Campbell, ed. *Sexual selection and the descent of man.* Chicago:Aldine Publishing Co.

Vorhies, C.R. and W.P. Taylor. 1922. Life history of the kangaroo rat *Dipodomys spectabilis*. Merriam. *USDA Bulletin No. 1091.*

Waser, P.M. 1976. *Cercocebus albigena*: site attachment, avoidance, and intergroup spacing. *American Naturalist* 110:911-935.

Waser, P.M. 1981. Sociality or territorial defense? The influence of resource renewal. *Behavioral ecology and Sociobiology* 8:231-237.

Waser, P.M. 1985. Does competition drive dispersal? *Ecology* 66:1170-1175.

Waser, P.M. 1987. "A model predicting dispersal distance distributions." In: B.D. Chepko-sade and Z. Halpin, eds. *Mammalian dispersal patterns: the effects of social structure on population genetics.* Chicago:University of Chicago Press, in press.

Waser, P.M. and W.T. Jones. 1983. Natal philopatry among solitary mammals. *Quarterly Review Biology* 58:355-390.

Waser, P.M. and R.H. Wiley. 1979. "Mechanisms and evolution of spacing in animals." pp. 159-224. In: P. Marler and J. Vandenberg, eds. *Social behavior and communication.* New York:Plenum Press.

Chapter 7

RESOURCES AND SOCIAL ORGANIZATION OF GROUND-DWELLING SQUIRRELS

Kenneth B. Armitage

Department of Systematics and Ecology
The University of Kansas 66045

I.   INTRODUCTION

Animals form many kinds of groups, ranging from temporary fish schools to highly integrated honeybee colonies (Wilson 1975). Much of the focus on the evolution of groups has centered on the costs and benefits of group living (Alexander 1974). Benefits of group living may include increased defense against predation and more effective exploitation of resources. Costs of group living include increased competition for resources, higher parasite loads and transmission, greater probability of misdirected parental care and infanticide, and increased conspicuousness to predators (Alexander 1974; Hoogland 1979a).

Costs and benefits are important only as they affect reproductive success. Thus increased parasite loads are not in themselves evidence of cost; the cost must be measured in terms of its effects on fitness, generally estimated from some measure of reproductive output. Likewise, increased vigilance behavior (Hoogland 1979b; Holmes 1984a) or modification of patch use (Holmes 1984a; Carey 1985) as antipredator strategies do not necessarily entail reproductive costs. A species or an individual's behavior may be modified by predation pressure, but the cost may be trivial compared to the benefit (death vs. survival). Many animals spend time apparently doing nothing (Herbers 1981); thus a simple demonstration that time is spent on an activity (e.g., vigilance behavior) does not demonstrate a cost. However, the reproductive suppression of some females by other females in the group (Wasser and Barash 1983; Armitage 1986a) does entail a cost; i.e., loss of reproductive output. Any analysis of the role of resources on social organization should focus on reproductive success. One may predict that an individual should be a member of a group when membership increases the probability of leaving reproductive descendants but should leave the group when membership decreases that probability.

Groups are so diverse in structure and function that it seems unlikely that, except in the broadest sense, the same pressures select for group living. For example, male groups of white-tailed deer may form because of the added vigilance of group living but female groups may form because of direct reproductive investment by mothers in daughters (Ozoga et al. 1982). In the former instance, the group may form because each individual has a lower probability of becoming prey; in a sense, the group is based on reciprocity and genetic relatedness is unnecessary for group formation to evolve. In the latter instance, an adult female forms a group only with daughters; genetic relatedness is a necessary condition for the evolution of the group.

This distinction among kinds of groups is critical when looking at groups of ground-dwelling squirrels. Ground-dwelling squirrels are typically described as living in colonies. The word colony connotes some type of group living. But in reality, many species are highly individualistic (Armitage 1981); five grades of sociality may be recognized (Michener 1983). The idea of colonial living stems from the aggregation of these species on patches of favorable habitat (Armitage 1981). Within these aggregations social groups may form. I have argued elsewhere that sociality occurs when body size combined with a relatively short active season results in reproductive maturity being delayed to age two or later (Armitage 1981). Sociality is viewed as a life-history trait in which females continue reproductive investment beyond weaning as a mechanism for increasing the probability of producing reproductive offspring.

It seems obvious that different selection pressures have acted on group formation and the development of sociality in these animals. Groups occur whenever resources are adequate. The interesting questions are what happens to sociality when resources on a patch are varied or when resources differ among patches and what is the likely role of kin selection in the formation of the more highly social groups.

In this paper I will examine the relationship of habitat variability to sociality in yellow-bellied marmots. Because sociality is part of a population-behavioral system (Armitage 1977), the effects of habitat variability on population processes will be considered. Second, I will review the effects of resource manipulation on the social systems of ground-dwelling squirrels. Third, I will discuss whether direct or indirect selection predominates in these systems. Parental investment is considered a component of direct fitness; indirect fitness (= kin selection) includes effects on non-descendant relatives (Brown 1980).

II. HABITAT QUALITY AND SOCIALITY IN MARMOTS

Yellow-bellied marmots (*Marmota flaviventris*) are large, diurnal, burrow-dwelling ground squirrels that occupy forest openings and the alpine of the Cascade, Sierra, and Rocky Mountains (Frase and Hoffman 1980). The populations are clumped on habitat patches consisting of an open area dominated by grasses and perennial forbs and in which talus, boulders, or rock outcrops occur (Svendsen 1974; Kilgore and Armitage 1978). Habitat patches range widely in size; larger patches that normally harbor three or more adults are designated colonial sites; smaller sites that typically support one or two adults are called satellite sites (Armitage and Downhower 1974; Svendsen 1974). This report will emphasize colonial sites.

The annual cycle of marmots is a circannual rhythm that characterizes the Marmotini (Davis 1976). Marmots in our study area in the upper East River Valley, Gunnison County, Colorado, elevation 2900 m, emerge from hibernation in early May. Mating occurs soon after and young are weaned usually by mid July. Body weights are minimal after emergence and all age classes gain weight during the summer (Armitage et al. 1976). Immergence begins in late August; young immerge last and rarely are animals active past mid September. Marmots, as well as all the hibernating ground-dwelling squirrels, are annual breeders. Reproduction must begin early in the year to ensure that young have sufficient time to achieve a body size and fat storage that will carry them through eight months of hibernation and that parous females have time to prepare for hibernation. Litters that are weaned late in the year have virtually no survivors through their first hibernation (Armitage et al. 1976). Yellow-bellied marmots are generalist herbivores (Frase 1983). Feeding preferences are unrelated to protein or water content of plants. Marmots use less than 4 percent of the above-ground primary production available to them (Kilgore and Armitage 1978). Population density apparently is not food limited; however, individual reproductive success may be related to food availability during pregnancy and lactation (Andersen et al. 1976).

Marmots form two kinds of social groups. Females establish matrilines of one or more closely related (mother:daughter or sister:sister) adults; males attach to one or more matrilines to form harems. Per capita reproductive output is unrelated to the number of females living in a matriline, but is inversely related to harem size (Armitage 1986a). The reduction in per capita output is attributable in part to reproductive suppression of females in one matriline by females in an adjoining matriline. The mechanism of suppression is unknown. Matrilines share space and members of a matriline may cooperate to defend against conspecific intruders (Armitage 1984, 1986a).

Although food usually comes to mind when thinking about social organization and resources, burrows are a critical resource for ground-dwelling squirrels. Burrows are usually of three types: (1) nest or home burrows where an animal lives during the active season; (2) auxilliary or flight burrows that usually are located near foraging areas and to which an animal can retreat when threatened (Armitage 1962); and (3) the hibernaculum, the burrow in which one or more animals hibernates. The hibernaculum may also be the nest burrow. Burrows in marmot habitats that are used as a hibernaculum are a small subset of the total number of burrows present. Unfortunately, no one has devised a means of evaluating burrow quality, but several lines of evidence suggest that burrows vary in quality. The same summer burrows are used year after year; some burrows have been used for each of the 24 years that I've studied marmots. Only a few burrows are used as hibernacula; the same hibernacula are used in successive years. Reproductive success may depend on access to a quality burrow. Reproductive sucess of the Wyoming ground squirrel (*Spermophilus elegans*) was related to maternity-burrow sites (Pfeifer 1982) and adult female *S. columbianus* may relinquish nest sites to the yearling daughters (Harris and Murie 1984). To conclude, burrows and food are the critical resources that are expected to affect social organization in marmots and other ground-dwelling squirrels.

A.   A Measure of Habitat Quality

Marmots prefer to use some parts of their habitat more than others (Armitage 1984; Frase and Armitage 1984). Because we have no objective way to evaluate burrow quality and other parameters of habitat quality are extremely difficult to measure (e.g., energy, nutrition, predator defense, avoidance of conspecifics), an indirect method of measuring habitat quality was developed. Because an animal's fitness depends on its reproductive success, marmots should use their habitat patches so as to maximize their reproductive success. Reproductive success is measured as:

$$RS = \frac{\text{number of yearlings}}{\text{number of young weaned}}$$

This ratio integrates overall habitat quality into one measure. For example, the production of young measures the ability of females to obtain needed energy and nutrients, to avoid predators, and to occupy a burrow in which the young can be nursed and sheltered. The number of yearlings measures the capacity of the habitat to provide to the young energy and nutrients for growth and hibernation, a burrow to serve as a hibernaculum, and defense against predators; e.g., flight burrows, good visibility. Because this measure is a ratio, it is independent of the size of the habitat patch and the number of females present.

RS was calculated for each year for each six habitats (= colonies). Yearly values, ranging from 0 to 1.0, were not normally distributed; therefore, differences among habitats were tested by the Kruskal-Wallis analysis of variance by ranks (Siegel 1956, p. 184). RS differed among habitats (Table I). The smallest habitat, Boulder, is characteristic of a satellite site and the lowest RS occurred there. The second-lowest RS occurred at the largest habitat patch. This relationship suggests that area is not related to RS; a rank correlation between area and RS was insignificant ($r_s = 0.03$, p > 0.1).

Habitat quality could affect survivorship (S) of adult females. Survivorship was measured as:

$$S = \frac{\text{number of adult females}}{\text{number of adult females the previous year}}$$

Annual survivorship ranged from 0 to 1.0; mean survivorship varied from 0.569 to 0.782 (Table I). Survivorship did not differ among habitats nor were RS and S correlated among habitats ($r_s = -0.31$, p > 0.1). However, RS and S were correlated within colonies in three habitats but not in three others (Table II). The significant values of r_s are not particularly large. These low correlations, the lack of significant correlations in three colonies, and the lack of significant correlations among habitats suggest that survivorship and reproductive success are not tightly coupled, but that there is some trend for habitats that favor survivorship to favor reproduction also. The areas where marmots live probably represent long-term selection for habitat choice by female marmots. Any tendency to occupy sites with low survivorship probabilities should receive strong negative selection. Doubtless we have chosen study sites

TABLE I. Mean values of reproductive success (RS) and survivorship (S) for adult female yellow-bellied marmots. Differences among colonies tested by Kruskal-Wallis one-way analysis of variance; for RS, $H = 9.38$, $0.1 > p > 0.05$; for S, $H = 0.6$, p > 0.9. N = number of years. Area in hectares.

Colony	N	RS	S	Area
River	23	0.477	0.569	2.73
Marmot Meadow	21	0.458	0.672	2.39
Cliff	8	0.467	0.713	1.88
Picnic	23	0.405	0.782	5.27
Boulder	22	0.170	0.673	0.02
North Picnic	20	0.329	0.688	7.24

TABLE II.   Spearman rank correlations between reproductive success (RS) and other population parameters within habitats (= colonies). Significance levels: 0.1*, 0.05**, 0.01***.

Colony (Habitat)	Adult Female			Per Capita RS
	Survivorship	Recruitment	Replacement	
River	0.331*	0.358**	0.029	0.930***
Marmot Meadow	0.489**	0.352**	0.390**	0.970***
Cliff	0.521*	0.428*	0.452*	0.911***
Picnic	0.167	0.143	-0.112	0.837***
Boulder	0.223	0.431**	0.101	0.987***
North Picnic	-0.048	0.326*	-0.059	0.973***

where marmots are successful; thus our results likely are biased toward the upper end of survivorship.   Calculations of survivorship of two additional satellite sites produced rates of 0.439 and 0.50.   These rates are lower than those of any of the long-term sites (Table I), but may be misleading.   Satellite females may seek a new habitat and be erroneously recorded as dead.   We have had females return to a satellite site after one or more year's absence; therefore, these survivorship values are minimal.   RS values for these two sites were 0.297 and 0.38, respectively. Both values are in the range of the two low values from the long-term study sites (Table I).   A tentative conclusion is that the places where female yellow-bellied marmots reside do not differ much in survivorship, but vary considerably in reproductive success.   Female marmots commonly live to five or more years of age; thus a female has several opportunities to breed and may be able to locate a better habitat patch in a subsequent year.   Our experience is that the better habitat patches are quickly occupied; thus many females have no alternative but to occupy a patch that promises little in the way of RS.   However, her fitness is likely to be higher if she reproduces, even if the probability of RS is low, than if she does not reproduce at all (RS = 0).   In effect, a female may have to make the best she can of a poor situation.

A female's fitness requires leaving reproductive offspring.   One estimate of fitness is recruitment, the retention of daughters in their natal area.   Recruitment was measured as:

$$\text{Recruitment} = \frac{\text{number of 2-year-old daughters added}}{\text{number of adult females resident the previous year}}$$

RS is significantly correlated with recruitment in all habitats except one.   RS and recruitment are highly correlated among colonies ($r_s = 0.943$, $p = 0.01$).   Both results are consistent with an analysis of recruitment that revealed that the production of yearling daughters was the single most important demographic factor affecting recruitment (Armitage 1984).

B.   Habitat Quality and Demography

The previous discussion suggests that habitat quality and demography may be related. Before considering the relationship between habitat and social organization, the various demographic patterns need to be described so that effects of habitat on demography can be distinguished from the relationship between habitat and social organization.

RS could be related to the density (= number) of resident adult females. Therefore per capita RS was calculated by dividing RS for each year by the number of resident adult (two-year-old or older) females. RS and per capita RS are highly correlated (Table II). This result is not surprising. When only one adult female is present, RS and per capita RS are identical; when either no young or no yearlings are produced, RS and per capita RS are identical. Only at Picnic is the number of females sufficiently variable to provide a good test of this relationship. The correlation is the lowest of the six habitats; however, it is quite high and suggests that density affects RS only in a minor way. This result is supported by the lack of any effect of the number of females living in a matriline on the per capita production of young or yearlings (Armitage 1986a). However, an increase in harem size is associated with a decrease in per capita production of both young and yearlings. The decrease in per capita production of offspring in harems was attributed to competition between matrilines. This effect probably accounts for the lower r_s between RS and per capita RS at Picnic than at other colonies, because at Picnic more than one matriline commonly occurred (Armitage 1984). I conclude that density is less important than social factors in determining RS in these populations.

Marmot populations are relatively stable (Armitage and Downhower 1974; Armitage 1975). There is no precise density-dependent regulation of density, but in general, losses of adults are replaced by recruits from within the population or by immigrants. RS and replacement (the addition of an adult female aged two or older to a population) could be related in that better habitat patches might be more quickly populated. Replacement (replacement is equal to the number of new adults added divided by the number of residents the previous year) and RS are highly correlated among colonies (r_s = 1.0, p < 0.01). However, replacement and RS are not highly correlated from year to year within habitat patches (Table II), nor is replacement related to survivorship (Table III). However, the number of females returning might determine replacement rather than survivorship (e.g., survivorship would be 0.5 when 1 of 2 and 2 of 4 females returned, but the number of females would be double). Within four of the six habitat patches, replacement is negatively correlated with the number of females returning (Table III). However, the correlations are relatively low; therefore, factors other than the number of females also affect replacement. One possible factor is area; residency might be more easily secured by a female where a large area could permit her to avoid antagonistic conspecifics. However, area of habitat and replacement are not correlated (Tables I and IV; r_s = 0.03, p > 0.1). Area of habitat is correlated with the mean number of adult

TABLE III. Spearman rank correlations between demographic parameters within colonies (= habitats). p = 0.1*, 0.05**.

Colony (Habitat)	Survivorship vs Replacement	Survivorship vs Recruitment	Number of Females Returning		
			vs Replacement	vs Immigration	vs Recruitment
River	-0.456**	0.116	-0.353**	-0.246	0.290
Marmot Meadow	-0.039	-0.115	-0.058	0.456**	0.045
Cliff	-0.256	-0.023	-0.673**	-0.083	-0.167
Picnic	0.025	-0.100	-0.474**	0.195	-0.406**
Boulder	-0.172	0.296	-0.340*	-0.084	0.294
North Picnic	0.179	0.220	-0.112	-0.094	0.319*

TABLE IV. Mean values for habitat and demographic parameters for six colonies of yellow-bellied marmots.

Colony (Habitat)	Number Females	Replacement	Recruitment	Recruitment as a Percentage of Replacement	Per Capita Production of Young
River	1.87	0.59	0.312	52.9	2.15
Marmot Meadow	1.76	0.390	0.295	75.6	4.76
Cliff	2.6	0.55	0.425	77.3	1.91
Picnic	5.35	0.314	0.244	77.7	1.74
Boulder	1.68	0.261	0.080	3.1	2.14
North Picnic	2.45	0.292	0.058	19.9	2.24

females (r_s = 0.771, p < .01). Therefore, the mean number of females could affect replacement; however, replacement and the mean number of females are not correlated among habitats (r_s = 0.257, p > 0.1).

Replacement has two components, immigration and recruitment. Within colonies, only three of 12 correlations between immigration or recruitment and the number of females returning were significant (Table III). These results suggest considerable variation among habitats. The correlations are low, again suggesting that factors other than population density affect replacement. Further evidence supporting the lack of simple density dependence are the low and insignificant correlations between survivorship and recruitment within colonies (Table III), and the low correlations among colonies between recruitment and the mean number of adult females (r_s = 0.581, p > 0.1), RS and the mean number of adult females (r_s = 0.257, p > 0.1), and RS and area of the habitat (r_s = 0.03, p > 0.1).

The production of young per female could be affected by the number of resident females. All rank correlations between per capita production of young and the number of resident females within colonies were insignificant (Table V). Among colonies the per capita production of young was negatively, but insignificantly related to the mean number of females (r_s = -0.6, p > 0.1). RS was not related to the per capita production of young (r_s = 0.029, p > 0.1), but adult survivorship was (r_s = -0.714, p ≅ 0.1). This result suggests a tradeoff between adult survivorship and the production of young. Those habitats of poor quality apparently cannot sustain both high productivity and survivorship. For the two colonies with the highest mean rates of RS, adult survivorship was significantly positively correlated with the per capita production of young (Tables I and V). As indicated previously, good habitat supports both a high RS and adult survivorship.

In conclusion, differences in reproductive success, survivorship, replacement, and recruitment are partly attributable to habitat differences, and partly to population density. But much of the variation

TABLE V. Spearman rank correlations between the per capita production of young and survivorship and the number of resident adult females within colonies, p = 0.1*, 0.05**.

Colony	Number of Resident Females	Survivorship
River	0.065	0.541**
Marmot Meadow	0.159	0.222
Cliff	-0.196	0.553*
Picnic	-0.022	0.256
Boulder	0.245	0.090
North Picnic	-0.092	0.159

remains unexplained. This unexplained variation could be a consequence of social organization.

III.   SOCIAL ORGANIZATION, REPRODUCTIVE SUCCESS, AND DEMOGRAPHY

Two demographic relationships suggest that social organization may be critical to population processes. First, among colonies, replacement is significantly correlated with recruitment (r_s = 0.943, p = 0.01). Because recruitment is the retention of yearling daughters in their natal colony, the high correlation between rates of replacement and recruitment suggests that the social system makes it possible for recruits to be preferred over immigrants as replacements. Second, the percentage of replacement that is recruitment (hereafter referred to as percent recruitment) varies widely among colonies (Table IV) and is not significantly related to replacement (r_s = 0.486, p > 0.1). But percent recruitment is significantly related to the mean number of resident females (r_s = 0.771, p = 0.1). Therefore, replacement is more likely to be recruitment when more adult females are present. This relationship can be interpreted to mean that females associating in a matriline can exclude immigrants and increase the probability that replacement will be a relative.

Percent recruitment is low at the two habitats where RS is low (Tables I and IV). However, the correlation among habitats between RS and percent recruitment is low (r_s = 0.486, p > 0.1), primarily because Picnic, with the highest percent recruitment, ranked fourth in RS and River, with the number one ranking in RS, ranked fourth in percent recruitment. The other habitats had identical rankings for both parameters. This relationship further suggests that habitat quality and demographic effects are influenced by some other factor, such as social organization.

Social organization varies among the habitats. There is little difference in the mean size of matrilines, but considerable variation in the mean number of matrilines among the six habitats (Table VI). None of

TABLE VI. Characteristics of matrilines among colonies.

Colony	Mean Number of Matrilines	Mean Size of Matrilines
River	1.5	1.31
Marmot Meadow	1.27	1.36
Cliff	1.3	1.92
Picnic	3.54	1.48
Boulder	1.09	1.56
North Picnic	2.06	1.05

the following habitat characteristics are significantly correlated among colonies with either the mean number or mean size of matrilines: RS, S, per capita production of young, replacement, recruitment, and percent recruitment (Table VII).

A. Habitat and Matrilineal Organization

The mean size of matrilines is negatively correlated with habitat area (Table VII). The mean number of matrilines is positively correlated with habitat area and the mean number of resident females. Thus, as habitat area increases, the number of resident females increases, but the higher number of females is subdivided into more matrilines rather than increasing the sie of individual matrilines. In smaller habitats, there is no space for matrilineal subdivision and all resident females must be compressed into a single matriline. Also, in larger habitats, immigrants may successfully establish a new matriline in the presence of residents, but immigration occurs in the smaller habitats only in the absence of residents.

The process of matrilineal formation and subdivision may be illustrated by comparing Picnic and Boulder, the habitats with the largest and smallest numbers of matrilines, respectively, but with similar mean size of matrilines (Table VI).

Only in 1975 and 1976 did Boulder have two matrilines. Each female immigrated into the habitat in the same year; one of the females emigrated and the other established a matriline continuing through 1985 (Fig. 1). By contrast, as many as six matrilines occurred in the same year at Picnic (e.g., 1965, 1968, Fig. 1) and in only six of the 22 years were there as few as two matrilines. At both habitats, new matrilines were initiated by immigrants, but matrilines persisted through time only by incorporating daughters. In both habitats, the average relatedness of members of a matriline usually was 0.5. When average relatedness decreased, matrilines generally subdivided (e.g., Picnic 1974, 1976, 1983).

TABLE VII. Spearman rank correlations between the mean number or mean size of matrilines and habitat characteristics, $p = 0.1*, 0.05**$.

Habitat Characteristics	Mean Number of Matrilines	Mean Size of Matrilines
Survivorship (S)	0.486	0.457
Reproduction success (RS)	0.143	-0.029
Per capita production of young	-0.314	-0.657
Replacement	0.143	-0.029
Recruitment	0.343	0.2
Percent recruitment	0.543	0.314
Area	0.886**	-0.771*
Mean number of females	0.829**	0.086

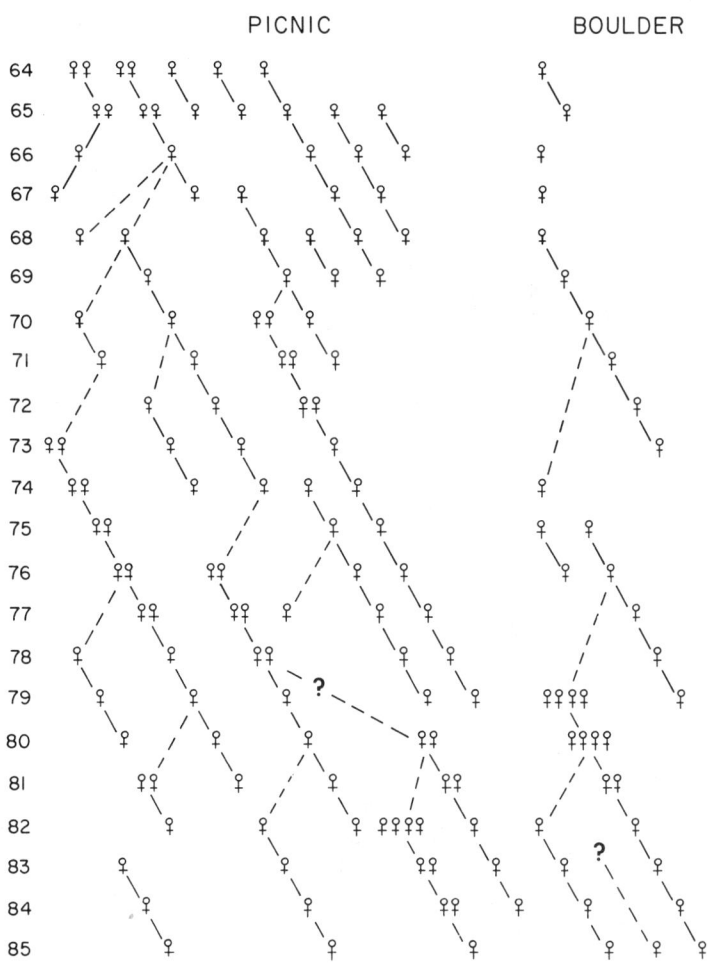

Figure 1. Matrilineal structure at Picnic and Boulder colonies. Each female symbol represents an individual adult female marmot. The female symbol is placed in the first year of residency. When a female is resident for more than one year, the female symbols are connected with a solid line and recorded on a diagonal. Recruits are recorded when they are two years old; a dashed line connects a recruit to her mother in the year of her birth. A dashed line with a ? indicates that the mother of the recruit could not be specified. This situation occurred when two or more females produced litters in the same burrow system and the young fully intermingled upon emergence. Two or more female symbols placed close together indicate that the females joined or formed a matriline in the same year. All females were residents in 1964 when the analysis was initiated.

Boulder has only one major burrow system and all residents share it. Because of competition between members of different matrilines (Armitage 1986a), it is unlikely that matrilines can subdivide and coexist. Matrilines larger than two occurred only three times in 22 years; at Picnic, matrilines larger than two occurred only four times in the same time span despite a three-fold greater mean numbr of residents. Thus, females tend to form a matriline with only one other female, a daughter or a sister. The larger area at Picnic includes many more burrow sites and foraging areas and provides the opportunity for females to dissociate themselves from individuals related by <.05 and to initiate their own matriline. The formation of a new matriline is characterized by the lack of overlap in space use between members of different matrilines (Armitage 1984). Furthermore, social behavior among females related by 0.5 is mainly amicable whereas social behavior among more distantly related females is characteristically agonistic (Armitage 1986b; Armitage and Johns 1982).

B.   Matrilines as a Mechanism for Increasing
      Direct Fitness

Matrilines are family units; specifically, mother-daughter units. In a sense, marmot sociality follows a familial pathway (Slobodchikoff 1984). Fifty-five matrilines can be distinguished in the six colonies during the period of study. Of these, 35 were established as mother-daughter matrilines and 20 as sister-sister matrilines. Of the 20 sister-sister matrilines, eight were derived from mother-daughter matrilines after the mother presumably died. Only 22 percent of the matrilines were formed in the absence of the mother; even in these 12 cases the mother usually was present the previous year, but died before her daughters became residents (e.g., Fig. 1, Picnic 1968, 1973; Boulder 1974).

The formation of matrilines is a form of parental investment. It is generally accepted that mortality is higher among dispersers than among residents (Gaines and McClenaghan 1980). A female's evolutionary fitness will be greater the larger her number of reproductive descendents. Therefore, she should act to increase the probability that her daughters will live long enough to be reproductive. One way to increase that probability is to retain daughters in their natal area. The natal area has proven successful; marmots have survived and reproduced there. The adult female can protect her daughters against conspecifics; daughters are highly philopatric (Armitage 1984). When a daughter reaches adulthood, she may form a coalition with her mother. A numerically dominant matriline can defend its space against conspecifics and can cause the numerically inferior matriline to abandon space (Armitage 1986a). Thus, both the mother and daughter can increase their potential fitness by gaining resources and cooperating in defense.

However, there are costs to the formation of a matriline. The reproduction of two-year-old females living with their mothers is significantly less than the population average (Armitage 1986a). These females fail to produce litters. The failure of these females to disperse in order to increase their likelihood of reproduction suggests that the survivorship of recruits is sufficiently greater than that of dispersers so that lifetime reproductive success compensates for any short-term annual loss.

Competition among females probably is the ultimate cause of matrilineal subdivision. A female will gain more in direct fitness by producing more daughters than by helping her daughters produce her grandchildren (Rubenstein and Wrangham 1980). The mother's offspring will be related to her by 0.5; her grandchildren likely will be related by 0.25. The mother can increase her direct fitness by being the only reproductive female and using her daughters(s) to help defend the matrilineal resources. Conversely, the daughter should try to maximize their direct fitness by producing offspring rather than maximizing their indirect fitness by assisting her mother. Thus, the daughter should avoid reproductive suppression either by becoming codominant with her mother or by forming an independent matriline. This competition is a clear example of parent-offspring conflict (Trivers 1974). The same arguments apply to sisters; they should remain together when the direct fitness of each is enhanced. But a sister whose direct fitness is lowered by competition with her sister should form her own matriline if that will increase her direct fitness. However, in many situations a subordinate sister may lose both direct and indirect fitness if the only choice she has is to emigrate. Therefore, she should make the best of her situation and gain as much inclusive fitness as possible.

In all of our sister-sister matrilines, one sister was dominant. The subordinate sister(s) were never known to disperse or to form a new matriline. One subordinate female, whose sister died, was forced to move by an immigrant to a marginal habitat at the edge of Picnic Colony where she produced one litter in six years and failed to recruit any daughters. In River Colony, a subordinate female moved to a peripheral location when an aggressive immigrant became resident, returned when the immigrant died, again became peripheral when new adults appeared, returned another year as the only adult present, but never reproduced. These examples suggest that subordinate females have little chance of success as immigrants to other habitats or to form and defend an independent matriline in the natal habitat. Only sisters of the same year class form a matrilineal group; full sisters from different year-classes and maternal half-sisters are treated as strangers; i.e., they are treated agonistically and space (with its food and burrow resources) is not shared.

The small size of matrilines, the subdivision of matrilines when average relatedness increases, the preponderance of mother:daughter matrilines, the patterns of cooperative and competitive behavior, and reproductive suppression (Armitage 1986a) indicate that female yellow-bellied marmots attept to maximize direct fitness. Post-weaning reproductive investment in which offspring are retained in their natal habitat for an additional growing season leads to the subsequent residency of some daughters and the formation of matrilines. Matrilines should be considered a result of direct reproductive investment and not a form of kin selection. Regardless of the resource level, the only sociality evidenced by yellow-bellied marmots is the matriline. Matrilines rarely form in small habitat patches where the quantity of resources does not typically allow more than one adult female resident.

IV. SOCIAL SYSTEMS OF GROUND-DWELLING SQUIRRELS

The fundamental social system of the ground-dwelling squirrels is based on female associations. Females must locate sufficient resources for growth, gestation, lactation, and fattening. These needs are especially critical during lactation (Kiell and Millar 1980). Thus, the primary issue of residency is set by the females whose fitnesses will be zero if they do not obtain sufficient resources to meet all their needs. Males associate with females in a variety of mating systems ranging from male-dominance polygyny in a resource-based lek in *Spermophilus beldingi* (Sherman and Morton 1984) to the highly territorial systems of marmots (Armitage 1974; Barash 1973, 1981) and prairie dogs (King 1955; Hoogland 1981a; Rayor 1985; also see reviewed by Michener 1983 and Dobson 1984). Males usually disperse from their natal areas the year before reaching reproductive maturity (Holekamp 1984). My analysis of behavior of male yellow-bellied marmots concluded that males seek out females. Where females are clumped, such as on habitat patches like Picnic or Cliff, a male may form a harem that is defended against intruding males. The harem frequently consists of more than one matriline (Armitage 1986). This mating system, from the male viewpoint, is female-defense polygyny (Emlen and Oring 1977).

On small habitat patches, such as Boulder, the male may be monogamous. Monogamy probably characterizes the mating system at widely separated satellite sites. However, we recently discovered through the use of radiotelemetry that some males may travel widely among two or more satellite sites and defend a dispersed harem (Van Vuren, pers. comm.). A population of 11 colonies of hoary marmots (*M. caligata*) in south-central Alaska was monogamous (Holmes 1984b). Food supported only one adult female and her young at each site and hibernacula were too far apart for a male to defend more than one. Litters were produced infrequently and movement of subadults to the periphery of the colony seemed to coincide with the appearance aboveground of a new litter of juveniles. Thus, resources would not allow the development of a social group consisting of at least two adult females.

Female kin associations occur among several species of *Spermophilus*. These associations, or kin clusters, occur in *S. tridecemlineatus* (Vestal and McCarley 1984), *S. beldingi* (Sherman 1981), *S. tereticaudus* (Dunford 1977), *S. richardsonii* (Michener 1979), *S. parryii* (McLean 1982), *S. columbianus* (Festa-Bianchet 1981; Harris and Murie 1984) and probably in *S. beecheyi* (Owings et al. 1977) and *S. armatus* (Slade and Balph 1974). It seems likely that female kin associations are universal in this genus.

Among *Cynomys*, female groups consist of descendent and collateral kin in *C. ludovicianus* (Hoogland 1985) and probably also in *C. gunnisoni* (Rayor, pers. comm.). For both species, the female assemblages are consistent with the interpretation that adult females increase their direct fitness by retaining daughters in their natal home range. Demography would then be expected to produce kin groups consisting of cousins, nieces, daughters, and sisters. The social organization of *C. leucurus* is poorly known, but females probably form kin clusters (Clark 1977).

Both *M. olympus* (Barash 1973) and *M. caligata* (Barash 1974) form female social groups that evidence sociality, i.e., several adult females with their offspring share burrow and foraging resources, although parous females tend to defend their burrows against conspecific intruders. It is very likely that these female groups are derived from mother-daughter matrilines. Among woodchucks (*M. monax*), some daughters gain access to a portion of their mother's home range (Meier 1985). This pattern is reminiscent of the kin clusters of spermophiles, but on an expanded scale. Home range of marmots is related to body size (Lindstedt et al. 1986); therefore, neighboring kin should be much further apart in the larger species. When adult females are not so dispersed, but share highly overlapping home ranges, sociality is indicated.

I conclude that there is a general trend among the ground-dwelling squirrels for some daughters to settle near their mothers. The tolerance of the adult for the juvenile is a form of parental investment that increases the likelihood that an adult will produce descendents that survive and reproduce. This association of kin may be expressed as sociality, especially in those species where reproductive maturity is delayed to age two or later and resources are clumped (Armitage 1981).

V. RESOURCE VARIABILITY AND SOCIAL ORGANIZATION

Previously we demonstrated that whether male *M. flaviventris* and *M. caligata* are monogamous or polygynous depends on the resources available to support females. If resources are too scattered or too sparse to support more than one adult female in a habitat patch, males are monogamous. Conversely, scanty resources limit female sociality; in the three well-studied species of montane marmots, matrilineal groups rarely or never occur on small habitat patches. However, the resources may be inherited by a daughter from her mother. In this section I will discuss the results of resource manipulation on ground squirrels and prairie dogs.

A. Natural Experiments

Several field studies have compared poulations living in habitats with observable differences in the abundance of food. Two populations of Gunnison's prairie dog (*C. gunnisoni*) were studied in south-central Colorado. One population occupied a dry, barren site and the other, a lush, irrigated site (Rayor 1985). No differences in social organization were observed. All differences were in life history traits; the prairie dogs in the lusher habitat were heavier, dispersed at an earlier age, reached sexual maturity more rapidly (some yearlings raised litters) and adult females produced larger litters. A young colony of black-tailed prairie dogs (*C. ludovicianus*) with access to abundant grasses evidenced earlier reproductive maturity, larger litters, a greater proportion of successful pregnancies, faster juvenile growth, and greater survivorship than an older colony whose habitat was characterized by plants of low palatability (Garrett et al. 1982). Again, all differences were in life history traits. No differences in social organization were reported.

Other evidence from field studies indicates that increased nutrition affects life history traits. Some female yearling *S. columbianus* reproduced in a favorable habitat (Festa-Bianchet 1981); normally, first reproduction is at age two. Better nutrition produced larger litters in *S. tereticaudus* (Reynolds and Turkowski 1972), *S. richardsonii* (Sheppard 1972), and *S. columbianus* (Murie et al. 1980). To summarize, all reports of differences in food resources indicate that individuals opt to initiate reproduction earlier or to produce more offspring. Each of these responses is a mechanism for potentially increasing the direct fitness of the individuals involved.

B.   Manipulative Experiments

Manipulation experiments involve either removing food resources, e.g., by mowing, or enhancing food by distributing some supplement in the habitat. In a colony of Gunnison's prairie dogs, mowing produced an increase in territory size and shift toward monogamy from the previous polygynous mating systems. Adding food to the mowed patches decreased territory size, but no change in monogamy (Slobodchikoff 1984). This response is similar to that of yellow-bellied or hoary marmots living on small patches which is equivalent to reduced resources. Apparently the experiment was not continued long enough to determine effects on reproduction and other traits.

Two populations of Columbian ground squirrels (*S. columbianus*) received food supplements whereas two additional populations served as controls (Dobson and Kjelgaard 1985a,b). Age at maturity decreased (yearling females successfully weaned litters) and litter size, reproductive effort, survival of young, and spring body weight increased in the food supplemented populations in comparison with the reference populations. There was no evidence for changes in social behavior or organization.

VI.   DIRECT FITNESS AND SOCIALITY

Throughout I have argued that sociality is a mechanism whereby a female ground squirrel attempts to maximize her direct fitness by increasing the probability that one or more daughters will survive to maturity. The retention of daughters in the natal area implies that survivorship is lower in dispersers than in residents. Female recruits may lose at least one year of reproduction because their reproductive capacity is suppressed by older females, often their mothers. This apparent willingness to forego reproduction in order to be a resident further suggests that the costs of dispersing are considerable. No female should disperse from her birth site unless her fitness would be increased over what it would be if she attempted to become a resident. A juvenile female should "assess" the probability of achieving reproductive success as a resident and "decide" whether to emigrate or remain. The proximal mechanisms initiating this decision are incompletely known (Gaines and McClenaghan 1980; Holekamp 1984), but adult residents probably play a major role. When all adults were removed from a colony of yellow-bellied marmots, none of the yearling females dispersed (Brody and Armitage 1985). Probably the proximal mechanisms are subtle; the mere presence of an adult who expresses dominant behavior may be sufficient to initiate dispersal of some juveniles. The benefits and costs of dispersal in comparison to recruitment must be determined if we are to understand the function of sociality as an evolutionary strategy for increasing fitness.

There is little evidence that resource manipulation changes the basic social structure of a species. Food supplementation has direct fitness consequences: the probability of producing reproductive offspring is increased by increasing litter size and decreasing the age of first reproduction. Both responses increase the lifetime reproductive output of an individual. This response has its physiological limitations; every species has an upper limit to the biomass of offspring a female can produce. In ground-dwelling squirrels, the age of first reproduction cannot be lower than one year of age because of the limitations imposed by hibernation or winter inactivity. If there are limits to maximizing fitness by increasing the probability of evolutionary success through the production of more offspring, then we would expect social mechanisms to evolve to increase that probability.

If sociality is such a successful mechanism for increasing fitness, why are not all ground-dwelling squirrels social? The answer seems to lie in mother-offspring conflict, in other words, in the costs of sociality. The costs are mediated through competition and expressed by loss of reproductive output or loss of space to distant kin ($r < 0.25$) that otherwise could be used by closely related kin ($r = 0.5$). Dispersal characteristically occurs in the summer before reproductive competition begins. This competition must make the costs of sociality too high for those species who are reproductively mature at age one. The establishment of separate home ranges in those species who do not form matrilineal groups is functionally similar to the subdivision of yellow-bellied marmot matrilines and the generally small size ($\bar{x} = 1.47$) of these matrilines. Females attempt to occupy a resource base and live in a social environment that will increase their direct fitness.

Prairie dogs are an interesting example of this process. The Gunnison and black-tailed prairie dogs are highly social, but the white-tailed is much less so (Hoogland 1981b). White-tailed prairie dogs breed at age one; most black-tailed prairie dogs breed at age two. A comparative study of the two species hypothesized that reduced predation may be the most important benefit of prairie-dog sociality (Hoogland 1981b).

However, I suggest that the benefits of sociality must be explained in terms of reproductive success. Various selective pressures can modify behavior without inducing sociality. Vigilance and alarm calling are common behaviors among ground-dwelling squirrels, but only a few species have developed sociality.

Sociality in prairie dogs can be explained in terms of reproductive benefits. In the not-too-distant past, black-tailed prairie dog populations extended for many miles and were (and still are) characterized by sparse vegetation (Koford 1958). The population is organized into female groups with one or two males called coteries (King 1955). Except for coteries on the periphery, all coteries are surrounded by competitors. Any disperser must enter hostile territory and the likelihood of finding a place to settle must be low. It does not matter whether a disperser would fail because of the inability to find a burrow site, or inability to obtain sufficient food, or by becoming prey; the key issue is that the likelihood of reaching reproductive maturity is very low. The probability of failure would be greater the younger and smaller the disperser. Thus, the probability of reproductive success would increase by retaining offspring in their natal area until they approach reproductive maturity. At the time of approaching maturity, decisions could be made. If mortality reduced coterie membership, the offspring could remain. If the coterie could not accept an additional member, the offspring could disperse. Because it would be older and larger, it would have some increased probability of survival.

It is critical to coterie success that new members be recruited. Coterie members defend their area against intruding conspecifics (King 1955; Hoogland 1981a). Small coteries may lose space (= resources) to larger, neighboring coteries. A female will maximize her fitness if all coterie members are her daughters. Demographic considerations suggest that situation would be difficult to achieve. In fact, coterie memberships may include half-sisters, full first cousins, half-aunt-nieces, etc. (Hoogland 1986). Thus, females probably are forced to accept less closely related kin in order to have assistance in coterie defense. However, coterie space is limited and only a few of the young that are produced can expect to achieve residency. A female should initiate behaviors that will increase the likelihood that her daughters will become residents rather than nieces, granddaughters, cousins, sisters, etc. One strategy is to kill offspring who would compete with one's own offspring for the limited space. Although female prairie dogs defend their burrows against conspecifics, infanticide is common (Hoogland 1985). Competition among female black-tailed prairie dogs is greatest during late pregnancy when the danger of infanticide is high (Hoogland 1986). Infanticide in this context is readily understood as an evolutionary strategy. Furthermore, amicable behavior among females is not distributed according to relatedness but varies inversely with competition (Hoogland 1986). Thus animals do not aid the reproduction of kin; they attempt to gain more by maximizing their direct fitness (see also Rubenstein and Wrangham 1980).

Sociality has both cooperative and competitive behaviors. Because social groups are frequently kin groups, too much attention has focused on cooperation under the assumption that individuals should assist kin. This assumption implies that indirect fitness is an important component of inclusive fitness. The blend of cooperation and competition suggest that ground-dwelling squirrels attempt to maximize direct fitness (see also Armitage 1986a). The indirect component to fitness is important only as it contributes to an animal's direct fitness. Thus, a female yellow-bellied marmot or a female black-tailed prairie dog tolerates a sister or a daughter, not because of indirect fitness benefits, but because these individuals contribute to her direct fitness. Black-tailed prairie dogs are highly agonistic toward non-coterie members, and amicable behavior within the coterie is always greater than amicable behavior with non-coterie members (Hoogland 1986). Obviously, inclusive fitness is greater when sociality involves kin groups rather than unrelated individuals. That is not the issue. What is critical is whether an animal's behavior is directed toward maximizing direct fitness.

I suggest that the behavior of ground-dwelling squirrels is directed toward maximizing direct fitness. Social behavior is both cooperative and competitive. In most species, "competitive decisions" are made in the first year of life; dispersal occurs at this time and most species, including white-tailed prairie dogs, do not form matrilineal groups. When the competitive decisions come at age two or later, matrilineal groups are formed. But competition becomes a way of life as each individual plays its game of attempting to maximize direct fitness.

I have sketched a scenario. This scenario can be verified or rejected only by focusing on the lifetime reproductive success of individuals of known relatedness. And these studies must include learning the fate of dispersers. Only by documenting the reproductive success of dispersers and residents can we construct a logical story of the relative importance of direct and indirect selection in animal sociality. I predict that direct selection will be by far the most important.

REFERENCES

Alexander, R.D. 1974. The evolution of social behavior. *Annual Review of Ecology and Systematics* 5:325-383.

Andersen, D.C., K.B. Armitage and R.S. Hoffman. 1976. Socioecology of marmots: Female reproductive strategies. *Ecology* 57:552-560.

Armitage, K.B. 1962. Social behaviour of a colony of the yellow-bellied marmot (*Marmota flaviventris*). *Animal Behaviour* 10:319-331.

Armitage, K.B. 1974. Male behaviour and territoriality in the yellow-bellied marmot. *Journal of Zoology, London* 172: 233-265.

Armitage, K.B. 1975. Social behavior and population dynamics of marmots. *Oikos* 26:341-354.

Armitage, K.B. 1977. Social variety in the yellow-bellied marmot: A population-behavioural system. *Animal Behaviour* 25:585-593.

Armitage, K.B. 1981. Sociality as a life history tactic of ground squirrels. *Oecologia* 48:36-49.

Armitage, K.B. 1984. "Recruitment in yellow-bellied marmot populations: Kinship, philopatry, and individual variability." pp. 377-403. In: J.O. Murie and G.R. Michener, eds. *Biology of ground-dwelling squirrels*. Lincoln:University of Nebraska Press.

Armitage, K.B. 1986a. "Marmot polygyny revisited: Determinants of male and female reproductive strategies." In: D.S. Rubenstein and R.W. Wrangham, eds. *Ecological aspects of social evolution*. In press. New Jersey:Princeton University Press.

Armitage, K.B. 1986b. Individuality, social behavior, and reproductive success in yellow-bellied marmots. *Ecology*, in press.

Armitage, K.B. and J.F. Downhower. 1974. Demography of yellow-bellied marmot populations. *Ecology* 55:1233-1245.

Armitage, K.B., J.F. Downhower and G.E. Svendsen. 1976. Seasonal changes in weights of marmots. *American Midland Naturalist* 96:36-51.

Armitage, K.B. and D.W. Johns. 1982. Kinship, reproductive strategies and social dynamics of yellow-bellied marmots. *Behavioral Ecology and Sociobiology* 11:55-63.

Barash, D.P. 1973. The social biology of the Olympic marmot. *Animal Behaviour Monographs* 6:171-245.

Barash, D.P. 1974. The social behavior of the hoary marmot (*Marmota caligata*). *Animal Behaviour* 22:256-261.

Barash, D.P. 1981. Mate guarding and gallivanting by male hoary marmots (*Marmota caligata*). *Behavioral Ecology and Sociobiology* 9:187-193.

Brody, A.K. and K.B. Armitage. 1985. The effects of adult removal on dispersal of yearling yellow-bellied marmots. *Canadian Journal of Zoology* 63:2560-2564.

Brown, J.L. 1980. "Fitness in complex avian social systems." pp. 115-128. In: H. Markl, ed. *Evolution of social behavior: Hypotheses and empirical tests*. Verlag Chemie.

Carey, H.V. 1985. The use of foraging areas by yellow-bellied marmots. *Oikos* 44:273-279.

Clark, T.W. 1977. Ecology and ethology of the white-tailed prairie dog (*Cynomys leucurus*). *Publications in Biology and Geology of the Milwaukee Public Museum* 3:1-97.

Davis, D.E. 1976. Hibernation and circannual rhythms of food consumption in marmots and ground squirrels. *The Quarterly Review of Biology* 51:477-514.

Dobson, F.S. 1984. "Environmental influences on sciurid mating systems." pp. 229-249. In: J.O. Murie and G.R. Michener, eds. *Biology of ground-dwelling squirrels*. Lincoln:University of Nebraska Press.

Dobson, F.S. and J.D. Kjelgaard. 1985a. The influence of food resources on population dynamics in Columbian ground squirrels. *Canadian Journal of Zoology* 63:2095-2104.

Dobson, F.S. and J.D. Kjelgaard. 1985b. The influence of food resources on life history in Columbian ground squirrels. *Canadian Journal of Zoology* 63:2105-2109.

Dunford, C. 1977. Social system of round-tailed ground squirrels. *Animal Behaviour* 25:885-906.

Emlen, S.T. and L.W. Oring. 1977. Ecology, sexual selection, and the evolution of mating systems. *Science* 197:215-223.

Festa-Bianchet, M. 1981. Reproduction in yearling female Columbian ground squirrels (*Spermophilus columbianus*). *Canadian Journal of Zoology* 59:1032-1035.

Frase, B.A. 1983. Spatial and behavioral foraging patterns and diet selectivity in the social yellow-bellied marmot. Ph.D. dissertation, University of Kansas.

Frase, B.A. and K.B. Armitage. 1984. Foraging patterns of yellow-bellied marmots: Role of kinship and individual variability. *Behavioral Ecology and Sociobiology* 16:1-10.

Frase, B.A. and R.S. Hoffman. 1980. *Marmota flaviventris*. *Mammalian Species* 135:1-8.

Gaines, M.S. and L.R. McClenaghan, Jr. 1980. Dispersal in small mammals. *Annual Review of Ecology and Systematics* 11:163-196.

Garrett, M.G., J.L. Hoogland and W.L. Franklin. 1982. Demographic differences between an old and a new colony of black-tailed prairie dogs (*Cynomys ludovicianus*). *American Midland Naturalist* 108:51-59.

Harris, M.A. and J.O. Murie. 1984. Inheritance of nest sites in female Columbian ground squirrels. *Behavioral Ecology and Sociobiology* 15:97-102.

Herbers, J.M. 1981. Time resources and laziness in animals. *Oecologia* 49:252-262.

Holekamp, K.E. 1984. "Dispersal in ground-dwelling sciurids." pp. 197-320. In: J.O. Murie and G.R. Michener, eds. *Biology of ground-dwelling squirrels*. Lincoln:University of Nebraska Press.

Holmes, W.G. 1984a. Predation risk and foraging behavior of the hoary marmot in Alaska. *Behavioral Ecology and Sociobiology* 15:293-301.

Holmes, W.G. 1984b. "The ecological basis of monogamy in Alaskan hoary marmots." pp. 250-274. In: J.O. Murie and G.R. Michener, eds. *Biology of ground-dwelling squirrels*. Lincoln:University of Nebraska Press.

Hoogland, J.L. 1979a. The effect of colony size on individual alertness of prairie dogs (Sciuridae:*Cynomys* spp.). *Animal Behaviour* 27:394-407.

Hoogland, J.L. 1979b. Aggression, ectoparasitism, and other possible costs of prairie dog (Sciuridae:*Cynomys* spp.) coloniality. *Behaviour* 69:1-35.

Hoogland, J.L. 1981a. "Nepotism and cooperative breeding in the black-tailed prairie dog (Sciuridae:*Cynomys ludovicianus*)." pp. 283-310. In: R.D. Alexander and D.W. Tinkle, eds. *Natural selection and social behavior*. New York:Chiron Press.

Hoogland, J.L. 1981b. The evolution of coloniality in white-tailed and black-tailed prairie dogs (Sciuridae:*Cynomys leucurus* and *C. ludovicicanus*). *Ecology* 62:252-272.

Hoogland, J.L. 1985. Infanticide in prairie dogs: Lactating females kill offspring of close kin. *Science* 230:1037-1040.

Hoogland, J.L. 1986. Nepotism in prairie dogs (*Cynomys ludovicianus*) varies with competition but not with kinship. *Animal Behaviour* 34:263-270.

Kiell, D.J. and J.S. Millar. 1980. Reproduction and nutrient resources of arctic ground squirrels. *Canadian Journal of Zoology* 58:416-421.

Kilgore, D.L., Jr. and K.B. Armitage. 1978. Energetics of yellow-bellied marmot populations. *Ecology* 59:78-88.

King, J.A. 1955. Social behavior, social organization, and population dynamics in a black-tailed prairie dog town in the Black Hills of South Dakota. *Contributions from the Laboratory of Vertebrate Biology of the University of Michigan* 67:1-23.

Koford, C.B. 1958. Prairie dogs, whitefaces and blue grama. *Wildlife Monographs* 3:1-78.

Lindstedt, S.L., B.J. Miller and S.W. Buskirk. 1986. Home range, time, and body size in mammals. *Ecology* 67:413-418.

McLean, I.G. 1982. The association of female kin in the arctic ground squirrel *Spermophilus parryii*. *Behavioral Ecology and Sociobiology* 10:91-99.

Meier, P.T. 1985. Behavioral ecology, social organization and mating system of woodchucks (*Marmota monax*) in southeast Ohio. Ph.D. dissertation. Athens:Ohio University.

Michener, G.R. 1979. Spatial relationships and social organization of adult Richardson's ground squirrels. *Canadian Journal of Zoology* 57:125-139.

Michener, G.R. 1983. "Kin identification, matriarchies, and the evolution of sociality in ground-dwelling sciurids." pp. 528-572. In: J.F. Eisenberg and D.G. Kleiman, eds. *Advances in the study of mammalian behavior*. American Society of Mammalogists Special Publication Number 7.

Murie, J.O., D.A. Boat and V.K. Kivett. 1980. Litter size in Columbian ground squirrels (*Spermophilus columbianus*). *Journal of Mammalogy* 61:237-244.

Owings, D.H., M. Borchert and R. Virginia. 1977. The behavior of California ground squirrels. *Animal Behaviour* 25:221-230.

Ozoga, J.J., L.J. Verma and C.S. Bienz. 1982. Parturition behavior and territoriality in deer: Impact on neonatal mortality. *Journal of Wildlife Management* 46:1-11.

Pfeiffer, S.R. 1982. Variability in reproductive output and success of *Spermophilus elegans* ground squirrels. *Journal of Mammalogy* 63:284-289.

Rayor, L.S. 1985. Effects of habitat quality on growth, age of first reproduction, and dispersal in Gunnison's prairie dogs (*Cynomys gunnisoni*). *Canadian Journal of Zoology* 62:2835-2840.

Reynolds, H.G. and F. Turkowski. 1972. Reproductive variations in the round-tailed ground squirrel as related to winter rainfall. *Journal of Mammalogy* 53:893-898.

Rubenstein, D.I. and R.W. Wrangham. 1980. Why is altruism toward kin so rare? *Zeitschrift für* Tierpsychologie 54:381-387.

Sheppard, D.H. 1972. Reproduction of Richardson's ground squirrel (*Spermophilus richardsonii*) in southern Saskatchewan. *Canadian Journal of Zoology* 50:1577-1581.

Sherman, P.W. 1981. "Reproductive competition and infanticide in Belding's ground squirrels and other animals." pp. 311-331. In: R.D. Alexander and D.W. Tinkle, eds. *Natural selection and social behavior*. New York:Chiron Press.

Sherman, P.W. and M.L. Morton, 1984. Demography of Belding's ground squirrels. *Ecology* 65:1617-1628.

Siegel, S. 1956. Nonparametric statistics. New York:McGraw-Hill.

Slade, N.A. and D.F. Balph. 1974. Population ecology of Uinta ground squirrels. *Ecology* 55:989-1003.

Slobodchikoff, C.N. 1984. "Resources and the evolution of social behavior." pp. 227-251. In: P.W. Price, C.N. Slobodchikoff and W.S. Gaud, eds. *A new ecology: Novel approaches to interactive systems*. New York:John Wiley and Sons.

Svendsen, G.E. 1974. Behavioral and environmental factors in the spatial distribution and population dynamics of a yellow-bellied marmot population. *Ecology* 55:760-771.

Trivers, R.L. 1974. Parent-offspring conflict. *American Zoologist* 14:249-264.

Vestal, B.M. and H. McCarley. 1984. "Spatial and social relations of kin in thirteen-lined and other ground squirrels." pp. 404-423. In: J.O. Murie and G.R. Michener, eds. *Biology of ground-dwelling squirrels*. Lincoln:University of Nebraska Press.

Wasser, S.K. and D.P. Barash. 1983. Reproductive suppression among female mammals: Implications for biomedicine and sexual selection theory. *The Quarterly Review of Biology* 58:513-538.

Wilson, E.O. 1975. *Sociobiology*. MA:Harvard University Press.

Chapter 8

SOCIAL SYSTEMS, RESOURCES, AND PHYLOGENETIC INERTIA: AN EXPERIMENTAL TEST AND ITS LIMITATIONS

Joel Berger

Department of Range, Wildlife, Forestry and Biology
University of Nevada, Reno 89512

and
Conservation and Research Center
Smithsonian Institution
Front Royal, Virginia 22630

THE ECOLOGY
OF SOCIAL BEHAVIOR

ACKNOWLEDGMENTS
REFERENCES

I. INTRODUCTION

Many behavioral ecologists and ethologists who study social systems attempt to explain them by identifying selection pressures and relying on adaptive explanations (see Crook 1970; Rowell 1979; Suzuki 1979; Thornhill and Alcock 1983; Lott 1984). Far fewer have considered whether social systems are the product of phylogenetic effects, resulting not from ecological conditions per se but rather historical ones (Eisenberg 1966, 1981; Clutton-Brock and Harvey 1979; Cheverud et al. 1985; Dobson 1985; Hardford and Mars 1985). The distinction between these two sets of factors is often blurred, especially when ecological conditions have changed little from the past to the present; in many cases it will not be possible to consider phylogenetic effects as alternatives to ecological ones because it is practically impossible to determine how an ancestral trait developed.

In this paper I want to deal with two issues concerning the study of social systems: whether or not species respond to a given set of ecological conditions in a predictable fashion *and* how phylogeny confuses such relationships. I first offer an overview of social systems by concentrating on ungulates, and then I suggest reasons why controlled experiments are necessary. Examples of field evidence are presented from some representative groups followed by experiments on two species of zebras in which resources are manipulated in an attempt to alter social systems. Finally, I point to problems of interpretation, to limitations of the data, and to areas where continued study may help resolve some of the issues brought forth.

II. ANALYSES OF SOCIAL SYSTEMS AND
SOME WORKING DEFINITIONS

"Social system" is a broad term with different meanings. It has been used in a variety of contexts often with an emphasis on general ways in which both species and individuals respond to local conditions. Descriptions of social systems have incorporated: 1) patterns of intraspecific spacing; 2) home ranges and daily use patterns; 3) demographic parameters; 4) resource distribution; 5) aggressive strategies of males, and less often of females; 6) female cohesiveness; and other variables (Eisenberg 1966; Fisler 1969; Crook 1970; Crook et al. 1976; Brown 1975; Wilson 1975; Clutton-Brock and Harvey 1977; Kleiman 1977; Waser and Wiley 1979; Michener 1983; Van Schaik and Von Hooff 1983; Lott 1984). These authors suggest that social *systems* of necessity must include descriptions of *mating and rearing the young,* and possible antipredator

behavior, while social *organization* is often reserved for broader categories which may also incorporate foraging (Eisenberg 1981). Discussions
of mating systems have traditionally focused on tactics used by individuals to improve reproductive success, but papers have also concentrated
on relationships between mating systems and sexual dimorphism (Alexander et. al. 1979; Clutton-Brock et al. 1980; Harcourt et al. 1981).

It is pertinent to review briefly some of the past approaches to the
study of social organization, and why these have led to the present experimental analysis. Because many factors can influence how populations and individual species respond to different conditions, it has not
been easy to determine which, if any, variables cause specific responses.
Significant progress was made by Peter Jarman's seminal (1974) paper on
antelope social organization. Jarman suggested that food resources and
body size explained much of the interspecific variation found in the
spacing and foraging patterns of African antelopes. More recent work
on a variety of mammalian taxa including lagomorphs (Swihart 1984),
sciurids (Armitage 1981; Murie and Michener 1984), carnivores (Bekoff
et al. 1981; Gittleman 1986a), and primates (Harvey and Clutton-Brock
1985) have affirmed the need to consider these variable and especially
body size when trying to understand factors which promote social relationships.

While many of the above analyses have been based on data collected
from field observations and provide cogent explanations for the observed
diversity of spacing, without experimental manipulations it has been difficult to determine the role that different factors have upon individual
species and how labile intraspecific patterns may be. Quite simply,
widespread experiments to determine the effects of variables X and Y
on pattern Z have not been possible. Although evolution has produced
different systems on a variety of continents, by necessity we must rely
primarily on observational data to explain patterns of diversity. Numerous difficulties arise when trying to account for variation that may be
due to phylogeny and body size (Harvey and Mace 1982; Jarman 1982).
Some of the problems concerning body size can be removed by holding
size constant through the use of multivariate statistics, but it is not always clear how to remove phylogenetic influences nor to separate cause
from effect (Clutton-Brock and Harvey 1984; Cheverud et al. 1985;
Gittleman 1986b).

Other problems are also encountered in trying to explain the interplay between social systems and local ecological conditions (LEC). A
traditional view has been that a species' typical social organization is a
direct response to its resource base, with environmental factors modifying to some extent resource-related behavior (e.g., group size, cooperation) (see Wittenberger 1981; Lott 1984; Woolfenden and Fitzpatrick
1984; Vehrencamp and Bradbury 1984 for reviews). Yet, Rowell (1979)
suggested that it may be difficult to detect whether social systems are
adaptive at all, and Jarman (1982) re-affirmed earlier views that social
systems are merely descriptions of how individuals are organized in
space and time, ostensibly responding to a set of conditions. Given the
evolutionary differences among certain lineages it may not be possible

to answer questions about adaptation (see also Lewontin 1978; Gould and Lewontin 1979). Harvey and Mace (1982) suggest some prospects for improving methods to study interspecific comparisons but we will face the dilemma of how to separate the multiple influences of factors which affect social systems.

One possible way to attempt to derive meaningful generalizations is to seek new species from which to examine predictions of specific models, recently done for marsupials (Lee and Cockburn 1985). If the predictions are powerful enough they should then explain the patterns found in the group under study. While in theory this is fine, there are at least two major problems.

First, most if not all field workers know that the study of social systems does not often proceed in this straightforward manner. Even when the most careful hypotheses are laid out in advance, the species, LEC, or data often fail to match the fit between expected and observed. There are probably many reasons for this failure. However, because a species' response is not in the predicted direction does not mean that the theory is wrong, or even that the data are inappropriate. It could be that the study was too short, the wrong variables were measured, the system was more complex than previously supposed, the age structure masked expected responses, or any other concoction of real or imagined factors.

Second, selecting the appropriate species (or taxonomic group) is especially important. This is because statistically independent points must be gathered for interspecific comparisons (Harvey and Mace 1982) but it is still not always clear at what level this occurs (species, genus, family, etc.) (Felsenstein 1985) or how it may vary among taxonomic entities. To illustrate this complication let's assume a genus with species all of equivalent size, all (presumably) monogamous, and all exploiting very similar food resources. Would phylogenetic rather than ecological factors be implicated as the major force producing such traits? Although such groups exist in many taxa (e.g. *Cephalophus* (duikers), *Madoqua* (dik diks), *Hylobates* (gibbons); Ralls 1976; Clutton-Brock and Harvey 1977), deciding how to weight or to tease apart ecological from phylogenetic effects on social systems remains a substantial challenge (Wilson 1975). Such difficulties are highlighted by ratites and tinamous. These flightless, relatively large, terrestrial birds offer striking instances of convergence in morphology and behavior, including male parental care despite mating systems which vary from monogamy to harem polygyny (Handford and Mares 1985). Among the more interesting creative ways to tease apart adaptive trends is the comparison of populations with long periods of evolutionary divergence in different habitats, (Rowe et al. 1986).

Before proceeding it is necessary to employ working definitions for several commonly used terms. I do not imply that my working definitions are any better than prior ones; only that to avoid some later confusion, I want to indicate more precisely what is meant here.

Phylogenetic Inertia: The occurrence of characteristics derived from common ancestors and shared by all members of a group (see Wilson 1975).

Local Ecological Conditions (LEC): Abiotic or biotic factors including food, water, energy, vegetation, precipitation, other species, etc., associated with any given site.

Social System: This term is often used synonymously with social organization, and both have been used for descriptions of intraspecific spatial relationships and group composition; some reserve social organization for accounts of intraspecific interactions (Van Schaik and Von Hoof 1983; see also Bekoff and Wells 1986). For my purpose, social systems will refer to descriptions of spacing and mating relationships while social organization will be used in a broader context which includes rearing of young, foraging as related to group structure, and demographic events (Wilson 1975).

Territory: Although many specific definitions are available (see Brown 1975; Wittenberger 1981), I prefer a rather broad (and admittedly simple) one; an exclusive area which is defended (see Burt 1943).

Harem: While the term "harem" has numerous connotations (Wrangham and Rubenstein 1986), I prefer the following: A group of females (with or without young) associating simultaneously for extended time periods with one (or occasionally more) male(s) and defended by their consort male(s) from interactions with other males.

Resources: Many factors including water, food, mates, etc. constitute resources (see Wiens 1984). I will restrict my operational definition to food and water.

III.   OVERVIEW OF UNGULATE SOCIAL ORGANIZATION
AND SOME ANOMALIES

Jarman (1974) suggested that in African antelopes a continuum of sizes and social/feeding classes exist. At one extreme small-bodied species such as duikers and dik diks associate in monogamous territorial pairs, foraging selectively on scattered and defendable food items. With enlarging body size, species' absolute food requirements increase and animals tend to occur in larger groups, feeding less selectively on grasses and forbs. Males are polygynous and group size may reach several hundred in such species as gazelles, kobs, and impala. At the opposite extreme are the largest species, eland and water buffalo, which comparatively speaking are less selective feeders, and vulnerable only to the largest predators. Their breeding systems are also polygynous. The relationship between body size and other socio-ecological variables in extant antelopes is so strong that it has been used to reconstruct the paleoecology of fossil ungulates (Janis 1982). Although more sophisticated quantitative analyses of social organization have followed for cervids (Clutton-Brock et al. 1980) and non-ungulate mammals (Bekoff et al. 1984; Armitage 1981), body size and diet specialization are still viewed as primary influences affecting species social organization (Eisenberg 1966, 1981; Western 1979).

Few, if any, analyses of ungulate social organization have considered how and where the family Equidae fits into the above scheme. This is an important omission; if broad ecological or evolutionary generalizations are to be accepted, they must be robust enough to include a variety of taxa. While a limited number of cases will always thwart general theory, and eloquent post hoc arguments can always be constructed for some groups, we are still left with the task of wanting to uncover the forces that produced certain systems.

In many respects, equids are interesting anomalies to the group; from which the above patterns have been described. Unlike bovids, cervids, giraffids, and antilocapids, equids and other groups (camels, pigs, tapirs) are not true ruminants (Langer 1973) since they lack a four-chambered stomach. In addition to being cecal digestors (Janis 1976) all equids are monomorphic in body size, they lack conspicuous secondary sexual characteristics, and some groups live in year-round harems (Berger 1986). Equids are not alone in their omission from contemporary analyses of ungulate social organization. Elephants, rhinos, tapirs, camels, and pigs have also not been included (but see Kiltle and Terborgh 1983; and Eisenberg 1981).

Lack of good data for these groups may have prevented their inclusion in existing theory, but there are also phylogenetic variables and evolutionary factors that may have rendered such groups as inappropriate choices for testing predictions. Groups which have been subjected to socioecological analyses (Geist 1974; Estes 1974; Jarman 1974; Clutton-Brock et al. 1980), cervids and bovids, are each categorized by species which range from small to large body size and have experienced relatively recent (Pliocene) radiations associated with the spread of grass-lands (Geist 1978; Cifelli 1981; Janis 1982). Conversely, rhinos, elephants, camels, tapirs, and equids are all represented by comparatively large-bodied species; these groups were formerly more successful and widespread than they now are (Maglio 1975; Webb 1977; Churcher and Richardson 1978). Whether or not these peculiarities are behind the exclusion of these groups from general treatments of social systems is not important. What is at issue is whether or not broadly derived generalizations will allow equids and other non-ruminant species to be placed within a cogent socioecological framework as well. My aim there is to tackle only a small part of this broader problem. I want to examine the evidence that equid social systems can be predicted in relation to their LEC by reliance on existing theory.

IV. EXPERIMENTAL ECOLOGY AND SOCIAL SYSTEMS

Much of the information on mammalian social systems stems from correlational approaches or from fortuitous natural perturbations. This is understandable. It is impractical, if not impossible, to simulate natural conditions under controlled settings, especially for large mammals. Since many of such species are becoming increasingly rare, experimental manipulations are logistically difficult and funding is often not available. Zoological and national parks with their conservation-oriented missions are rightfully not the appropriate place to perform these manipulations if the subject might experience detrimental effects simply for the sake of basic science.

Most experimental studies of mammalian social organization have used smaller species, usually rodents, as models in which variables such as sex ratios, habitats, food resources, age structures or densities are manipulated. Nevertheless, the dearth of experimental work on large mammal social systems places a strong reliance on observational data. As a result, the testing of numerous hypotheses is based on inference and arguments of design (see Williams 1966; Symons 1978) rather than on controlling the influences of specific variables.

Lack of experimentation has resulted in the acceptance of several assumptions. It is often supposed that ungulate aggregations are responses to highly dense food resources, rather than clumping to avoid potential predators. Unfortunately though, it is difficult to disentangle the influence of one variable from the other. The interaction between potential predators and food resources could possibly be separated using two experimental settings, one in which predation pressure was held constant and the other in which food resource quality and availability were constant. In the first situation, ungulate group sizes should be larger in areas where food density is greatest if food rather than predators was responsible for increased grouping. In the second scenario, group sizes should be largest where potential predators are common. Some evidence in support of the former and latter exist (Hirth 1977; Berger et al. 1983) but without controlling some of the possible confounding variables a clean picture of cause and effect cannot emerge. It is clear that food resources affect in some general way spacing and grouping patterns but finding methods to quantify specific influences has not proved easy. The South American marsh deer, a forest browser living in small herds, is a good example. Herd formation may be the result of any number of factors including flooded marsh habitats, low carrying capacities in which animals group around isolated food pockets, and unforeseen heavy predation pressure, or it could possibly be a recent event brought about by increased human hunting pressure (Eisenberg 1981; Redford 1985). In this case, as in many others, it will take additional work to uncover the causes. Most important, however, is that based on the existing data one could construct plausible arguments in any number of directions. Without a detailed look at the existing *data* and conditions under which they were collected, one could be lured into supporting chosen arguments when the data are inconclusive.

V. RESOURCES AND GROUPING AS INFLUENCED
BY PHYLOGENY

There is no simple or straightforward way to determine the level at which phylogeny does not have a major effect on a pattern. Clutton-Brock and Harvey (1977) suggested that one way to detect when phylogenetic effects are minimal is to use nested ANOVAS to identify the lowest taxonomic level at which the maximum variance in the trait is expressed. Yet, if species from two seemingly independent groups are compared and show divergence or convergence with regard to some trait, it may not be due to ecological factors, though we commonly presume this to be the case. For instance, macropod kangaroos form larger groups in open habitats than in closed ones (Kaufman 1974) and so do African ungulates. This could be due to differences in body size of respective groups, or because predator pressures vary between the two continents, although it is logical to assume that common resource-related factors operate on each group (Jarman 1982). However, if we were to select New World deer (though the analysis to my knowledge has not been done) and compare them to Old World antelope, it is not clear how we might interpret any of the resulting similarities or differences. Small forest-dwelling deer such as pudu are probably not monogamous while several species of small antelope dwelling in dense cover are (Ralls 1976; Kleiman 1977). Reasons for such differences are yet to be clearly explained.

Some fruitful between-species constrasts of social systems stem from comparisons of similar-sized members of the same genus. For illustration, I compare below information based on the literature in two carnivore and two herbivore lineages. In each case the species are close enough in body size and genealogy that interspecific matings have occurred under natural conditions, or in captivity. Because genetic differences between such species are relatively small, any observed behavioral traits might result from ecological differences rather than phylogenetic ones. However, this idea suffers because one can also argue that selective pressures on individuals within species have favored local traits. In essence, without information on the origins of ancestral traits it is not possible to distinguish between adaptation and inheritance. Thus, when possible I have relied on intraspecific comparisons and outline whether resource-related differeces produce lability in social systems.

A. Carnivore Examples

For some canids, inter- and intraspecific differences arise due to food size. Coyotes form packs where prey size is large and carrion is defendable, but they are in smaller groups or pairs when only small food items are available (Bowen 1982; Bekoff and Wells 1986). It remains unclear whether similar facultative responses occur in wolves (Zimen 1976; Allen 1979). Perhaps these closely related species of presumed similar "niches" respond in common fashion to varied resource bases. If this is the case, support for an ecologically induced social system hypothesis would be bolstered.

The same argument could be constructed for two members of the genus *Panthera*, tigers and lions. Tigers are primarily solitary hunters selecting large but scattered prey in thickly vegetated habitats (Sunquist 1981). In contrast, lions in a variety of areas cooperate socially to harvest and defend their food (Schaller 1972; Bertram 1978; Packer 1986). While lions show a marked degree of plasticity based on local ecological conditions (Van Orsdol et al. 1985), it is still unknown whether tigers would respond to congregated food resources by becoming more social. Tigers in mixed hardwood forest in the USSR had home ranges about ten times the size of those in an Asian lowland grassland; equally notable was that females had exclusive home ranges in Nepal but had overlapping ones in the USSR (see Sunquist 1981). Thus it is clear that tigresses exhibited some degree of social tolerance for one another due to ecological conditions.

Whether male tigers exhibit flexibility in their social tolerance for others is unclear, but some anecdotal accounts of male associations with both provisioned food and under natural conditions lend some support to this idea (Schaller 1967; Sunquist 1981). Like the canids in the above case it may be that phylogeny has not been at the root of differences in tiger and lion social systems. Variation in the type and distribution of food bases, brought about by habitat differences, may be all that is needed to explain the observed variation in the social systems of these two felids. Such examples highlight how tempting it may be to conclude that interspecific differences in social systems were induced by ecological conditions. Clearly, without experimental manipulations it will not be possible to rule out a variety of alternative explanations.

B. Herbivore Examples

In ungulates, congenerics such as white-tailed and mule deer are sympatric in the wild in a variety of areas, hybridize and have similar social systems (Geist 1981). They respond almost identically to ecological conditions, forming fairly large groups in open areas when resources are clumped and forming smaller groups in dense cover when resources are scattered (Hirth 1977; Kucera 1978). Bighorn sheep in desert and mountain environments also show consistent group size fluctuations with varying plant productivity (Fig. 1) (r_s = 0.93; p < 0.001). While these examples illustrate the very unsurprising relationships that intra-and interspecific grouping varies with underlying resource distributions, they say little about social systems or whether these social systems are directly shaped by either ecology or phylogeny.

In several primates, a greater complexity occurs in trying to sort out how taxonomy and ecology might influence social structure. Entire families or subfamilies can be characterized by the same social groupings, as indicated by the monogamous gibbons (Hylobatidae) (Chivers 1972; Tenaza 1975; Tilson 1981 or the Pitheciinae of the Cebidae, which include owl (*Aotus*) and titi (*Callicebus*) monkeys and sakis (*Pithecia*)

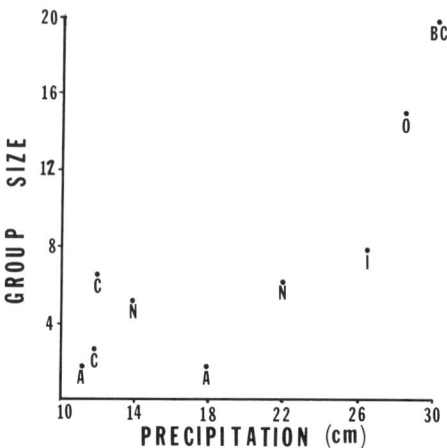

Fig. 1. Relationships between mean size of bighorn sheep ewe groups and mean annual precipitation based on the literature (from Berger unpub.). Ewe group sizes consist of ewes, young rams, juveniles and yearlings after peak lambing periods (lambs excluded). A - Arizona; BC - British Columbia; C - California; I - Idaho; N - Nevada; O - Oregon.

(Robinson et al. 1986). Because species within these different taxonomic groups all share the same social systems and family members all live in similar habitats exploiting what appear superficially to be the same kind of resources (but see Leighton 1986 and Rodman this volume), it is again difficult to say whether phylogeny or ecology produced such behavioral traits.

The situation is complicated to a much greater extent when considering two congeneric, arboreal and folivorous primates: the black and white, and the red colobus monkeys. They are sympatric in the Kibale Forest where red colobus occur in troops of about 50 members with several adult males while their black and white relatives live in single adult male troops numbering about eight animals (Struhsaker 1975; Oates 1977; Leeland et al. 1984). While numerous examples of clear correlates between primate ecology and social organization are available (see Eisenberg et al. 1972; Wilson 1975) there are also many exceptions (Clutton-Brock and Harvey 1977; Crockett and Eisenberg 1986.

VI.   RESOURCES AND LIVING IN YEAR-ROUND HAREMS AS INFLUENCED BY PHYLOGENY

Extending upon the above theme, I now want to examine further how a behavior different from casual grouping, that of living in year-round harems, may be influenced by phylogeny. Few groups of mammals are characterized by individuals who live in harems throughout the year. Marmots (Downhower and Armitage 1971), Gelada and Hamadryas baboons (Dunbar 1984; Kummer 1968), some langurs and other columbine monkeys (Hrdy 1977), some populations of African lions (Schaller 1972; Bertram 1978; Packer and Pusey 1983), dwarf mongooses (Rood 1978, 1980), and some populations of guanacoes, and vicunas (Franklin 1983) are among the notable examples. Among equids, feral and Przewalski horses, and Mountain and Common zebras (Klingel 1972; Berger 1986; Rubenstein 1986) are the species. As might be expected, factors promoting such long-term associations conceivably vary among species and ecological conditions.

In trying to account for year-round harems unitary explanations generally fail dismally; most arguments favoring a multitude of factors are post hoc and not amenable to experimental verification. Further, explanations can be constructed to favor phylogeny as a factor promoting these unusual stable groups. In the above, lions, marmots, and Gelada and Hamadryas baboons are without congenerics living in harems; within the same genera of columbines, camelids, equids, and mongooses, there are species living in year-around harems. By inference, phylogeny has had an influence, but because these groups (e.g., primates, carnivores, rodents, and ungulates) differ from one another in their phylogenetic history, we must conclude that year-round harem living has evolved independently several times. Ecological factors can also be implicated to some extent. In guanacoes and vicunas, plasticity in harem living is evident. Populations in different areas vary in their propensity to live in such groups and Franklin (1983) suggested that resource-related variation produces these contrasting social systems.

Even when local ecologies can be identified as sources for variability in social systems, it may not be evident why some species live in year-round harems. Among members of the genus *Papio*, there are baboons living in year-round harems with a single male, in multi-male groups, and in more casual assocations (Altmann 1980; Kummer 1968; Packer 1979; Stammback 1986). Some of these species are similar in body size and in methods of habitat exploitation, and some are even sympatric. The same is true of several equids. Grevy's and Common zebras overlap in areas of Kenya (Keast 1965), sometimes feeding in mixed groups, yet in Common zebras males defend harems year-round while this is untrue of Grevy's zebras (Rubenstein 1986). Among feral horses and burros, which are sympatric in a few areas of the Great Basin Desert of the USA, horses are in harems year-round whereas burrows are not. Because each of these equids tend to exploit graminoids when available and differ in body size by not more than about 15 to 20 percent, it is not obvious why their social structures differ so dramatically.

VII.   OBLIGATE AND FACULTATIVE SOCIAL SYSTEMS IN EQUIDS

Among equids there are six extant native and two feral species. These, their social systems, and the number of studies that reported social groupings in a variety of ecological zones are summarized in Table

I. Two basic types exist: year-round harem and casual mixed groups with mothers and non-sexually mature young forming the only permanent social unit (Klingel 1972).

Since all of the non-haremic equids derive from the asinus lineage of Equidae and the harem-dwellers are all caballine derivatives (see Groves 1974 for equid taxonomy), might these different social systems arise simply as a product of phylogenetic intertia? A problem in implicating phylogeny is that ecological factors also correlate nicely since the harem dwellers all live in relatively mesic areas whereas the non-harem dwellers are from more xeric environments. In accordance with Brown's (1964) formulation of economic defensibility, it has been postulated that males in xeric areas cannot mobilize the reserves necessary to support the costs of year-round harem defense (Klingel 1972). While this idea is certainly plausible, it suffers because phylogeny and ecology covary as all non-harem dwelling equids live in xeric areas. What is needed is an experimental test.

Data collected by McCort (1980) offer opportunities to examine directly how ecological factors affect social systems in feral asses. Whereas prior studies of feral asses in Death Valley (Moehlman 1974) and the Sonoran Desert (Woodward 1979) found burros to be organized into loosely structured groups with females and their young forming the permanent associations, McCort found strikingly different patterns. On the relatively lush, subtropical Ossabau Island situated off the Georgia Coastline, feral asses were in year-round harems (McCort 1980). His

Table I. Summary of the number of studies reporting different species of equids to be in either year-round harems OR casual, loosely organized (e.g. non-harem) groups.

	Horses		Zebras		Hemiones		Asses	
	Feral	Prze-walski	Moun-tain	Common	Grevy	Kulan	Native	Feral
Harem	18	?	3	6	0	0	0	1
Non-Harem	1	?	0	0	3	3	2	6

References: Berger (1977, 1983b, unpubl.); Grubb (1981); Hoffman (1983); Joubert (1974); Kaseda (1981); Keiper (1976); Klingel (1967, 1968, 1972, 1972*, 1975, 1977); McCort (1980*); Miller (1981); Moehlman (1974*); Monfort and Monfort (1978); National Research Council (1980); Norment and Douglas (1977); Penzhorn (1984); Rubenstein (1981*); Salter and Hudson (1982); Seegmiller (1977); Smuts (1975); Solmatin (1973); Welsh (1975); Woodward (1979); Zicardi (1970).

*Referenced in this paper; for others see Berger (1986).

data demonstrate that at least one species of feral equid possess the social resiliency to alternate between year-round harems and other less permanent types of social groupings when ecological conditions vary.

Despite this one apparent test for feral asses, the correlation between occupancy of relatively mesic habitats and year-round harem living makes it difficult to suggest whether phylogeny or resources have promoted such social systems. Perhaps ass habitats do not differ to the extent needed to disentangle the influences of the above two variables. To tease these apart, data are needed on species differing in social systems but living in identical habitats or on the social systems of populations living in extremely different ecological areas.

Some data exist that meet this former criterion, and they support a phylogenetic argument. Both feral horses and burrows are sympatric as pointed out above, but horses still live in harems while burros do not (see Berger 1986). Similarly, Common and Grevy's zebras overlap in several areas in Kenya but their social systems show no convergence (Klingel 1972).

In summary, limited success has been achieved in trying to sort out which of many variables influence equid social systems. In part, problems exist with the questions themselves, both because data cannot be easily collected which will directly address the issues and because equid phylogeny cannot be manipulated beyond what it already is. Other difficulties also occur: five of the six native equids are now endangered species and it is not possible to conduct natural experiments in the field.

Fortunately, possibilities for manipulations of resources are available under certain conditions, and in 1984 with these thoughts in mind I set about trying to find ways in which to analyze experimentally which variables were most important because of unique characteristics associated with a large endangered species preserve.

VIII.   EXPERIMENTAL ANALYSES OF EQUID SPACING SYSTEMS

A.   Overview

Based on prior literature (Rubenstein 1981, 1986), I assumed that equid social organization is a direct response to resources, with female distribution being explained by food and water distribution, and males tracking females (Emlen and Oring 1977). If these assumptions are correct, then similar patterns of movement and social cohesion would be expected of two zebra species when resources are identical. As a test of this prediction, I examined the movement patterns and social use of space of Grevy's and mountain zebras under conditions in which food resources are experimentally altered.

B.   Study Site and Methods

Canyon Colorado Equid Sanctuary is a privately owned 40,000 acre+ (ca. 160 km^2) preserve (Grunerwald pers. comm.) at about 1900 m on the high plains of northern New Mexico (Fig. 2). It straddles the western

Fig. 2 (A) Overview of Canyon Colorado Equid Sanctuary and (B) Grevy's zebras females in Park (A).

boundary of the Canadian River Gorge, a deeply cut drainage of the Canadian River, and is situated east of the Sangre de Cristo Mountains near the town of Wagon Mound. Some areas are fenced but a connecting gorge, Canyon Colorado, serves as a barrier to zebra movements to the east.

The reserve was established as a breeding center for rare and endangered equids, and three species (kulans, and mountain and Grevy's zebra) occurred there during the time the study was performed, January until August, 1984. With the exception of mountain zebras, the other animals originated from zoos. Twenty-two Mountain zebras were captured in the wild in Namibia and brought to the United States in late 1982. All animals at the preserve were offered artificial water and food (primarily alfalfa and grain), using the latter most heavily in late fall and winter when native grasses were less available.

Because of concern over the potential deleterious effects of inbreeding, Canyon Colorado animals were separated when possible from one another on the basis of their genetic relationships or the number of offspring sired by particular unrelated males. Overrepresented males as well as fathers and sons were segregated from others, sisters, and other close kin. Unrelated animals were maintained in enclosures which varied in size from several acres to about 6.4 km^2. The resultant associations provided the bases for tests of how food resources influenced several aspects of equid behavior.

C.   Experimental Perturbation and Data Collection

For each species two replicates were performed. The approximate size of available area and sex ratios of species for each are given in Table II. All treatments (A, B, C, and D) occurred in parks which had similar habitat components--spacious grasslands, pinyon-juniper forests, and deeply-cut drainages and rimrock. Artificial water sources were provided and these were not changed over the duration of the study. In treatments A-C, several adult males were present, but for mountain zebras in D, only one male occurred with the females (Table II). This unequal sampling treatment creates problems in trying to understand how potential social competition might influence spacing in relation to resources and there is no simple way to dismiss this bias. I will argue below that, although the problem is real, it is still possible to make some deductions about zebra social groupings and resources.

Table II. Approximate size of parks available to zebras and adult sex ratios.

Species	Park	Approximate Area (km^2)	Adult Sex Ratio
Grevy	A	5.1	3:6
Grevy	B	4.2	3:5
Mountain	C	4.0	3:4
Mountain	D	2.7	1:5

Zebras were identified on the basis of distinguishing stripe patterns. Grevy's zebra ages were known by birthdates and records from originating zoos. Unfortunately, ages of the mountain zebras were not determined at capture and they remained unknown. Grids of approximately 100 x 100 m were drawn on aerial photographs and the number used per unit time allowed estimation of home range size. Observations were made from jeeps or foot by Carol Cunningham, Michael Henry, and myself. The data reported here are based on 661 hours of observations.

Zebra foraged on native plants and roamed freely within each park. Food manipulations were performed as follows. All animals were provisioned with a protein supplement (grain) at the location where they had also received it for the prior several months. Patterns resulting from this situation will be referred to as baseline data. Experiment I involved the movement of grain and provisioning to a flattened and visible area in the approximate center of the cumulative home range of all animals within a given park (Fig. 3). Experiment II involved the movement and provisioning of grain to areas outside the 90 percent frequency-use polygons of at least 70 percent of the members of each population. I assumed that animals would have to search widely and beyond their traditionally used areas if they were to feed on grain. Each experimental perturbation followed the prior one sequentially by approximately six weeks, and began about six weeks after baseline data had been collected in each park. The manipulations were not synchronized among all parks, but they never differed by more than eight days.

D. Size of Areas Used and Their Changes

Areas of use were calculated by using frequency-use polygons for 90 percent and 50 percent of the most clustered observations; these areas

Fig. 3. Experimental trough (lower left) and Grevy's zebra male and females in an open area at Canyon Colorado Equid Sanctuary.

were called home ranges and core area, respectively (see Berger 1986). Mean sizes of these regions for females in each park and resultant increases with food manipulations are shown in Table III. Both core areas and home ranges increased significantly in size from prior measurements. It was evident that the frequented areas diminished in size once animals discovered the new locations of grain if the grain remained in that area without additional short-term changes in location. Contrasts between home range size three weeks before and three weeks after the food manipulations within each park, and for each experimental treatment, showed that size differed significantly (p < 0.05 for each contrast within A and B using Mann-Whitney U Tests, and p < 0.001 for within C and D contrasts using T tests). Such changes suggest that females were tracking their food resources, but they do not reveal whether the females sought out the food resources by themselves, or in groups.

E.   Social Cohesiveness and Food Manipulations

Does the degree to which females associate with other females (their cohesiveness) change in relation to resource distribution? Since the same home ranges could be exploited by all members of each park,

Table III.   Mean home range core area sizes (in acres) used by female zebras during different treatments.   For experiments 1 and 2 numbers refer to percent increases.

Treatment	Species/Park		Home Range	Core Area
Baseline				
	Grevy	A	610	240
	Grevy	B	630	210
	Mountain	C	575	375
	Mountain	D	540	275
Experiment 1				
		A	6*	35***
		B	10**	32***
		C	7*	7*
		D	5*	11*
Experiment 2				
		A	40***	66***
		B	55***	60***
		C	29***	24***
		D	15**	18**

Asterisks refer to comparisons made with the prior treatment. *, p < 0.05; **, p < 0.01; ***, p < 0.001; Mann-Whitney Test.

but usage could vary temporally (at different times of the day or night), it may be that group cohesiveness was lacking. To show that members of each park occurred in cohesive groups requires the demonstration that home ranges overlapped temporally and spatially.

During the collection of baseline data, Grevy's zebras occurred in bands only 15 percent (Park A) and 9 percent (B) of the time. These were associations where all females were within 50 m of the same male. In contrast mountain zebra bands were found during 97 percent (C) and 99 percent (D) of the scan samples. During Experiments I and II the following results were obtained concerning the percentage of observations in bands: (I) Grevy's zebras: (A) 18 percent, (B) 17 percent; mountain zebras: (C) 99 percent, (D) 100 percent; (II) Grevy's zebra: (A) 6 percent, (B) 7 percent; and mountain zebras: (C) 99 percent (D) 99 percent. For any given trial, differences in band cohesion were not significant within a species (two-way Z test; Sokal and Rohlf 1981), but differences between species were significant ($p < 0.001$, two-way Z tests) for all paired comparisons between species for the baseline data and two follow up experiments. Additionally, mountain zebra bands remained intact despite food manipulations, and no changes over time occurred. In contrast, when grain was provided outside the normal home ranges of Grevy's zebras, the degree of cohesiveness dropped from 18 to 6 percent in A and from 17 to 7 percent in B ($p < 0.05$ for each; two-way Z tests).

These data suggest that the two zebra species had consistent but dissimilar grouping responses to experimental manipulations of protein supplements. Food resources mediated changes in home ranges and movement patterns for all replicates of both species and during all control and experimental perturbations but mountain zebras were still found in bands, whereas Grevy's zebras were so far less frequently. Hence, the data further suggest that protein manipulations did not dictate whether zebra females would be found in harems since resources were changed in similar fashions for all replicates.

F.   Exclusive Use Areas by Males

Because active defense of many areas did not occur, I prefer to report only on areas of exclusive use. The size of exclusively used regions was determined by plotting home ranges and calculating the size of areas where other same-sexed individuals did not overlap. The results are shown in Fig. 4. Grevy's males in the replicates and during experimental manipulations showed considerable variability in the size of exclusive areas, but throughout all cases these areas were sigificantly larger than those shown by mountain zebra males in Park C ($p < 0.01$; Mann-Whitney U test). The data are consistent with the available literature (Klingel 1967), that Grevy's males establish territories, whereas mountain zebra males do not. One limitation of these data is that no mountain zebra replicate was available since in Park D only one male existed (Table II).

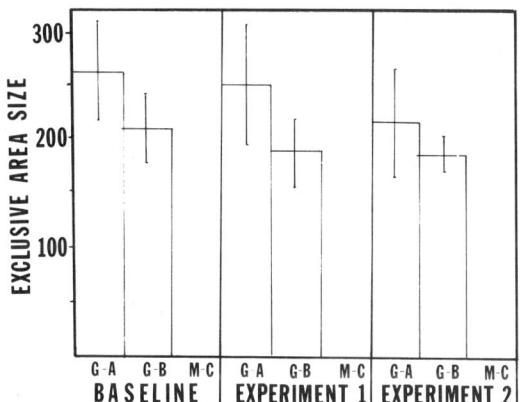

Fig. 4.   Relationships between mean size of areas (in acres) used exclusively by males and treatment in two species of zebras. Vertical lines – standard deviations; G – Grevey's M – Mountain zebras; (A), (B), and (C) refer to parks (see Table 2).

G.   Resources and Phylogeny in Zebras

The above results point to two interesting relationships. First, each zebra species responded in similar fashion to unpredictably distributed food; increased home range size resulted when protein supplements were unpredictably distributed, and home ranges diminished after the grain remained in areas for several weeks. Thus, with regard to food distribution it appears that females did indeed track food. Second, despite identical food manipulations in four parks the two zebra species differed consistently in their social systems. The conclusion that phylogenetic inertia produced these contrasting social response is tempting.

H.   Potential Biases Arising from the Experiment

A phylogenetic inertia hypothesis is bolstered by the above results, but the above experimental analyses are far from conclusive. Given the logistical difficulties of working with large mammals, and especially endangered species such as mountain and Grevy's zebras, it may never prove easy to tease apart variables which produced zebra social systems. Although I suspect that phylogenetic inertia plays a large role in explaining why many, if not most, species exist in harems defended year-round by males, the above experimental work faces several shortcomings.

(1) The time period over which the manipulations were performed may have been too short to alter species social systems. While numerous species have the ability to shift social systems in rather short periods of time (Lott 1984) such plasticity may not be possible in zebras. It may be that Grevy's zebras would have shifted to harems, or mountain zebras to more loosely organized associa-

tions, but the time period over which data were collected was simply too short. Further, the past environments of the two species might have differed, and produced the social systems which exist today. Yet, by examining putative adaptive consequences of extant social systems we may be creating an error: the failure to distinguish between selection pressures responsible for the origin of a trait and its current use (see Gould and Vrba 1982; Kiltie 1985).

(2) Resource levels were not harsh enough to cause shifts from one social system to another. In essence, the above work might not have addressed whether species converge in social systems when resources are depressed below a threshold. In the manipulation experiments it may be that resources were not sufficiently depressed so that it would no longer benefit mountain zebra stallions to continue to defend females. If this was true, energetic arguments could account for the differences in social systems rather than purely phylogenetic ones. Of course, these need not be distinctly different alternatives.

(3) The resources within each of the replicates, while being qualitatively similar, could have differed in properties that my coworkers and I failed to detect. Because even seemingly homogeneous equid home ranges differ in plant quality (Berger 1986), subtle differences in underlying vegetation components cannot be ruled out as possible determinants of the differences between the two zebra species.

(4) The age structure of the groups could have produced subtle variations in male behavior, including subsequent patterns of spatial defence. While this might have been veiwed as an unlikely explanation for variation in social systems several years ago, recent works now suggests that male age strongly influences individual and population responses. In wild horses old males are more likely to escalate the intensity of fights than are younger males, and the former range more widely (Berger 1986). Also, pronghorn social systems may switch between harem defense and territoriality based on male age structure (Byers and Kitchen 1987).

(5) The ecological setting does not represent the species' true habitats. This argument probably has little to do with why the two zebra species responded differently in the resource manipulation experiments. In areas where a variety of different species have been transplanted to new habitats, or raised in captive environments, they still show characteristic patterns of behavior and social organization. Further, despite the maintenance of Grevy's and mountain zebras in areas with fencing, these species still ranged over relatively large areas, and regions larger than their home ranges were available for additional movements.

The above potential pitfalls should not diminish the importance of the experimental results. What they should do is underscore the difficulties that one faces when dealing with broad questions about evolutionary antecedents at the interface of ecology, behavior, and phylogeny, regardless of taxonomic group. In most cases it will not be possible to control for demographic, ecological, and behavioral sources of variation. This does not mean that questions in these areas should not be pursued or that true insight cannot be gained: only that rigorous examination of the data must be exercised when trying to understand how different factors have influenced social systems. Because historical factors underline all living systems, it seems prudent to expect that explanations for why systems evolved will take into account presumed past as well as present selection pressures. Nevertheless, finding ways to evaluate hypotheses dealing with historical effects remains difficult in the areas of evolutionary and behavioral ecology (Harvey et al. 1983; Conner and Simberloff 1986).

IX.   INFERENCES ABOUT ECOLOGICAL AND PHYOGENETIC EFFECTS

Numerous approaches have been adopted in trying to tease apart roles that specific factors have had in shaping individuals and species' responses to ecological conditions. The most promising have employed multivariate analyses where specific variables are held constant (Harvey and Mace 1982; Clutton-Brock 1985), but even these face problems; for, many variables still cannot be compared directly. Below, I illustrate some of the possible inferences that might be made by relying on direct pair-wise contrasts when closely related groups of equivalent body size are compared.

Consider for example which of two factors, ecology or phylogeny, might be implicated under the following eight scenarios. The first four are comparisons of hypothetical congeneric species (of similar body size) when social systems are either the same or different and in habitats which may be similar or different. The second four comparisons are contrasts between hypothetical populations of the same species, again differing in potential social systems and habitat. For the sake of simplicity, only phylogeny or ecology should be implicated (or checked) as possible determinants of social systems in the spaces below.

Quite obviously interpretations will vary depending upon the groups which serve as sources for comparisons. I am not confident that any of the above comparisons offered conclusive demonstrations to support either ecological or phylogenetic effects. Clearly, comparisons (1) and (2) are inconclusive since phylogeny might mask any secondary influence of ecology. Scenarios (5) and especially (6) offer little support for either factor. Phylogeny could plausibly be a greater cause than ecology for similar social systems in comparison (3) while ecological OR phylogenetic influences could produce the different social system highlighted by

Treatments	Social Structure	Implicate Phylogeny	Implicate Ecology
within genus			
1) cf spp in same habitats	if similar	---------	---------
2) cf spp in same habitats	if differ	---------	---------
3) cf spp in unlike habitats	if similar	---------	---------
4) cf spp in unlike habitats	if differ	---------	---------
within species			
5) cf pop in same habitats	if similar	---------	---------
6) cf pop in same habitats	if differ	---------	---------
7) cf pop is unlike habitats	if similar	---------	---------
8) cf pop is unlike habitats	if differ	---------	---------

comparison (4). Only comparisons (7) and (8) go a bit further in separating ecology from phylogeny with (7) implicating the latter while (8) does the former.

While the above exercise may be of heuristic value, it is fraught with problems. In the real world it is often unclear what constitutes anything except grossly similar and dissimilar habitats. Moreover, since habitats often differ in underlying distribution and quality of food (or predator and competitor densities) populations vary in size. And, as demography influences social behavior and social organizations (Woolfenden and Fitzpatrik 1984), differences detected in paired comparisons might not be due to any ecological factor at all. As a result of such difficulties, it may not be possible to refine more cleanly which, if any, specific factors influence social organization.

Perhaps one of the most promising approaches entails long-term studies of known individuals in the same habitats. Recent work at the National Bison Range in Montana is illustrative. Because of peculiarities of the study site, densities of pronghorn have remained fairly constant over a 12-year period (Byers and Kitchen 1987). Only the age structure has changed, in part the result of a catastrophic winter die-off of most adult males. Comparison of social systems before and after the male die-off demonstrate that pronghorn could be categorized as either harem dwelling or territorial (Byers and Kitchen 1987). This study underscores the importance of long-term research in sorting out how ecology alone may influence social systems. And, although it is certainly not possible for most researchers to carry out 12-year studies and benefit from natural perturbations, such fortuitous events may prove to be the only way that we refine our knowledge of how ecological and behavioral factors interact. This still leaves unresolved how phylogenetic inertia, rather than a multitude of other factors, might be implicated as a producer of the social systems that we see today.

The hypothesis that phylogeny may be responsible for social systems has been examined statistically by Clutton-Brock and Harvey (1977) (see also Felsenstein 1985), by inference (Eisenberg 1966, 1981), and by cladistics (Dobson 1985). In taxa with a large number of species such comparative approaches are essential, but for taxa with only a few species, of monophyletic origin, and with little variation in body size such as camels, zebras, or tapirs, the power of such methods is more limited. Nevertheless, if our knowledge about factors which produce such systems is to improve, additional work must focus on taxa which have received less attention and experimental analysis (Row et al. 1986).

X.   SUMMARY AND CONCLUSIONS

Much of the early work concerning social systems and ecology relied on adaptationist paradigms in explaining the observed diversity between a species' behavior and its environmental setting. More recent work has drawn attention to alternative explanations, especially historical factors and models which predict individual rather than species' responses.

At the species level a problem lies in trying to recreate the selective forces responsible for the origin of what has been termed "species-specific social organization." And, while natural selection may account for the mechanism by which animals adapt to their surroundings, it predicts little about the form of fit between organisms and their ecological settings. Because there is more than one way to approach a problem and because species adopt different solutions to the same suite of problems, there is no reason to believe that all species will adopt the same social pathways in responding to ecological conditions. This is unfortunate because it merely underscores the fact that a priori knowledge may be insufficient to predict how species respond to local conditions, and that phylogeny may complicate our models to a greater extent than previously believed.

The present study examined the effects of ecological conditions on individual behavior in two zebra species with different social organizations and divergent evolutionary pathways. Systematic manipulations of food resources were performed to test the hypothesis that social systems are a response to resource distribution. It was predicted that different species would have the same social systems given identical food perturbations. Because the species did not conform, phylogeny was suggested as a factor to explain behavioral differences. However, numerous biological pitfalls underlie experimental studies of this type. It is practically impossible to provide enough evolutionary time to determine whether adaptive responses might occur, and many experiments cannot distinguish between adaptation and phylogenetic inertia.

On a broader level, and when considering species of different size, morphology, and history, problems are further heightened. Robert MacArthur once said, "the structure of the environment, the morphology of the species, the economics of species behavior, and the dynamics of population changes are the four essential ingredients of all interesting biogeographical patterns. Any good generalization will be likely to build on all these ingredients, and a bird pattern would only be expected to look like that of Parmecium if birds and Parmecium had the same morphology, economics, and dynamics, and found themselves in environments of the same structure" (Macarthur 1972:1). Clearly, the expectation that species from distantly related groups will conform in biological patterns, including social systems, can at times be very naive. Among the challenges facing behavioral ecologists today will be the incorporation of the above factors to develop robust models to predict how social systems operate under varying ecological settings and how we might treat exceptions that fail to conform to expected generalizations.

ACKNOWLEDGMENTS

Permission to work at Canyon Colorado Equid Sanctuary was provided by William Gruenerwald. The comments and critical insight of Gary Bateman, Carol Cunningham, Rich Kiltie, Con Slobodchikoff, Peter Stacey and an anonymous reviewer are greatly appreciated.

REFERENCES

Alexander, R.D., J.L. Hoogland, R.D. Howard, K.M. Noonan and P.W. Sherman. 1979. "Sexual dimorphism and breeding systems in pinnipeds, ungulates, primates, and humans." pp. 402-435. In: N.A. Chagnon and W. Irons, eds. *Evolutionary biology and social behavior: An anthropological perspective.* North Scituate, Massachusetts:Duxbury Press.

Allen, D.L. 1979. *Wolves of Minong.* Boston:Houghton-Mifflin Co.

Altmann, J. 1980. *Baboon mothers and infants.* Cambridge:Harvard University Press.

Armitage, K.A. 1981. Sociality as a life-history tactic of ground squirrels. *Oecologia* 48:36-49.

Bekoff, M., J. Diamond and J.B. Mitton. 1981. Life history patterns and sociality in canids: Body size, reproduction and behavior. *Oecologia* 50:386-390.

Bekoff, M., T.J. Daniels and J.L. Gittleman. 1984. Life history patterns and the comparative social ecology of carnivores. *Annual Review of Ecology and Systematics* 15:191-232.

Bekoff, M. and M.C. Wells. 1986. Social ecology and behavior of coyotes. *Advances in the Study of Behavior* 16:251-338.

Berger, J. 1979. Weaning, social environments, and the ontogeny of social behavior in bighorn sheep. *Biological Behavior* 4:363-372.

Berger, J. 1986. *Wild horses of the Great Basin: Social competition and population size.* Chicago:University of Chicago Press.

Berger, J., D. Deneke, J. Johnson and S.H. Berwick. 1983. Pronghorn foraging economy and predator avoidance in a desert ecosystem: Implications for the conservation of large mammalian herbivores. *Biological Conservation* 25:193-208.

Bertram, B. 1978. *Pride of lions.* New York:Scribner.

Bowen, W.D. 1982. Home range and spatial organization of coyotes in Jasper National Park, Alberta. *Journal of Wildlife Management* 46:201-216.

Bowyers, T. 1985. Correlates of group living in southern mule deer. Ph.D. dissertation. Ann Arbor:University of Michigan.

Brown, J.L. 1964. The evolution of diversity in avian territorial systems. *Wilson Bulletin* 76:160-169.

Brown, J.L. 1975. *The evolution of behavior*. New York:Norton.

Burt, W.H. 1943. Territoriality and home range concepts as applied to mammals. *Journal of Mammalogy* 24:346-352.

Byers, J.A. and D. Kitchen. 1987. Mating system shift in pronghorn. Typescript.

Cheverud, J.M., M.M. Dow and W. Leutenegger. 1985. The quantitative assessment of phylogenetic constraints in comparative analyses: Sexual dimorphism in body weight among primates. *Evolution* 39:1335-1351.

Chivers, D.J. 1972. "The siamang and the gibbon in the Malay Peninsula." pp. 101-135. In: D.M. Runbaugh, ed. *Gibbon and Siamang*. Basel: Karger.

Churcher, C.S. and M.L. Richardson. 1978. "Equidae." pp. 379-422. In: V.J. Maglio and H.B.S. Cooke, eds. *Evolution of African mammals*. Cambridge:Harvard University Press.

Cifelli, R.L. 1981. Patterns of evolution among the Artiodactyla and Perissodactyla (Mammalia). *Evolution* 35:433-440.

Clutton-Brock, T.H. and P.H. Harvey. 1977. Primate ecology and social organization. *Journal of Zoology* 183:1-39.

Clutton-Brock, T.H. and P.H. Harvey. 1979. "Home range, population density, and phylogeny in primates." pp. 201-214. In: I.S. Bernstein and E.O. Smith, eds. *Primate ecology and human origins*. New York:Garland STPM Press.

Clutton-Brock, T.H. and P.H. Harvey. 1984. "Comparative approaches to investigating adaptation." pp. 8-29. In: J.R. Krebs and N.B. Davies, eds. *Behavioural ecology: An evolutionary approach*. Sunderland, Massachusetts:Sinauer.

Clutton-Brock, T.H., S.D. Albon and P.H. Harvey. 1980. Antlers, body size, and breeding group size in the cervidae. *Nature* 285:565-567.

Conner, E.F. and D. Simberloff. 1986. Competition, scientific method, and null models in ecology. *American Scientist* 74:155-162.

Crockett, C.M. and J.F. Eisenberg. 1986. "Howlers: Variation in group size and demography." pp. 54-68. In: B.B. Smuts, D.L. Cheney, R.M. Seyfarth, R.W. Wrangham and T.T. Struhsaker, eds. *Primate societies* Chicago:University of Chicago Press.

Crook, J.H. 1970. *Social behaviour in birds and mammals*. London:Academic Press.

Crook, J.H., J.E. Ellis and J.D. Goss-Custard. 1976. Mammalian social systems: Structure and function. *Animal Behaviour* 24:261-274.

Dobson, F.S. 1985. The use of phylogeny in behavior and ecology. *Evolution* 39:1384-1388.

Downhower, J.F. and Armitage K.B. 1971. The yellow-bellied marmot and the evolution of polygyny. *American Naturalist* 105:355-370.

Dunbar, R.J.M. 1984. *Reproductive decisions: An economic analysis of Geleda baboon social strategies*. Princeton:Princeton University Press.

Eisenberg, J.F. 1986. The social organization of mammals. *Handbook of Zoology*. Band 8, Lieferung, 39(10):1-92.

Eisenberg, J.F. 1981. *The mammalian radiations: An analysis of trends in evolution, adaptation, and behavior*. Chicago:University of Chicago Press.

Eisenberg, J.F., N. Muckenhirn and R. Rudran. 1972. The relationship between ecology and social structure in primates. *Science* 196:863-874.

Emlen, S.T. and L.W. Oring. 1977. Ecology, sexual selection, and evolution of mating systems. *Science* 197:215-223.

Estes, R.D. 1974. "Social organization of the African bovidae." pp. 166-205. In: V. Geist and F. Walther, eds. *The behavior of ungulates and its relation to management*. Morges, Switzerland:IUCN Publications #24.

Felsenstein, J. 1985. Phylogenies and the comparative method. *American Naturalist* 125:1-15.

Fisler, G.F. 1969. Mammalian organizational systems. *Los Angeles County Museum Publications* 167:1-32.

Franklin, W.F. 1983. "Contrasting socioecologies of South America's wild camelids: The vicuna and the guanaco." pp. 543-629. In: J.F. Eisenberg and D.G. Kleiman, eds. *Advances in the study of mammalian behavior*. Lawrence, Kansas:American Society of Mammalogists, Special Publication 7.

Geist, V. 1974. On the relationship of social evolution and ecology in ungulates. *American Zoologist*. 14:205-220.

Geist, V. 1978. *Life strategies, human evolution, environmental design: Toward a biological theory of health*. New York:Springer-Verlag.

Geist, V. 1981. "Behavior: Adaptive strategies in mule deer." pp. 157-224. In: O.C. Wallm, ed. *Mule and blacktailed deer of North America*. Lincoln:University of Nebraska Press.

Gittleman, J.L. 1986a. Carnivore brain size, behavioral ecology and phylogeny. *Journal of Mammalogy* 67:23-36.

Gittleman, J.L. 1986b. Carnivore life history patterns: Allometric, phylogenetic, and ecological associations. *American Naturalist* 127:744-771.

Gould, S.J. and R.C. Lewontin. 1979. The spandrels of San Marco and the Panglossian Paradigm: A critique of the adaptationist programme. *Proceedings of the Royal Society of London B*. 205:581-598.

Gould, S.J. and E.S. Vrba. 1982. Exaptation - a missing term in the science of form. *Paleobiology* 8:4-15.

Groves, C.P. 1974. *Horses, asses, and zebras in the wild*. Hollywood, Florida:Curtis Books.

Handford, P. and M.A. Mares. 1985. The mating systems of ratites and tinamous: An evolutionary perspective. *Biological Journal of the Linnaen Society* 25:77-104.

Harcourt, A.H., P.H. Harvey, S.G. Larson and R.V. Short. 1981. Testis weight, body weight, and breeding systems in primates. *Nature* 293:55-57.

Harvey, P.H. and G.M. Mace. 1982. "Comparisons between taxa and adaptive trends: Problems of methodology." pp. 343-362. In: King's College Sociobiology Group, eds. *Current problems in sociobology.* Cambridge:Cambridge University Press.

Harvey, P.H., R.K. Colwell, J.W. Silvertown and R.M. May. 1983. Null models in ecology. *Annual Review of Ecology and Systematics* 14:189-211.

Harvey, P.H. and T.H. Clutton-Brock. 1985. Life history variation in primates. *Evolution* 39:559-581.

Hirth, D.H. 1977. Social behavior of white-tailed deer in relation to habitat. *Wildlife Monographs* 53:1-55.

Hrdy, S.B. 1977. *The langurs of Abu.* Cambridge:Harvard University Press.

Janis, C. 1976. The evolutionary strategy of the equidae and the origins of rumen and caecal digestion. *Evolution* 30:757-774.

Janis, C. 1982. Evolution of horns in ungulates: Ecology and paleoecology. *Biological Review* 57:261-318.

Jarman, P.J. 1974. The social organization of antelope in relation to their ecology. *Behaviour* 48:215-267.

Jarman, P.J. 1982. "Prospects for interspecific comparisons in sociobiology." pp. 323-342. In: King's College Sociobiology Group, eds. *Current problems in sociobiology.* Cambridge:Cambridge University Press.

Jourbert, E. 1972. The social organization and associated behavior in the Hartmann zebra. *Equus Zebra Harmannae Madoqua* 6:7-56.

Kaufmann, J.H. 1974. Habitat use and social organization of nine sympatric species of macropod marsupials. *Journal of Mammalogy* 55:66-80.

Keast, A. 1965. Interrelationships of two zebra species in an overlap zone. *Journal of Mammalogy* 46:53-66.

Kiltie, R.A. and J. Terborgh. 1983. Observations on the behavior of rain forest peccaries in Peru: Why do white-lipped peccaries form herds. *Zeitschrift für Tierpsychologie* 62:241-255.

Kiltie, R.A. and J. Terborgh. 1985. Evolution and function of horns and hornlike organs in female ungulates. *Biological Journal of the Linnean Society* 24:299-320.

Kleiman, D.G. 1977. Monogamy in mammals. *Quarterly Review of Biology* 52:39-69.

Klingel, H. 1967. Soziale organization und verhalten freibender steppen zebras (*Equus guagga*). *Zeitschrift für Tierpsychologie* 24:580-624.

Klingel, H. 1972. Social behavior of African equidae. Zoology of Africa 7:175-184.

Kucera, T.F. 1978. Social behavior and breeding systems of the desert mule deer. *Journal of Mammalogy* 59:463-476.

Kummer, H. 1968. *Social organization of Hamadryas baboons*. Chicago: Chicago University Press.

Langer. 1973. Stomach evolution in the Artiodactyla. *Mammalia* 38:295-314.

Lee, A.K. and A. Cockburn. 1985. *Evolutionary ecology of marsupials*. Cambridge:Cambridge University Press.

Leighton, M. 1986. "Hornbill social dipersion: Variations on a monogamous theme." pp. 108-130. In: D.I. Rubenstein and R.W. Wrangham, eds. *Ecological aspects of social evolution*. Princeton:Princeton University Press.

Leland, L., T.T. Struhasaker and T.M. Butynski. 1984. "Infanticide by adult males in three primate species of Kibale Forest, Uganda: A test of hypotheses." pp. 151-172. In: G. Hausfater and S.B. Hrdy, eds. *Infanticide: Comparative and evolutionary perspectives*. New York:Aldine Publishing.

Lewontin, R.C. 1978. Adaptation. *Scientific American* 239:156-169.

Lott, D.F. 1984. Intraspecific variation in the social system of wild vertebrates. *Behaviour* 88:266-325.

MacArthur, R.H. 1972. *Geographical ecology*. New York:Harper and Row.

Maglio, V.J. 1975. "Pleistocene faunal evolution in Africa and Eurasia." pp. 419-476. In: K. Butzer and G. Isaac, eds. *After the australopithecines*. The Hague:Mouton Publications.

McCort, W.D. 1980. The behavior and social organization of feral asses (*Equus asinus*) on Ossabau Island, Georgia. Ph.D. dissertation. University Park:Pennsylvania State University.

Michener, G.R. 1983. "Kin identification, matriarchies, and the evolution of sociality in ground-dwelling sciurids." pp. 528-572. In: J.F. Eisenberg and D.G. Kleiman. *Advances in the study of mammalian behavior*. Lawrence, Kansas:American Society of Mammalogists, Special Publication 7.

Moehlman, P. 1974. Behavior and ecology of feral asses (*Equus asinus*.) Ph.D. dissertation. Madison:University of Wisconsin.

Murie, J.O. and G.R. Michener. 1984. *The biology of ground-dwelling sciurids*. Lincoln:University of Nebraska Press.

Oates, J. F. 1977. The social life of a black and white colobus monkey, *Colobus guereza*. *Zeitschrift für Tierpsychologie* 45:1-60.

Packer, C. 1979. Inter-troop transfer and inbreeding avoidance in *Papio anubis*. *Animal Behaviour* 27:1-36.

Packer, C. 1986. "The ecology of sociality in felids." pp. 429-451. In: D.I. Rubenstein, R.W. Wrangham. *Ecological aspects of social evolution*. Princeton:Princeton University Press.

Penzhorn, B.L. 1979. Social organization of the Cape mountain zebra, *Equus zebra*, in the Mountain Zebra National Park. *Koedoe* 22:115-156.

Penzhorn, B.L. 1984. A long term study of social organization and behavior of Cape mountain zebras *Equus zebra*. *Zeitschrift für Tierpsychologie* 64:97-146.

Ralls, K. 1976. Mammals in which females are larger than males. *Quarterly Review of Biology* 51:245-276.

Redford, K.H. 1985. Emas National Park and the plight of the Brazilian cerrados. *Oryx* 19:210-214.

Robinson, J.G., P.C. Wright and W.G. Kinzey. 1986. "Monogamous cebids and their relatives: Intergroup calls and spacing." pp. 44-53. In: B.B. Smuts, D.L. Cheney, R.M. Seyfarth, R.W. Wrangham and T.T. Struhsaker, eds. *Primate societies*. Chicago:Chicago University Press.

Rood, J.P. 1978. Dwarf mongoose helpers at the den. *Zeitschrift für Tierpsychologie* 48:277-288.

Rood, J.P. 1980. Mating relationships and breeding suppression in the dwarf mongoose. *Animal Behaviour* 28:143-150.

Rowe, M.P., R.G. Coss and D.H. Owings. 1986. Rattlesnake rattles and burrowing owl hisses: A case of acoustic Batesian mimicry. *Ethology* 72:53-71.

Rowell, T.E. 1979. "How would we know if social organizations were not adaptive?" pp. 1-22. In: I.S. Bernstein and E.O. Smith, eds. *Primate ecology and human origins*. New York:Garland STPM Press.

Rubenstein, D.I. 1981. Behavioral ecology of island feral horses. *Equine Veterinary Journal* 13:27-34.

Rubenstein, D.I. 1986. "Ecology and sociality in horses and zebras." pp. 282-302. In: D.I. Rubenstein and R.W. Wrangham, eds. *Ecological aspects of social evolution*. Princeton:Princeton University Press.

Schaller, G.B. 1967. *The deer and the tiger*. Chicago:Chicago University Press.

Schaller, G.B. 1972. *The Serengeti lion*. Chicago:Chicago University Press.

Sokal, R.W. and F.J. Rohlf. 1981. *Biometry*. San Francisco:W.H. Freeman.

Stammbach, E. 1986. "Desert, forest, and montane baboons: Multilevel-societies." pp. 112-120. In: B.B. Smuts, D.L. Cheney, R.M. Seyfarth, R.W. Wrangham and T.T. Struhsaker, eds. *Primate societies*. Chicago:Chicago University Press.

Struhsaker, T.T. 1975. *The red colobus monkey*. Chicago:Chicago University Press.

Sunquist, M.E. 1981. The social organization of tigers (*Panthera tigris*) in Royal Chitawan National Park, Nepal. *Smithsonian Contributions in Zoology* 336:1-98.

Sukuki, A. 1979. "The variation and adaptation of social groups of chimpanzees and black and white colobus monkeys." pp. 153-173. In: I.S. Bernstein and E.O. Smith, eds. *Primate ecology and human origins*. New York:Garland STPM Press.

Swihart, R.K. 1984. Body size, breeding season length, and life history of lagomorphs. *Oikos* 43:282-290.

Symons, D. 1978. *Play and aggression: A study of Rhesus monkeys*. New York:Colombia University Press.

Tenaza, R.R. 1975. Territory and monogamy among Kloss's gibbons (*Hylobates klossi*) in Sieberut Island, Indonesia. *Folia Primatology* 24:60-80.

Thornhill, R. and J. Alcock. 1983. *The evolution of insect mating systems.* Cambridge:Harvard University Press.

Tilson, R. 1981. Family formation strategies of Kloss's gibbons. *Folia Primatology* 35:259-287.

Van Orsdol, K.G., J.P. Hanby and J.D. Bygott. 1985. Ecological correlates of lion social organization (*Panthera leo*). *Journal of Zoology, London* (A) 206:97-112.

Van Schaik, C.P. and J.A.R.A.M. Van Hooff. 1983. On the ultimate causes of primate social systems. *Behaviour* 85:91-117.

Vehrencamp, S.L. and J.W. Bradbury. 1984. "Mating systems and ecology." pp. 251-278. In: J.R. Krebs and N.B. Davies, eds. *Behavioural ecology: An evolutionary approach.* Sunderland, Massachusetts: Sinauer Associates.

Waser, R.M. and R.H. Wiley. Mechanisms and evolution of spacing in animals. *Handbook of Behavioral Neurobiology* 3:159-223.

Webb, S.D. 1977. A history of savanna vertebrates in the New World, Part 1. North America. *Annual Review of Ecology and Systematics* 8:335-380.

Western, D. 1979. Size, life history, and ecology in mammals. *African Journal of Ecology* 17:185-204.

Wiens, J.A. 1984. "Resource systems, populations and communities." pp. 397-436. In: P.W. Price, C.N. Slobodchikoff and W.S. Gaud, eds. *A new ecology: Novel approaches to interactive systems.* New York: Wiley.

Williams, G.C. 1966. *Adaptation and natural selection: A critique of some current evolutionary thought.* Princeton:Princeton University Press.

Wilson, E.O. 1985. *Sociobiology: The new synthesis.* Cambridge:Harvard University Press.

Wittenberger, J.E. 1981. *Animal social behavior.* Boston:Duxbury Press.

Woodward, S.L. 1979. The social system of feral asses (*Equus asinus*). *Zeitschrift für Tierpsychologie* 49:304-316.

Woolfenden, G.E. and J.W. Fitzpatrick. 1984. *The Florida scrub jay: Demography of a cooperative-breeding bird.* Princeton:Princeton University Press.

Wrangham, R.W. and D.I. Rubenstein. 1986. "Social evolution in birds and mammals." pp. 452-470. In: D.I. Rubenstein and R.W. Wrangham, eds. *Ecological aspects of social evolution.* Princeton:Princeton University Press.

Zimen, E. 1976. On the regulation of pack size in wolves. *Zeitschrift für Tierpsychologie* 40:300-341.

IV. ECOLOGY OF SOCIAL BIRDS

Chapter 9

IDEAL FREE COLONIALITY IN THE SWALLOWS

W.M. Shields[1]
J.R. Crook[2]
M.L. Hebblethwaite[3]
S.S. Wiles-Ehmann[3]

Department of Environmental and Forest Biology
State University of New York
College of Environmental and Forest Biology
Syracuse, New York 13210

I.   INTRODUCTION
II. STUDIES OF GROUP LIVING IN THE BARN SWALLOW
   A.   Potential Benefits: Antipredator and Energetic Efficiency
   B.   Potential Costs: Increased Competition and Ectoparasitism
   C.   The Net Value of Group Living in the Barn Swallow
   D.   Exclusionary Conclusions Lead to Unsatisfactory Solutions
III. DO BARN SWALLOWS AGGREGATE OWING TO SHORTAGES OF IDEAL HABITAT?
   A.   Traditional Sites and Ideal Swallow Nesting Habitat
   B.   Reusing Old Nests and Ideal Swallow Nesting Habitat
   C.   Traditional Aggregations in Ideal Habitat: Predictions and Evidence
IV. THE EVOLUTION OF COLONIALITY IN THE SWALLOW FAMILY
V.   SUMMARY AND CONCLUSIONS: A PROSPECTIVE PERSPECTIVE ON THE PAST

[1]*Supported by the State University of New York, the Cranberry Lake Biological Station, and National Science Foundation grant BSR 83-19559.*
[2]*Present address: Department of Biology, William Paterson College, Wayne, New Jersey. Supported by Sigma Xi, a Chapman grant from the American Museum of Natural History, a Fuertes award from the Wilson Ornithological Society, and a Herbert and Betty Carnes Award from the American Ornithologists Union.*
[3]*Supported by Sigma Xi.*

189

ACKNOWLEDGMENTS
REFERENCES

I. INTRODUCTION

Questions about why some animals live or breed in groups while others are more or less solitary have been, and remain, central to behavioral and evolutionary ecology. An interest in developing both descriptive and explanatory "rules for group living" has resulted in numerous reviews and attempts at synthesis for a variety of taxa (Lack 1954, 1968; Wynne-Edwards 1962; Hamilton 1964; Crook 1965; Orians 1969; Brown and Orians 1970; Lin and Michener 1972; Alexander 1974; Wilson 1975; Brown 1975; Emlen and Oring 1977; Pulliam and Caraco 1984; Slobodchikoff 1984; Wittenberger and Hunt 1985).

Comparative studies of group living in the birds have, since the beginning, focused on a limited number of "social" or "colonial" taxa. One of the most prominent has been the swallow family (Hirundinidae) which usually has been classified as breeding colonially (Crook 1965; Lack 1968; Wittenberger and Hunt 1985). These analyses, as well as ours, have been well informed by the many general reviews of the family's taxonomy, life histories, and social systems (Bent 1942; Mayr and Bond 1943; Mayr and Greenway 1960; Thompson 1964; Hall and Moreau 1970; St. Peters 1973; Bryant 1985).

The swallow family (consisting of 75–80 species) has a cosmopolitan distribution, with member species breeding on every continent save Antarctica. All are aerial insectivores. Temperate breeders migrate in large gregarious flocks to the tropics where they overwinter, often in large mixed-species groups. Family members have been classified generically, in part, on the basis of distinct and consistent nesting differences among species (Mayr and Bond 1943). All species nest in more or less protected sites. One group of species nests in cavities as secondary tenants in tree cavities, in burrows dug in the ground by other organisms, or in rocky niches. Members of a second group modify their sites, either by digging burrows or building their own mud nests. Such mud nests are situated on walls or ledges or in niches which are situated on or in cliffs, bridges, culverts, caves, and in many areas, buildings. The tradition shift to human artifacts as nest sites has resulted in increased populations in species that have switched from, or added to, traditional sites (reviewed in Bent 1942; Erskine 1979; Speich et al. 1986).

The swallows are probably better suited to exploring sociality than their usual colonial classification would imply. All members of the family are morphologically similar, do capture insects on the wing, usually while foraging over open fields or water, and nest in more-or-less inaccessible sites. Rather than all being colonial, however, swallow sociality varies widely among genera, species, and even among populations within species (for review, see Bent 1942; Shields and Crook 1987).

The suitability of swallows for comparative study of sociality was and is widely recognized. Pioneering studies of many swallow species often included sections comparing their results with results on other populations or species breeding in different areas or in different sized groups. Many made explicit reference to the costs and benefits of colonial breeding (Stoner 1936; Moreau and Moreau 1939; Kuerzi 1941; Bent 1942; Edson 1943; Hosking and Newberry 1946; Allen and Nice 1952; Emlen 1952, 1954; Veitinghoff-Riesch 1955; Chapman 1955; Petersen 1955; Lunk 1962; Mayhew 1958; Samuel 1971; Lohrl and Gutscher 1973; Bryant 1973, 1975, 1978).

Others were designed to document the factors expected to influence group living in the swallows. In North America, these included studies of the solitary tree swallow (e.g., Sheppard 1977; Lombardo 1984, 1985; Robertson and his group, Leffelaar and Robertson 1984, 1985; Muldal et al. 1985; and Stuchbury and Robertson 1985) and the colonial bank swallow (e.g., Emlen and Demong 1975; Hoogland and Sherman 1976; and Beecher and Beecher 1979) and cliff swallow (e.g., Wilkinson and English-Loeb 1982; Brown 1984, 1985; Brown and Brown 1986).

The semicolonial barn swallow attracted even more attention, even though some of the more extensive studies can only be accessed through unpublished or underpublished theses. Such studies have been done in North America (e.g., Snapp 1973, 1976; Meyers 1977; Meyers and Waller 1977; Lohoefener 1977, 1980; Barrentine 1978; Wolinski 1981; Ball 1982, 1983; and most recently by our group, Shields 1984a,b; Shields and Crook 1987; Crook 1984; Crook and Shields 1985, 1987) as well as in Europe (e.g., M ller 1983, 1984, 1985a,b, 1987a,b).

It is our intention here to use this extensive data base to review traditional hypotheses about the ecological and evolutionary causes and consequences of group living in the barn swallow (Section II). Our assertion that such hypotheses are inadequate is intended as a prelude to our own hypothesis about the causes of barn swallow sociality (Section III). We currently believe that group living in the barn swallow is the unselected effect of an otherwise adaptive response to a shortage of safe and inexpensive "housing." We believe that barn swallows aggregate because *ideal* nesting habitat is patchily distributed and in short supply (Section IIIA). After exploring the evidence available to test our hypothesis in the barn swallow (Section IIIB), we expand the domain of the hypothesis to explore its value for the rest of the swallow family (Section IV) and for the evolution of group living more generally (Section V).

II.   STUDIES OF GROUP LIVING IN THE BARN SWALLOW

Snapp's (1973, 1976) pioneering studies of barn swallow sociality in central New York have stimulated a great deal of research. Among their most direct descendants are the studies focused on coloniality by Lohoefener (in Kansas, 1977; and in Mississippi, 1980), Barrentine (in Washington State, 1978), Wolinski (in Michigan, 1978), M ller (1983, 1984, 1985a,b in Norway; 1987a,b in Denmark), and ourselves (Shields

1984a,b; Shields and Crook 1987; Crook and Shields 1985, 1987, in the Adirondacks of New York). Following Snapp's lead, all were designed, at least in part, to measure and explain the reproductive costs and benefits of group living. Unfortunately, many were reported in unpublished or underpublished theses (e.g., Snapp 1973; Lohoefener 1977, 1980; Barrentine 1978; Wolinski 1978). Because they contained important dates and ideas, however, we decided to include them in our review.

While they differed in significant ways, the study areas and populations were also markedly similar in size and social structure. In each, the barn swallows nested semicolonially, with some solitary pairs (5-15% of the focal pairs), many (15-33%) observed in small colonies (2-8 pairs), and with most pairs (54-80%) breeding in moderate colonies (9-35 pairs). Within colonies, nests were usually clumped rather than regularly or randomly distributed (Lohoefener 1980; Shields 1984b; M ller 1987a). These colony sizes and nest distributions were similar to those reported for birds nesting in ancestral sites (caves, cliffs, Speich et al. 1986) and on human artifacts elsewhere (Bent 1942; Veitinghoff-Reisch 1955).

Nesting phenology and life history did differ in minor ways among these populations (Shields and Crook, 1987). The duration of the incubation, brooding, and nestling stages of the breeding cycle varied little. In contrast, the initiation and termination of breeding, the mean clutch size for all broods, and the number of broods per season varied, but in the expected directions, given the latitude and climate differences among areas. On balance, we are confident that none of the study area populations were sufficiently different to invalidate using all of them in a review of barn swallow sociality.

A. Potential Benefits: Antipredator and Energetic Efficiency

Barn swallow nests were generally inaccessible to nonflying, nonhuman predators. Despite a low *a priori* expectation of predation, the synchronous breeding, contagious alarm calling, and large mobbing groups observed in barn swallows were thought to be consistent with the hypothesis that sociality reduces predation. Numerous studies have shown that colonial swallows *were* likely to detect potential predators more quickly than solitary nesters. In addition, the size of mobbing groups, and presumably their deterrence or distractive value, increased with colony size as well (Smith and Graves 1978; Shields 1984a; M ller 1984). These patterns are consistent with coloniality functioning to reduce predation.

The remainder of the data, however, argue against this hypothesis. In the best documented life history studies of the barn swallow, estimates of the impact of nest predation ranged from less than 1 percent to a maximum of 8 percent of the total egg and nestling mortality. Most documented events, in turn, were attributed to nest competitors like house sparrows (*Passer domesticus*) or predators like humans and domestic cats (Bent 1942; Mason 1953; Samuel 1971a; M ller 1987a; Shields and Crook 1987). Predation on adults and fledglings occurred more frequently (M ller 1987a; Shields and Crook 1987) and could have favored coloniality, but only if group living reduced the per capita probability of being taken, owing to dilution or swamping effects (Hamilton 1971; Wilkinson and English-Loeb 1978). *Every* study with direct evidence, however, documented either a constant (Snapp 1976; Shields and Crook 1987) or an increasing (Lohoefener 1977; M ller 1987a) per capita predation rate on nests, fledglings, and adults with increasing colony size.   While coloniality should have obvious consequences for antipredator behavior and predation rates, reduced predation could be ruled out as an ultimate explanation for the origin or maintenance of barn swallow coloniality (Snapp 1973, 1976; M ller 1987a; Shields and Crook 1987).

Barn swallow foraging efficiency might benefit from group living, if centralization of nests allowed for more efficient central place foraging (Horn 1968; Wittenberger and Hunt 1985). Consistent with this hypothesis, the flying insects barn swallows hunt were distributed patchily in time and space and can occur in sufficiently large numbers to swamp an individual or pair's ability to use or defend them economically. Arguing against the hypothesis were the many observations that individual swallows often limited their foraging to small areas near their own nests (Snapp 1976; M ller 1987a; Hebblethwaite pers. obs.). If swallows were central-place foragers, individuals should share common colony foraging grounds, but they do not.

Colonial barn swallows could still benefit energetically if per capita foraging efficiency increased owing to social facilitation or information transfer about ephemeral prey (Ward and Zahavi 1973; Waltz 1983; Wittenberger and Hunt 1985). Arguing against these hypotheses, however, barn swallows were rarely observed foraging in large social groups during the breeding season.   Within colonies, individual arrivals and departures appeared to occur randomly, both temporally and in direction (M ller 1987a). In addition, no one has yet reported that barn swallows followed nonmate conspecifics to common foraging grounds (needed for information transfer) or that foraging swallows recruited group members into the small foraging groups that were observed (needed for social facilitation) (Snapp 1973, 1976; M ller 1987a; Shields and Crook 1987). The unanimous conclusion here was that energetic benefits were also improbable as ultimate factors favoring barn swallow coloniality (Snapp 1973, 1976; M ller 1985a, 1987a; Shields and Crook 1987).

B.   Potential Costs: Increased Competition and Ectoparasitism

Living in groups increases the spatial proximity of potential competitors in the barn swallow. This could result in increases in the frequency or intensity of competition for limiting resources like nests, mates, or food. It was the case that guarding behavior, including nest and mate guarding, as well as general levels of individual aggression, increased with increasing colony size in the barn swallow (Crook 1984; M ller 1985a, 1985b). This could have been a response to the increases in attempted extrapair copulations observed in larger colonies (Barrentine 1978; Wolinski 1978; Lohoefener 1980; M ller 1985b). Similarly, our (Crook and Shields 1985) discovery of sexually selected infanticide by unmated males as a competitive mating tactic was confirmed by M ller (1987a). In Denmark, though not in New York, infanticide occurred more frequently on a per-nest basis in larger swallow colonies. M ller (1987b) also documented barn swallows dumping eggs in the nests of other colony members. Since the primary brood parasites were neighbors, the probability of victimization by egg dumping also increased with colony size (M ller 1987b).

Many studies have reported on the occurrence of different kinds of ectoparasites in or on the nests and nestlings of barn swallows. Hematophagous fly larva (genus *Protocalliphora*) were reported as major pests in Europe (Boyd 1935, 1936) and in North America (Mason 1936; Shields and Crook 1987). Other nestling parasites, including hippoboscids (*Stenepteryx* spp.), fleas, and mites were also known to affect nestling barn swallows adversely (Veitinghoff-Riesch 1955; M ller 1987a). In every study with appropriate data, both the probability and severity of parasitism increased with colony size (M ller 1987a; Shields and Crook 1987).

C. The Net Value of Group Living in the Barn Swallow

In any evolutionary analysis, the bottom line is net reproductive success. Most ornithological studies estimate fitness by measuring the number of nestlings fledged by a pair per nest attempt or per season. Snapp's (1973, 1976) study continues to generate interest because she failed to document statistically significant differences in the number of nestlings fledged as a function of colony size. Her results cannot be blamed on a paucity of data, since her sample sizes were large (ranging from 18 to 81 pairs for each colony size class). In addition, there was no suggestion of a monotonic increase or decrease in fledging success with changing colony size (Snapp 1973, 1976). In fact, if any trend existed, it hinted at an intermediate optimum colony size, since the minimum fledging rate was associated with solitary nests *and* those in the largest colonies (Snapp 1973). Snapp (1976) concluded that there were no demonstrable colony size effects on barn swallow reproduction in central New York.

Unlike Snapp, Lohoefener (1977, 1980) studied a nonrandom sample of colony sizes during his M.S. and Ph.D. thesis research. He specifically searched for solitary nests (one active nest in a culvert) and compared them to colonial nesters (pairs breeding in a sample of focal culverts containing six or more active nests). In Kansas (Lohoefener 1977), solitary pairs fledged significantly more young than colonial pairs. The major costs of coloniality were: (1) a reduction in the proportion of eggs hatched successfully, which he attributed to social competition, and (2) increased per-nest predation rates. The increased predation included observations of a bobcat (*Felis rufus*) destroying in one season 34 active nests and the 78 nestlings and 34 eggs they contained. These losses occurred in a *single* 25-pair colony (many pairs lost both renests and first nests). He also noted that even if this exceptional incident was excluded from the analysis, solitary pairs still did better than the colonial breeders. He concluded that, in Kansas, group living resulted in a net loss for barn swallows.

In Mississippi, Lohoefener (1980) reported that colonial breeders were *more* successful than solitary pairs. Based on their earlier breeding and larger clutches, he concluded that on average, colonial breeders were older than solitary breeders. Relative hatching success and nestling survivorship did not differ with clutch size or between solitary and group breeders. He concluded that it was the larger clutch of the older colonial birds that was responsible for their greater fledging success. Since his population was only partially marked and he studied it for only two years, his conclusion remained tentative. He did conclude that in Mississippi, group living had little or no net effect on reproductive success in the barn swallow.

During our long-term study of a completely marked population of barn swallows in the Adirondacks of New York, we found no significant fitness differences between solitary pairs and those living in small groups (Shields and Crook 1987). We did document a significant reduction in the net reproductive success of pairs living in larger colonies and subcolonies. In our population there was a trend for younger birds to be found more often in the larger colonies. By examining different fitness components, we showed that clutch size, hatching success, and nestling survivorship were reduced in the larger groups, independent of breeder age (Shields and Crook 1987). We also showed that the primary factor reducing fitness in the larger groups was the higher probability and severity of parasitism. Like all of our predecessors, then, we concluded that coloniality provided no obvious benefits, and generated a net cost for those living in large groups in our population (Shields and Crook 1987).

D.   Exclusionary Conclusions Lead to Unsatisfactory Solutions

Snapp was the first to study most of the "factors" that might have explained the evolution of colonial breeding in the barn swallow. In her study, all of Alexander's (1974) positive factors (predation, foraging efficiency, breeding synchrony and ultimately reproductive success) failed to vary significantly or in the predicted directions with colony size. She followed Sherlock Holmes' dictum that, "When you have eliminated the impossible, whatever remains, however improbable, must be the truth." She suggested that: "Coloniality in the barn swallow does not appear to occur as a response to predation pressure, food shortage, or social stimulation. Rather, nest-site availability seems to be of greatest importance in determining the size of a colony" (Snapp 1973, p. 154). She concluded that "colonies represent passive aggregations of breeding birds and do not actively recruit additional pairs" (Snapp 1976, p. 479). She provided no data to test her nest-site limitation hypothesis, despite an assertion that "the number of pairs in a colony rose in parallel with the increase in the size of the building and/or the number of entrances to the building" (Snapp 1976, p. 479).

Lohoefener (1977, 1980) attemtped to test this part of Snapp's thesis directly by looking for size and suitability differences between colonial and solitary nest sites. In both Kansas and Mississippi, colony culverts averaged larger (primarily longer), but there was much overlap in culvert size (height, width, length, and volume). Very small culverts never hosted colonies, but many culverts with solitary nesters were as large as, or larger than, many culverts that did host colonies. Other potentially important factors (e.g., culvert orientation, the presence of standing or running water, the number of projections for supporting nests) did not differ between solitary and colonial sites. The similarity between colonial and solitary sites and the occurrence of empty culverts falsified Snapp's thesis, at least in its simplest form. Despite this falsification, however, Lohoefener's easy exclusion of all of the obvious alternatives led him to conclude "that colonialism in Barn Swallows is the result of passive recruitment as proposed by Snapp (1976)" (Lohoefener 1980, p. 64).

We (Shields and Crook 1987) also had data that were not consistent with the simplest version of Snapp's thesis. In our population, colony size was not associated with consistent physical differences in site or site size. In fact, the number of breeding pairs at each site varied over time (Shields 1984a; Shields and Crook 1987). One boat house at our main focal colony actually ran the gamut, hosting 4, 2, 0, 1, 2, 6, and 7 pairs in each year from 1979 to 1985. Similarly, in every year there were apparently suitable buildings near active colonies that were unused by barn swallows. Thus we could not demonstrate any shortage of suitable nest sites. Nevertheless, we could demonstrate that all of the other obvious factors (predation, foraging effects, breeding synchrony) failed to explain the actually costly coloniality observed in our population. Thus, our tentative, and ultimately unsatisfying, conclusion was, "we are forced to agree with Snapp (1976), that the simplest explanation for why some barn swallows are colonial is that some aspect of their nesting or nest site resources is limited and so forces them to breed in groups" (Shields and Crook 1987).

The same dissatisfaction we felt about our conclusion that colonial swallows were aggregating at limiting nest sites forced M ller (1987a) to a less-than-satisfying conclusion. He had ruled out nest site limitation as a factor in swallow sociality on *a priori* logical grounds. He noted that at any site "a doubling of breeding pairs between two seasons is not uncommon"; and that colonial nests are aggregated, implying that "late-arriving birds thus tend to settle near other swallows rather than away from them" (M ller 1987a, p. 821-822). He argued that the value of coloniality must be considered for different classes of individual swallows (older vs. yearling male breeders vs. older vs. yearling female breeders vs. unmated males). His analysis of individual costs and benefits, however, showed that while some classes of male could benefit, no class of female increased and most females actually lost by living in groups.   His data were similar to everyone else's and his conclusion just as unsatisfactory.   None of Alexander's (1974) obvious positive factors could explain the observed coloniality in female swallows, and yet there was no *obvious* shortage of nest sites or nesting habitat that could have forced the observed aggregations.

Our consensus conclusion was unsatisfactory because it was exclusionary.   It may have seemed that it was *all* that was left, but we could not rule out the possibility that we had overlooked the correct alternative.   More tellingly, our conclusion was not fully consistent with the data.   The logic behind rejecting Snapp's conclusion was simple enough. For example, Holroyd (1975) noted that in his study area "population size remained relatively constant despite an apparent abundance of nest sites.   The increased use of the ruins in 1970, after the closing of the barn, indicates that there were unused nest sites in previous years.   Thus there was always an excess of nest sites, and some factor other than nest-site availability limited the size of the breeding population of Barn Swallows."   Similarly, Wittenberger and Hunt (1985, p. 6) noted that nest site limitations often were unlikely as explanations of coloniality because: "Neighboring habitat islands remain unused while one becomes crowded, or only a small portion of an island will be occupied by a very dense colony . . . swallows and swifts could often disperse their nests over cliff faces, bluffs, or other substrates instead of nesting colonially.   Thus, an analysis of coloniality must include a consideration of why nests or roosts occur at higher densities than habitat clumping dictates." In order to develop a convincing argument that barn swallow colonies were passive aggregations, then, we felt that we would need to develop an explicit model about what exactly was in short supply.

III.   DO BARN SWALLOWS AGGREGATE OWING TO SHORTAGES OF IDEAL HABITAT?

Suitable barn swallow nesting habitat has been defined as possessing three critical resources: (1) foraging sites; open areas usually over water or fields, (2) near a nest site(s); that includes one or more microsites with vertical or horizontal structure and surfaces that permit nest attachment, usually inside an enclosure, and always directly under and close to some sort of cover (roof or overhang), and (3) a nearby source of mud for nest building (Bent 1942; Samuel 1971b; Lohoefener 1977; Wolinski 1978).   Jackson and Burchfield (1975) noted that "ideal" habitat was also likely to contain utility wires or other perch sites.   The notion that suitable habitat, nest site, or nest microsite shortages, as defined, could have led to group living was contradicted by the abundance of each throughout the range of the barn swallow.

Variation in site or microsite quality could have effects on the settling decisions of individual birds, and thus on the occurrence of group living.   We agree with M ller (1987a) that exploring the decisions of individual barn swallows in the context of optimality models, like those developed for habitat choice in patchy environments (Fretwell and Lucas

1970; Fretwell 1972; Whitham 1980), or for mate choice (Orians 1969; Emlen and Oring 1977; Murray 1984) might shed light on swallow sociality.

For example, the Verner-Willson-Orians model for the evolution of polygyny states that a female should choose an already mated male if the resources on his site (or that he controls) offer her a greater net gain (after paying the cost of sharing that site and male with another female) than choosing an unmated male at a lower quality site. Similarly, Fretwell's ideal-free model of habitat choice assumes that animals settle in the patch that yields the local maximum net gain. Individuals are expected to settle first in the highest quality patch. As others settle in the highest quality patch, competition or other crowding effects will reduce the per capita gain. Then a patch of lower absolute quality could yield a higher relative gain. Only then would an animal choose to settle in the lower quality patch. In the end, and in the absence of aggressive exclusion by conspecifics, the number of females or settlers (i.e., local density in a male's territory or in a habitat patch) is expected to be positively correlated with the absolute quality of sites.

If barn swallow nesting habitat or nest sites vary enough in absolute quality, then similar assumptions and an analogous model could help to explain their nesting dispersion. In this case, swallows should settle first in patches of ideal nesting habitat (those with the highest absolute quality), rather than settling in any suitable nesting habitat. We predict that some individuals will gain by settling in already settled patches of "ideal" habitat, thereby creating colonies, and in spite of the costs of crowding. This would occur if the absolute quality of some patches were great enough to insure a net gain, relative to breeding in unsettled but otherwise suitable patches of lower intrinsic quality. Only when a high quality patch was so crowded that the net gain for settling in a group would be less than for settling in an empty patch, would swallows nest solitarily. In this sense, swallow coloniality would be one side effect of adaptive habitat choice and so nonadaptive (sensu Williams 1966; Murray 1978). As Snapp (1976) suggested, such colonies would exemplify "passive aggregations" and would result from a shortage of habitat or nest sites. The shortage, however, would be of high-quality patches or *ideal* habitat, rather than the more abundant *suitable* habitat that remains empty.

A. Traditional Nest Sites and the Ideal
    Swallow Nesting Habitat

The use of traditional nesting habitat and sites by many organisms is often discussed under the rubric of philopatry and site tenacity (reviewed in Baker 1978; Shields 1982; Greenwood and Harvey 1982; Rowley 1983; Waser and Jones 1983). Breeder site tenacity is well known in the barn swallow (Shields 1984b) and in other swallow species (Freer 1979). The pattern, which is found in birds generally, is that successful breeders that survive tend to return to the same area, habitat, nest sites, microsites, and even nests in subsequent breeding seasons. Pairs that fail, in contrast, have a higher probability of dispersing away from the area and sites in which they have failed (Shields 1984b).

The usual explanation of this pattern is that site-faithful individuals may gain through familiarity with the location and phenology of local resources and dangers including the social milieu provided by conspecific neighbors. In addition, their own prior success is direct evidence that the site in question does not harbor any hidden but fatal flaws (e.g., is physically unstable or subject to catastrophes like flooding or regular visitation by predators or parasites). Failed breeders are thought to do better by dispersing if their failure was caused by site- or habitat-specific factors that are likely to remain dangerous in subsequent attempts or seasons (e.g., a local abundance of predators or parasites, or a physically unsafe site). Thus, the site tenacity hypothesis relies on the notion that experience at particular sites will benefit the individuals that *reuse* those sites.

We do not dispute the value of this thesis in explaining adult site tenacity, and in fact Shields (1984b) used it in his explanation of barn swallow dispersal.   The thesis, however, has also been advanced to explain juvenile philopatry. In many organisms, juveniles fail to disperse before breeding, or in migratory species return to, or near to, their natal ranges for their first breeding attempts. Ghiselin (1974) provided an eloquent rationale for why juveniles might forgo dispersal in his discussion of seal philopatry:

> Seals can hardly be expected to know, through abstract reasoning, when to seek a better place for giving birth. Rather they ought to go where they were born, where everyone else goes, or both. That mother succeeded in a given place is clear and sufficient evidence that daughter may."

Shields (1982, p. 22) criticized this version because it failed to provide a reason for why daughter's success was contingent upon settling where *mother* did. As stated the principle does nothing more than state that philopatry is heritable. If mother was philopatric and succeeded, and daughter's existence guarantees the success, then daughter should be philopatric and should succeed as well. The statement, however, does not provide an explicit reason for why daughter does better to stay than disperse. The circularity of the argument can, perhaps, best be illustrated by making the same statement about hypothetical seals that *always* disperse.

> Seals can hardly be expected to know, through abstract reasoning, when to seek a better place for giving birth. Rather they ought to disperse like their mother did, like everyone else does, or both. That mother succeeded by dispersing is clear and sufficient evidence that daughter may.

A complete explanation of site tenacity or dispersal would have to provide explicit ecological or genetic factors that penalized or favored dispersing or staying at home (Shields 1981 1987).

While we remain convinced that Ghiselin's dictum is flawed as an explanation of juvenile philopatry, we are also convinced that it contains a very important idea that has not been given due consideration in discussions of habitat quality or settling behavior. It is this kernel that makes Ghiselin's hypothesis, in its many incarnations (reviewed in Shields 1982), so seductive, and in part so true. The notion is simply that *any* living, and especially successfully reproducing, conspecific could be an indicator of habitat or site quality. Rephrasing Ghiselin's argument one last time:

> Seals can hardly be expected to know, through abstract reasoning, where to seek a place for giving birth. Rather they ought to settle where some seal has bred before. That *any* other seal has succeeded is clear and sufficient evidence that they may succeed.

Note that this rule neither implies nor excludes juvenile or adult philopatry or dispersal. It does imply that an animal should settle in a habitat patch that shows evidence of having been used, or even better, successfully used, by conspecifics. Thus, it is not a hypothesis about individual site tenacity but rather concerns the existence of traditional sites. This traditional-site hypothesis implies a habitat choice system that includes evidence of conspecific use, and preferably successful breeding, as one proximate cue triggering settling behavior (for an actual example, see Stamps 1987).

The primary prediction of this hypothesis would be continual use of the same habitat patches and/or sites in an area. It is irrelevant whether the tenants of a traditional site are the same site-faithful individuals each year or a parade of constantly changing individuals. Juvenile settlers at such traditional sites might have been born there or elsewhere. Adult settlers might have bred there or elsewhere in prior attempts. All that the traditional-site hypothesis demands is that when an individual chooses a site new to itself, it will tend to settle where conspecifics are currently breeding or have bred before.

Of course this traditional-site hypothesis depends on many assumptions. It could only apply to organisms with breeding habitat, sites, or microsites that were temporally stable or predictably changeable, so that prior success would be predictive of current or future prospects. It would not apply to species that totally consume local resources or irrevocably damage their breeding sites by using them. It would also be unlikely to apply to species whose breeding requirements vary either so unpredictably or with such long renewal cycles in space or time that individuals were forced to wander widely in search of suitable breeding conditions (nomadic species, Andersson 1980). We would expect the hypothesis to apply, and the rule to be of most value, in species using habitat or site types that show little temporal variation, but much unpredictable or nonobvious spatial variation in site quality. Thus, intrinsically better and poorer sites should retain their relative value, regardless of temporal changes in their absolute values.

These are the same conditions expected to favor adult site tenacity in successful breeders and dispersal in failed breeders (Freer 1979;

Greenwood and Harvey 1982; Redmond and Jenni 1982; Dow and Fredga 1983; Shields 1984b; Blancher and Robertson 1985). Thus, we would predict that the traditional-site strategy would be found *only* in species that display adult site tenacity after successful breeding. If a species uses the rule--reuse a site if successful (Win-Stay) and disperse if unsuccessful (Lose-Switch)--then the presence of breeders *would* imply success and thus suitability. In fact the more breeders at a site, the higher the implied quality.

Empty, but apparently suitable sites, could be just that. They might, however, be poor sites that are empty because of nonobvious problems like greater susceptibility to environmental disturbance or predators. The value and strength of the traditional-site strategy, then, would depend on the relative abundance of high- and low-quality empty sites. If many empty sites were poor and *a priori* assessment of quality difficult, then settling in proven sites with conspecifics could be favored. If most or even many of the empty sites were of equal quality to those in use, then if there were *any* crowding costs, settlement of empty sites might be favored.

We believe that safety and quality variation characterize barn swallow nesting habitat, nest sites, and microsites. While there is lots of vertical substrate under an overhang or roof, not all of it is equally suitable. While some humans will allow swallows to build their nests on some of their artifacts, not all buildings are equally safe. When a swallow attempts to build above a cherished auto or boat or over an often-used walkway, it is likely to return to a nest that has been torn down by its human host who has rebelled at the thought or sight of nestling fecal deposits. The risk associated with non-artifact sites is also likely to vary among otherwise suitable sites. Thus, different sea caves and cliffs are likely to vary in their susceptibility to flooding or rockfalls. In addition, some sites are likely to be more susceptible to parasites, predators, or competitors than others, if for no other reason than because some have been discovered while others remain hidden.

Since barn swallows display breeding site tenacity when successful and are more likely to disperse after nest failure (Shields 1984b), the presence of breeders is indicative of a safe site and of one that may be more suitable than nearby empty sites. Since yearlings rarely settle in their natal colonies, yet often settle at traditional sites (Shields 1984b), a swallow site tradition would not require juvenile nondispersal. Thus, adult site tenacity could combine with a traditional-site preference and result in barn swallows breeding in groups rather than solitarily.

Barn swallows would be attracted to other swallows, not to gain social benefits, but because the presence of conspecifics would be a good indicator of site safety and quality. The clumped distribution of breeding swallows that resulted would exemplify a passive aggregation in limited habitat as Snapp (1976) suggested. Swallow dispersion, however, would not be constrained by shortages of potentially suitable habitat, which actually occurs in abudance. Rather, swallows would be attracted in numbers to high-quality sites that have proven their quality, and continue to demonstrate it by hosting successful and therefore site-faithful conspecifics. A traditional site without current breeders, in contrast, might actually deter settlement because evidence of past breeding (e.g., old nests in the swallow) in the absence of current tenants could imply recent breeding failure rather than success. In this way, successful traditional sites, which always would be in short supply, would attract greater numbers of breeders until some upper limit was reached.

In the barn swallow, the upper limit should depend on (1) the absolute size of a traditional site, (2) the number of suitable microsites at a site, (3) the degree to which residents resist immigration (defend a site), and/or (4) the increasing costs of crowding with increasing colony size at a site. Thus, a prospecting swallow could not settle at a traditional site with room enough for a single occupied nest, or at a larger site with dominant breeders at every suitable microsite. A potential immigrant also could be deterred from settling by the aggressive resistance of current residents, even if a site contained empty microsites. Resident aggression, in turn, would be favored when any critical limiting resource, especially food, nest microsites, or nests (see below) were economically defendible (Brown 1964, 1969). Similarly, residents should resist immigration aggressively when the addition of a pair at their site would raise the cost of crowding above the cost of aggression.

Alternatively, potential settlers could avoid previously settled sites, rather than being excluded. For example, a potential settler could decide that increased ectoparasitism or other crowding costs would make the effort of settling at a crowded traditional site too risky. If a prospector failed to recognize the crowding at a particular site as pathological, and was not excluded by resident aggression or an absence of microsites, its subsequent breeding failure would raise its probability of dispersing from that site before its next breeding attempt. Whether prospectors were deterred by residents, avoided settling, or dispersed after failing at a particular traditional site, they would be expected to look for a less crowded traditional site, or failing that, to settle at a new site. Thus, there would be a natural cap on colony size at each traditional site, probably as a result of more than one of these interacting factors. The final result would be a dynamic balance with some swallows breeding alone, some in small groups, and some, at a few high-quality or especially large traditional sites, in moderately large groups.

Alternatively, potential settlers could avoid previously settled sites, rather than being excluded. For example, a potential settler could decide that increased ectoparasitism or other crowding costs could make the effort of settling at a crowded traditional site too risky. If a prospector failed to recognize the crowding at a particular site as pathological, and was not excluded by resident aggression or an absence of microsites, its subsequent breeding failure would raise its probability of dispersing from that site before its next breeding attempt. Whether prospectors were deterred by residents, avoided settling, or dispersed after failing at a particular traditional site, they would be expected to look for a less crowded traditional site, or failing that, to settle at a new site. Thus, there would be a natural cap on colony size at each traditional site, probably as a result of more than one of these interacting factors.   The final result would be a dynamic balance with some swallows breeding alone, some in small groups, and some, at a few high-quality or especially large traditional sites, in moderately large groups.

B.   Reusing Old Nests and Ideal Barn Swallow Nesting Habitat

Barn swallows modify their nest sites by building mud pellet nests. These are usually sufficiently strong, sheltered, and inaccessible (to most ground dwellers except humans with ladders) that they can remain useable indefinitely.   In Europe, Veitinghoff-Reisch (1955) reports on individual nests in use, off and on, for up to 48 years.   Vansteenwegen (1982) estimates that the average nest in his French study area lasted for eight years.   In our study area, a few nests have been in continuous use since 1979 and others have seen intermittent use since at least 1974 (Shields 1984b).   The refurbishment, usually by adding a small layer of fresh mud to the rim and cup, and reuse of old nests is commonplace in the barn swallow throughout its range.   When old nests are available, from 40 to 90 percent of the pairs in any area refurbish one, rather than build anew (Table I) (Boyd 1935, 1936; Bent 1942; Veitinghoff-Reisch 1955; Samuel 1971b; Lohoefener 1977, 1980; Barrentine 1978; Wolinski 1981; Vansteenwegen 1982; Hill 1982; Ball 1983; Barclay in press).

The consensus explanation for nest recycling is that nest repair requires less time and energy than building a nest from scratch.   In our population, and others, birds refurbishing old nests require from 1/2 to 1/10 the time for nestbuilding as those that build new nests (Table I) (reviewed in Boyd 1935, 1936; Bent 1942; Kuzniak 1967; Hill 1982; Barclay in press).   They also require many fewer mud gathering trips and so conserve the considerable energy required to fly from 1000 to 2000 mud pellets back to a nest site (reviewed in Withers 1977; Collias and Collias 1985).   If the usual time and energy constraints on breeding in migratory passerines (Lack 1954, 1968) hold for the swallow, then in the absence of balancing costs, the savings associated with nest reuse should be translatable into increased reproductive success.

In addition, the traditional-site hypothesis could easily be focused down to a traditional-microsite or even traditional-nest hypothesis for the barn swallow.   A microsite used successfully has demonstrated the suitability and safety of its precise location (e.g., it was not rained on, flooded, or predated) and its support structure and attachment surfaces (e.g., the nest did not fall down).   Similarly, a nest used successfully has demonstrated its strength and suitability for breeding.   Thus, it might be safer, as well as more time and energy efficient, to reuse a proven microsite and the nest it contains, rather than gamble on building a new one even in the same safe site.   In the swallow, successful adults do tend to use the same microsite(s), even building a new nest if the old one has disappeared, and an old nest or nests when available.   Failed breeders, in contrast, change their microsites and nests more frequently (Shields 1984b).

TABLE I. Variation in breeder age, nesting phenology, and components of seasonal reproductive success (mean (N)) as a function of nest age in the barn swallow at Cranberry Lake, New York (1979-85).[1]

Breeding Factor	Nest Age				Signif.
	Old		New		
% of total nests	82	(154)	18	(34)	NT
Duration of building	4.8	(45)	11.0	(15)	**
Incubation date	151	(94)	157	(24)	*
% ectoparasitized	67	(128)	38	(29)	**
% in small colonies	32	(96)	64	(25)	**
% yearling females	60	(96)	64	(25)	NS
% yearling males	41	(96)	56	(25)	*
Number of eggs/pair	6.9	(90)	6.0	(25)	NS
Number hatched/pair	5.2	(90)	5.3	(25)	NS
Number fledged/pair	3.0	(90)	3.2	(25)	NS
% successful doublebrood	6	(90)	0	(25)	NS

[1]Nests are placed into two classes: old = older nests built during or before the most recent prior nesting attempt either across or within seasons; new = nests built completely during a current attempt. Incubation date is given as day of the year (e.g., 1 June = 152). Reproductive success includes the total eggs, hatchlings, and fledglings produced per pair for a season as a function of whether their first nest was old or new. Statistical comparisons made using T-tests with corrections for heterogeneous variance when required to compare means and Chi-squared or Fisher's Exact Tests for frequencies; NS = a nonsignificant comparison; * = significance at an alpha level of $P < 0.05$; ** = $P < 0.001$; and NT = no test was performed.

Nest reuse in the swallow is likely to carry potential costs as well as potential benefits. For example, aging nests, and especially adherent nests attached only at a vertical surface, should have a higher probability of falling as they weather (Vansteenwegen 1982). Nest falls can be a significant source of nest mortality in the barn swallow (Shields and Crook 1987). If a swallow could estimate nest age or structural stability it should prefer to reuse younger or more stable nests. If the only nests available were very old or decrepit, it might pay a settler to expend the time and energy to build a new nest at a preferred site rather than risk a fall.

Since swallow nests are sheltered, they often provide an ideal microclimate and habitat for ectoparasites (reviewed in Bent 1942; Brown 1985; Shields and Crook 1987; Barclay in press). As ectoparasites can be a major mortality factor for nestling barn swallows (M ller 1987a;

Shields and Crook 1987), it should pay to minimize their impact. Continuous use of the same nest might allow ectoparasite populations to grow continuously. Using an old nest intermittently might provide reduced opportunities for the ectoparasites. If a swallow could tell whether a nest had been used in a recent attempt, it might reduce its ectoparasite load by choosing one that had lain fallow during a prior attempt or season (Shields 1984b; Barclay in press). In this way, it could gain the time and energy benefits and still reduce one of the risks of nest reuse. If no fallow nests were available, it might gain by expending the time and energy on a new nest at its preferred site, rather than reusing a recently used nest known or suspected to harbor a larger ectoparasite load.

When combined, the traditional-site and nest-reuse hypotheses imply a hierarchy of quality classes for barn swallow habitat. The highest quality, and therefore preferred, areas would contain traditional sites with fallow nests available for reuse and low densities of conspecifics to minimize potential crowding costs. A traditional site nearer saturation density or with only continuously used old nests available for reuse might offer the same safety benefits as the highest quality sites, but would be more likely to suffer increased risks of ectoparasitism, nest fall, or crowding effects. It would then be less suitable than the highest quality sites.

Any site with old nests would, by definition, be a traditional site, but not all traditional sites need possess old nests available for new settlers. For example, if all old nests at a site were controlled by returning site-tenacious adults, none would be available for subordinate immigrants. In addition, traditional sites could lose their old nests every year (e.g., humans do remove them each winter from many sites in Norway, M ller 1985a; some sites could be scoured clean by annual floods, landslides, or winds, Barclay in press). As long as such catastrophes occurred outside the breeding season, they might not affect breeding success enough to change the pattern of site tenacity.

In both cases, returning birds and potential immigrants would be forced to build new nests and so would lose the time and energy savings associated with nest reuse. As long as they built at traditional sites and microsites (using nest scars to locate both), the nests might prove safer than those sited randomly. Such traditional but nestless sites, then, might not rank as high as those with old nests. In the absence of spatial saturation or crowding effects, however, they might still be preferred to untried sites, which by definition lack reusable nests and a detectable history of success.

C. Traditional Aggregations in Ideal Habitat:
Predictions and Evidence

The traditional-site and nest-reuse hypotheses, in combination with other aspects of barn swallow life history and ecology, offer a comprehensive, but still tentative, working hypothesis for the origin and control of group breeding in the species. This traditional aggregation, or TAG, hypothesis falls within the realm of Alexander's (1974) resource localization or limitation hypothesis, but is more explicit than previous attempts to explain swallow sociality on that basis (Snapp 1976; M ller 1987a; Shields and Crook 1987). Since we synthesized this hypothesis recently, and it differs from our earlier working hypotheses, we have only just begun to gather data and design experiments to explicitly test its many predictions. For now, therefore, we feel that we can best stimulate additional discussion and research by listing *some* of its, and its component hypotheses', predictions--especially those that can distinguish among alternative hypotheses.

1. The TAG hypothesis predicts that groups of breeding swallows will be found at traditional sites and *only* at such sites. All of the alternative explanations of sociality, including simpler forms of habitat shortage, do not predict such a limitation. If conspecifics are drawn to one another for direct social benefits, or to a general resource in short supply, there is no reason that a novel site should not be settled by many pairs simultaneously, as does occur in some colonial nomads (Crook 1965; Lack 1968).

The TAG hypothesis, however, predicts that single pairs would usually settle novel sites. Only after such pioneers were successful, and had left evidence of their success (e.g., an old nest or nest scar), or had returned to such a site, would the site be expected to attract additional pairs. Thus, groups regularly settling at novel sites would falsify a strict TAG hypothesis. The value of using a traditional site might still apply, but additional factors would be needed for a complete explanation. In our study, and in every other study of the barn swallow with appropriate data of which we are aware, large breeding groups are found only at traditional sites. Sites were assigned to the traditional class if their histories were known or the site contained old unused nests and/or nest scars.)

2. The TAG hypothesis predicts that the maximum colony size at a traditional site will increase with the absolute size of the site, with the number of microsites it contains, or both. Larger sites can host more pairs, because nests would be less crowded and would suffer lower crowding costs than the same number of nests at a smaller site. Even a large site, however, could not host a colony in the absence of suitable microsites. For barn swallows, the appropriate measure of suitable site size might be the linear extent of vertical and horizontal attachment surfaces directly under and close to cover. This could be estimated by measuring the length of the walls, rafters, and beams directly under roofs, eaves, and overhangs on or within buildings, bridges, culverts, or caves. Alternatively, the number of old nests and nest scars could also provide an estimate of the minimum number of microsites, or the number of improved or proven microsites, available at a traditional site. As predicted, maximum colony size does increase with the amount of attachment surface available at traditional sites at Cranberry Lake, despite the availability of empty microsites in otherwise similar nontraditional sites nearby (Fig. 1).

3. The traditional-site subhypothesis predicts that pairs building new nests at novel sites will be less safe, and ultimately less successful, on

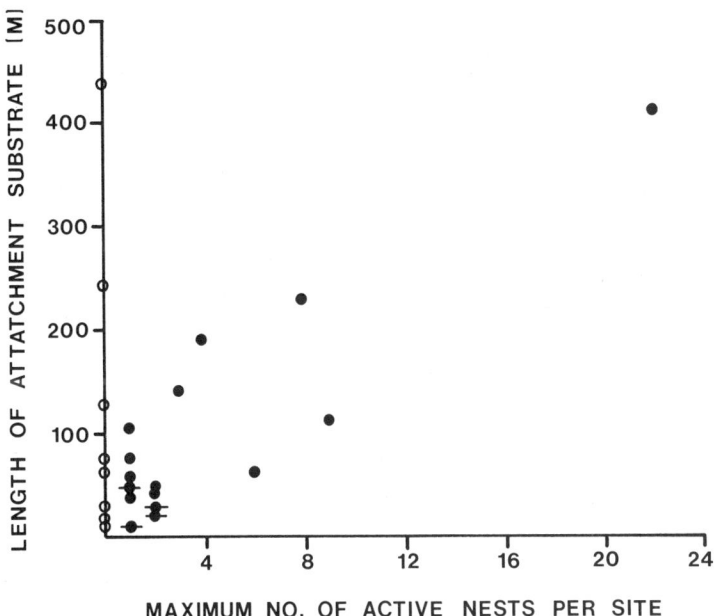

Figure 1.   Relationship between the size of a suitable nest site in the form of the linear extent of suitable attachment surface under overhead cover available to barn swallows at Cranberry Lake at traditional (open circles) and nontraditional (closed circles) sites and the maximum colony or subcolony size observed from 1979 to 1985.

average, than those building new nests at traditional sites.   Since our focal populations occurred only at traditional sites, we have no evidence on this point.   While Lohoefener (1980, p. 16) did not make a distinction between traditional and novel sites, he did note that, in Mississippi, more solitary (36%) than colonial (13%) nests were built anew.   This difference is consistent with an assumption that in this population solitary nests were located more often in new culvert sites, and colonies in traditional culverts exclusively.   If true, then his report (Lohoefener 1980, p. 48) that "no colonial sites ever flooded to the point that nests were dislodged.   Many of the solitary sites were subject to flooding and many nests were lost . . . (27% of the solitary sites, none of the colonial sites)" would be consistent with this prediction.   Data gathered or reanalyzed in light of the novel-traditional site distinction will be required for a more rigorous test.

4. The traditional microsite corollary predicts that even within traditional sites, pairs building new nests at traditional microsites should be safer, and ultimately more successful, than pairs building new nests at novel microsites.   At our traditional focal sites at Cranberry Lake, six of twelve new   nests   built   at   novel   microsites fell during their first

season, which resulted in egg or nestling losses for three of the six pairs. In contrast, and as predicted, only one nest fall resulting in mortality was observed in the 20 new nests built at traditional microsites (Fisher's Exact Test, P = 0.006).

5. At its simplest, the nest-reuse subhypothesis predicts that the time and energy saved by refurbishing an old nest should result in increased reproductive success. Nest reuse was associated with higher probabilities of successful double brooding in our study (Table I) and in Pennsylvania (Hill 1982), but not in Manitoba (Barclay in press). In New York, the *only* pairs that raised two broods were those that reused old nests for both attempts (Table I). Such double-brooded pairs fledged more young per season than any other class of swallow, so nest reuse was associated with the maximum potential payoff as predicted (Fig. 2). Unexpectedly, and contradicting the prediction, the *mean* number of young fledged per season did not differ between pairs that reused nests and those that built new ones (Table I, Fig. 2). This resulted primarily from a greater probability of complete failure balancing the higher probability of double brooding in old nests (Fig. 2). In Hill's (1982) study, in contrast, and in complete agreement with the prediction, both the maximum and mean payoff per season were higher for swallows that reused nests than for those that built new ones. Thus, the data are supportive, but still equivocal, with respect to the simplest version of this prediction.

As the benefits of reusing old nests should be the same everywhere, spatial or temporal differences in net payoffs would be expected only as a result of spatial or temporal differences in the costs of reuse. The hypothesis does predict that crowding around old nests should increase costs enough to favor building new nests. This would occur when all microsites with old nests at a traditional site were defended vigorously by dominant residents, or when crowding at, and/or continuous use of, a site increased the levels of social interference, predation, or ectoparasitism. In line with this, in New York, nest reuse was associated with an increase in ectoparasitism (Table I). In addition, younger, and presumably more subordinate, birds built proportionately more new nests than did older birds (also reported in Pennsylvania, Hill 1982) and they usually located them in less crowded sites (Table I). This is consistent with the prediction that birds build new nests to avoid the costs of reuse in crowded sites.

If nest reuse in Pennsylvania were not associated with increased crowding costs like increased ectoparasitism, then the higher net payoff observed there would be expected. If some of the new nests in Pennsylvania had been built at novel sites, then their reduced success could also lead to a higher payoff for old nests. Unfortunately, Hill (1982) did not distinguish between traditional and novel sites or gather data on ectoparasitism or other crowding costs, so this explanation of the difference between New York and Pennsylvania remains speculative (see also Barclay in press).

An optimal nest-reuse subhypothesis, then, makes different predictions. In the absence of balancing costs, reusing old nests at traditional sites should result in increased payoffs. When the costs are higher, for

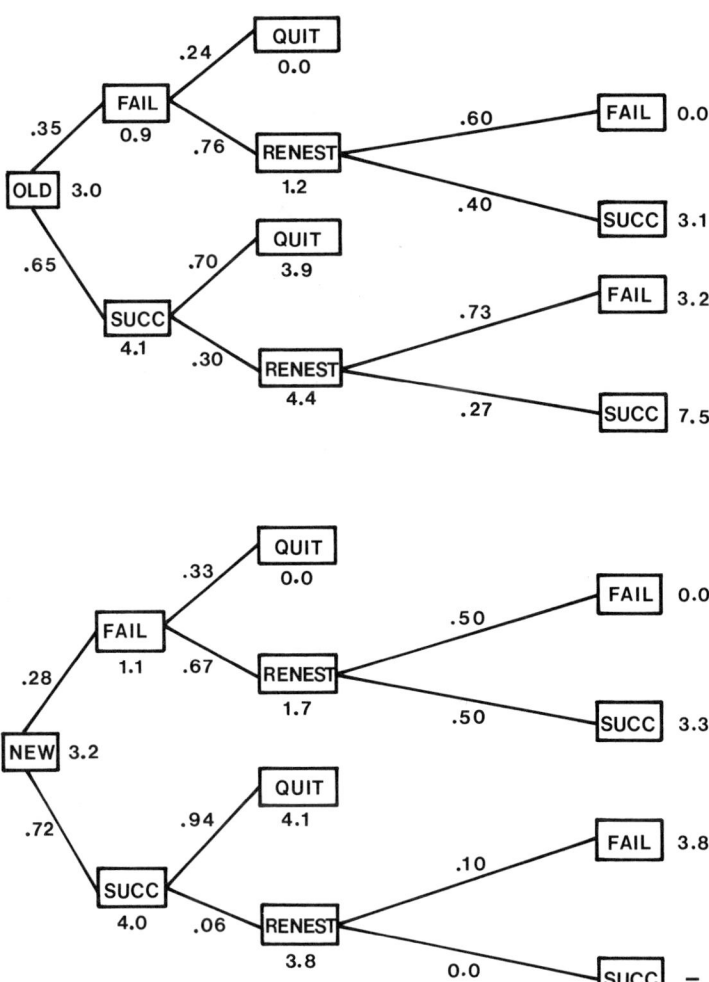

Figure 2. Decision and event tree illustrating the empirically derived probabilities of particular decisions, events, and outcomes along with their expected payoffs (the mean number of young fledged per year for individuals following particular paths) as a function of whether for their first attempts of the year, swallows chose to refurbish a nest (OLD) or build a new nest from scratch (NEW).

example due to increased parasitism, then building new nests in less crowded areas should result in increased payoffs. Ideally, then, if settlers have enough information and aggression does not force them to choose suboptimally, the fitness of reusers and builders should be equal as they settle in an ideal-free fashion. Alternatively, despots could prevent individuals from making the optimal choice and their building would result in lower payoffs following an ideal despotic distribution. At Cranberry Lake, the mean payoffs were equal, implying that the choice may have been ideal free (Fig. 2). In addition the variance in fitness was much greater for the reuse tactic, implying that it might be more suitable for risk-prone than for risk-averse individuals that might prefer the lower variance tactic of building anew (Fig. 2) (for a discussion of variance effects on optimization, see Caraco 1981a,b; Yoshimura and Shields 1987).

6. Putting the traditional-site and nest-reuse subhypotheses together generates a final critical prediction for the TAG hypothesis. All other factors being equal, traditional sites with old nests should be preferred and host larger colonies than traditional sites without such nests. The prediction can be tested experimentally by documenting the settlement pattern at matched traditional sites and then removing old nests from some, adding old nests to the others, or both. One then would document the settlement of immigrants at the two site types (ignoring site-tenacious adults because their behavior might not be changed by the protocol). Factors requiring control include the absolute size of the sites, the number of microsites available in them, and the size of the breeding group using them (Prediction 2 above).

Though this experiment has not been done, humans may be performing analogous natural experiments. For example, M ller (1985a) notes that in Denmark, "most nests were destroyed by the farmers overwinter." He also noted that barn swallow colonies are generally smaller in Scandinavia than they are in Germany or North America. We predict that human hosts in Germany and North America do not remove nests from traditional sites overwinter at the same frequencies they do in Scandinavia. We would also predict that the larger-than-average swallow colonies observed in Denmark occur at traditional sites where old nests are left undisturbed.

IV. THE EVOLUTION OF COLONIALITY IN THE SWALLOW FAMILY

Snapp (1976) was the first to note that "the development of coloniality in the swallow family paralleled the development of the ability to build nests." She modestly cited Mayr and Bond (1943) for this observation, but our reading of their paper failed to unearth a discussion by them of any association between nesting habits and sociality. Rather, Mayr and Bond discussed using differences in nesting habit as a taxonomic character. In contrast, Snapp explicitly noted that "Burrow-excavators of the genus *Riparia* and mud-nest builders of the genus *Petrochelidon* often breed in large colonies. On the other hand, the genera that utilize naturally-occurring cavities for nest sites, e.g., *Iridoprocne* and *Stelgidopteryx*, breed in a dispersed fashion, perhaps because of the erratic distribution of such cavities." We concur with her more explicit hypothesis that differences in nesting habit might control the variation in nesting dispersion observed among swallow species.

Our model suggests that nesting habit differences could interact with the dispersal behavior and nesting habitat selection strategies common to all swallow species to control nesting dispersion. The major assumptions of this model are (1) *all* swallow species are cavity nesters, (2) the quality of potential nest sites (cavities) varies with the factors assumed important by the traditional aggregation hypothesis (e.g., history of use, degree of site improvement, and local breeder density), and (3) the abundance and dispersion of breeding swallows is controlled primarily by the abundance, dispersion, size, and especially variation in the quality of suitable nests (cavities) and nest sites.

Cavity Nesting.   Traditionally, swallow species have been classified as nest or nest site borrowers, burrowers, or builders (Fig. 3) (Bent 1942; Mayr and Bond 1943).   As secondary cavity users,   the borrowers (e.g., *Tachy cineta* spp., *Progne* spp., *Stelgidopteryx* spp., or *Psalidoprocne* spp.) (Fig 3,   Table II)   have   always   been   classified   as   hole   or   cavity

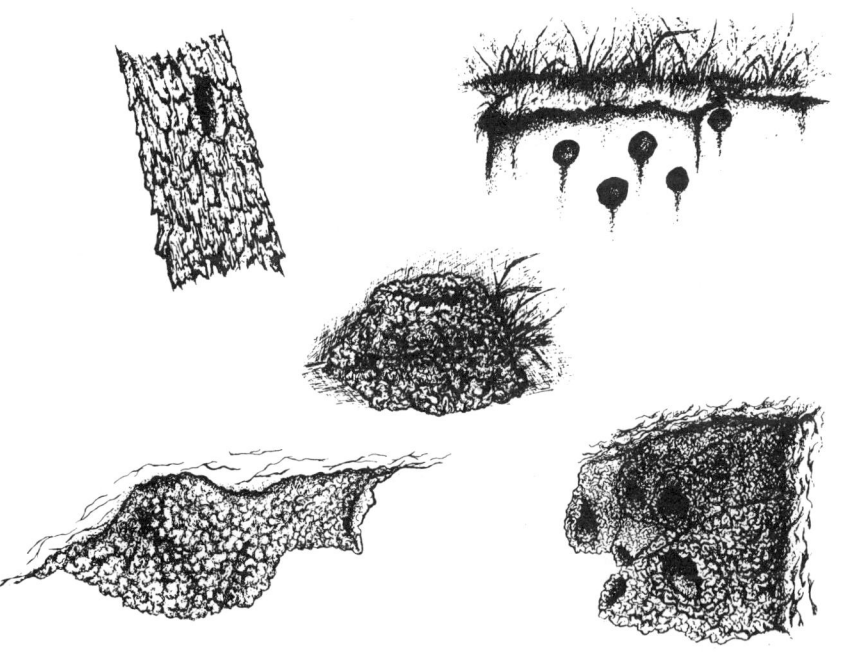

Figure 3. Typical nest and site types used by members of the swallow family (Hirundinidae). Clockwise from the upper left: tree cavities, e.g., T. bicolor; burrows in vertical banks, e.g., R. riparia; closed mud spheres with attached entrance spout, e.g., H. pyrrhonota; attached mud retort with an attached tunnel entrance, e.g., H. daurica; and in the center of an open cup mud nest, e.g., H. rustica.

TABLE II. Associations between swallow (Hirundinidae) nesting habits and social behavior.

Genus	Spp. or Super-Species Group	No. of Species	Nest Site	Modify Site	Nest Type	Nest Material	Nest or Site Reused	Breeding Sociality	Feeding Sociality
Tachycineta	spp.	8	T,R,H	N	OC	V	Y	S	S
Progne	spp.	3-5	T,R,H	N(R)	OC	V(M)	Y	S	S
Phaeoprogne	tapera	1	T,R,H,G	N	OC	V	Y	S	S
Neochelidon	tibialis	1	T,G,	N	-	-	-	S	-
Atticora	spp.	2	R	N	OC	V	Y	S	-
Pygochelidon	spp.	4	R,H,B	N	OC	V	Y	S	S
Phedina	borbonica	1	R,H	N	OC	T,V	-	S	-
Stelgidopteryx	spp.	2	B,H	N(R)	OC	V	Y	S	S
Psalidoprocne	spp.	3-10	B,G,H	N(R)	OC	V	-	S	-
Hirundo	griseopyga	1	G,B	Y	OC	M,V	Y	S	-
H.	atrocaerula grp.	2	G,H	Y	OC	M,V	Y	S	S
H.	nigricans	1	T	Y	OC	V,M	Y	SC	-
H.	rupestris grp.	3-4	R,H	Y	OC	M,V	Y	SC	S
H.	smithii grp.	2	R,H	Y	OC	M,V	Y	SC	-
H.	dimidata grp.	3	R,H	Y	OC	M,V	Y	SC	-
H.	rustica grp.	5-7	R,H	Y	OC	M,V	Y	SC	S
H.	daurica grp.	3-4	R,H	Y	AR-T	M,V	Y	SC	S
H.	cecropis grp.	3-4	R,H	Y	CR-T	M,V	Y	SC-C	-
H.	petrochelidon grp.	6-9	R,H	Y	CS-S	M,V	Y	C	C
Delichon	spp.	3	R,H	Y	CS-S	M,V	Y	C	S
Phedina	brazzae	1	B,G	Y	OC	V	Y	SC	-
Pseudochelidon	spp.	2	G	Y	OC	V	Y	C	-
Cheramoeca	leucosterna	1	B	Y	OC	V	Y	C	C
Riparia	spp.	4	B	Y	OC	V	Y	C	C

The binomial classification used in this table comes from Hall and Moreau 1970, which was based on Mayr and Bond 1943 and Mayr and Greenway 1960. Recent changes in the A.O.U. Checklist 1985 are reflected in the western hemisphere species. Y = yes, it is reported. N = no, it is not reported, and N(R) = reported but rare. Nest Site column: T = tree cavities; R = rocky substrates including niches, ledges, or walls; H = human artifacts, including buildings, bridges, culverts, pipes, or nestboxes; G = cavities in level ground, including wells, mines, holes, animal dens; B = cavities in vertical soil substrates, including banks, animal dens, termite mounds, and road cuts. Nest Type column: OC = open-cup nest lining or nest; AR-T = retort-shaped nest attached to over-head substrate with a tunnel entrance; CR-T = closed retort attached to vertical substrate with a tunnel en-trance; CS-S = closed sphere attached to vertical substrate with a spout entrance. Nest Material column: V = vegetation pad or mat, usually lining a harder cavity; T,V = a twig nest lined with vegetation; M,V = a mud nest lined with vegetation; V,M = vegetation lining in a tree cavity, with mud used to reduce entrance hole. Breeding Sociality column: S = solitary; SC = semicolonial; C = colonial, defined explicitly in text. Feeding Sociality col-umn: S = only solitary foraging reported; C = group foraging reported.

The tabled entries were compiled from numerous regional field guides, life history summaries, and individ-ual species acounts. General sources included Thompson 1964; Hall 1964; Hall and Moreau 1970; St. Peters 1973; and Bryant 1985. Local sources included, Africa: Moreau 1939, 1940, 1947; White 1947, Macworth-Praed and Grant 1955, 1973; McLachlan and Liversidge 1963; Snell 1969; Hall and Moreau 1970; Benson et al. 1973; Harri-son 1982. Asia: Delacour 1947; Whistler and Kinnear 1949; Dement'ev 1968; Ali and Ripley 1972; King and Dick-inson 1975; Hails 1984; Meyer de Schauensee 1984. Australia: Gould 1967; Anon. 1978; Pizzey 1980. Europe: Dement'ev 1968; Bannerman and Bannerman 1971; Harrison 1982; Pforr and Limbrunner 1982. North America: Stoner 1936; Bent 1942; Friedmann et al. 1950; Lunk 1962. South and Central America including the Caribbean: Bent 1942; Skutch 1952, 1960, 1961; Meyer de Schauensee 1964; Haverschmidt, 1968; Ridgely 1981; Dyrcz 1984; Meyer de Schauensee and Phelps 1984; Turner 1984; Wetmore et al. 1984.

nester. They nest in relatively inaccessible and enclosed spaces in tree cavities, especially old woodpecker holes, in burrows in vertical banks, in holes in level ground, in aerial termite nests, in rocky niches, or in holes or cavities provided by human artifice (e.g., in pipes, cornices, nestboxes). They are classed as borrowers because they simply move into cavities provided geologically, or more often through the toil of other species. They spend little or no time or energy modifying or improving the sites. They do construct a nest pad or line the cavity with vegetation and other soft material, especially feathers (Table II). As borrowers (beggars) they are not very choosy, and many species use broader arrays of nest substrates than do the burrowers or builders (Table II).

Burrowers build their own enclosed cavities, in relatively inaccessible sites, by digging burrows in level ground (e.g., *Pseudochelidon* spp. which use sandbars in rivers) or more commonly in vertical banks (e.g., *Cheramoeca leucosterna* and *Riparia* spp.) (Fig 3, Table II). Like secondary cavity nesters, they construct nest pads or linings out of vegetation and feathers (Table II). Because they dig their own burrows, they are defined as having improved their nest site by modifying it (Table II). As they require relatively soft soil or sand for digging, they are more limited in their choice of substrates than secondary cavity users. Since their old burrows are often used by secondary cavity nesters, including other swallows, few would argue against their classification as primary cavity nesters.

Builders species (e.g., *Hirundo* spp., *Delichon* spp.) also nest in relatively inaccessible sites, usually on rocky substrates (cliffs and caves) or on or in similar human artifacts (e.g., bridges, culverts, buildings), though a few nest in trees or in holes in level ground (Table II). We classify the builders as primary cavity nesters for the same reasons as we did the burrowers. They do modify their nest sites extensively (Table II). Rather than digging, however, they use mud pellets to build more or less enclosed brick nest cavities to their own specifications (Fig. 3).

Those that build open-cup mud nest structures (e.g., most of the genus *Hirundo*, especially the barn swallow allies) (Table II) enclose them by attaching them to vertical or horizontal surfaces that are always directly under and near overhead cover provided by the nesting substrate (under roofs, eaves, or overhangs) (Fig. 3). Those that build completely enclosed spheres or retorts (e.g., the *H. petrochelidon* group, or *Delichon* spp.) (Fig. 3, Table II) attach their nest structures to vertical or horizontal surfaces, but not always under or close to overhead cover. Like the the rest of the swallows, then, the builders construct nest pads and linings with vegetation and feathers (Table II). For all builder species, the result is a more or less inaccessible cavity nest offering similar or greater protection from the elements and potential predators as the tree or ground cavities borrowed, or the burrow cavities dug, by other swallow species.

Traditional Aggregation Hypothesis. Positive evidence for the reuse of nest cavities or nest sites within and across seasons, and the continuous use of traditional breeding sites has been reported, at least anecdotally, for almost every swallow species (Table II). For no species is there any direct evidence for an absence of site and nest reuse.   In addition, all studies with marked individuals have reported dispersal from natal sites and settlement at traditional sites for juveniles combined with strong adult site tenacity, especially of successful breeders, and regardless of nesting habits (Chapman 1955; Lunk 1962; Freer 1979; Shields 1984b; Sikes and Arnold 1984).

That all swallow species might suffer crowding costs at high densities, which could act to limit group size, is illustrated by the reports of nest ectoparasites in all of the well-studied groups (Table II).   Many studies have estimated the costs of ectoparasitism and have documented increased costs at higher breeding densities (Moss and Camin 1970; Chapman 1973; Hoogland and Sherman 1976; Brown and Brown 1986; M ller 1987a; Shields and Crook 1987).   Many have also documented the increased social competition expected in larger breeding groups, at least in the semi- and fully colonial species (e.g., bank swallow, Hoogland and Sherman 1976; cliff swallow, Brown 1985; barn swallow, M ller 1985a; Shields and Crook 1987).

Abundance, Size, and Dispersion of Nest Sites.   The availability of nest sites is expected to limit reproduction in most secondary cavity nesting bird species (reviewed in von Haartman 1957; Lack 1968; Collias and Collias 1984).   Like similar species in other families, then, the abundance and dispersion of tree cavity or burrow borrowing swallows should be controlled by the abundance and dispersion of nest cavities.   Since most of the cavities and burrows they use result from the efforts of breeding woodpeckers, kingfishers, trogons, and other normally solitary species, such swallows normally breed solitarily as well (Table II).

That the local abundance of nest sites can limit their numbers is illustrated by the ease with which local densities of secondary cavity-using swallows can be increased by providing additional cavities in otherwise suitable habitat (e.g., nestboxes for tree swallows, Chapman 1955; DeSteven 1978; plastic or pipe burrows for rough-winged swallows, Lunk 1962; reviewed in Bent 1942; Holroyd 1975).   Indeed, by providing grouped or multichambered nestboxes or apartment houses, breeding aggregations of normally solitary species can be induced (e.g., purple martins, Allen and Nice 1952; tree swallows, Sheppard 1977; Muldal et al. 1985).

Thus, the secondary cavity nesting swallows might best be considered facultatively colonial.   If the nest cavities they find in nature are dispersed regularly or randomly, which they usually are, they nest solitarily.   If the available cavities are clumped, however, they will aggregate at such concentrations of an otherwise limiting resource and end up breeding in groups.   Since it is primarily the secondary cavity users that are classified as solitary in the swallows (Table II), the evidence is quite consistent with Snapp's (1976) suggestion that they usually nest alone owing to the "erratic" dispersion of their nest sites.

Normally, secondary cavity-nesting swallows breed in groups only when humans manipulate cavity availability and distribution by providing clumps of nestboxes or artificial burrows.   The cavity-making swallows, including both burrowers and builders, operate under the same rules, and with similar results. The essential difference is that they are less limited by what the environment, including humans, has to offer. By building their own cavities, they are no longer limited by the occurrence of "natural" cavities. Rather, they are limited by the abundance and distribution of high-quality sites at which they build their own nestboxes or cavities. The size of their breeding aggregations, in turn, is expected to be controlled by the characteristic abundance and size of the nest sites they require.

For example, for builders of open-cup mud nests, each site must have enough attachment substrate under and close enough to overhead cover to protect the nest. Because the amount of cover at a single site is often limited, and traditional sites are always limited in number, open-cup nesters would be expected to aggregate if sites were large enough. Since most of the sites with cover are usually of limited size, however (e.g., caves, cliff faces under large overhangs, on buildings under eaves, or in open buildings near roofs), small or moderate-sized groups would be expected to occur most often. Only at large traditional sites with large amounts of cover would large colonies be expected. The fact that most open-cup mud nest builders are classified as semicolonial breeders is consistent with this hypothesis (Table II).

The closed sphere or retort builders and burrow diggers, in contrast, are less limited by the types or sizes of sites they can use. Because the cavities they build are enclosed, neither group *requires* overhead cover. Both groups, however, do require specialized sites with suitable properties for digging or nest attachment. They also should prefer traditional sites. Because of their emancipation from overhead cover, the size of a group at any site could be larger than for similar open-cup builders. The result would be larger breeding groups at traditional sites, even when otherwise suitable but nontraditional sites remain unused nearby. Consistent with this prediction, it is the burrow diggers and covered nest builders that nest in the largest groups and that are normally classified as colonial breeders (Table II).

The available data, then, are consistent with the hypothesis that an ecological factor, namely the dispersion of nests or nest sites, determines the size of breeding groups and hence the level of coloniality in all of the swallows. As Mayr and Bond (1943) noted, however, nesting habits are also indicative of phylogenetic relationships within the family. Thus, it is at least possible that the patterns of coloniality observed are the result of descent without modification. Arguing against this is the strong association between coloniality and the building or digging of covered nests in a phylogenetically diverse group of swallows, including the most primitive (*Pseudochelidon* spp.) and perhaps the most advanced (*Delichon* spp.) genera (Table II).

The few apparent exceptions to the ecological hypothesis actually strengthen its case. For example, one tree cavity nester, the Australian tree martin (*H. nigricans*) is reported to nest semicolonially despite its status as a secondary cavity user (Table II). Its phylogenetic affinities to the mud builders are not in doubt and are indicated by its habit of using mud to decrease the size of the entrance holes to its nest cavities. The cavities result from natural rots in knots in eucalyptus trees. The tree martin's semicoloniality could result from its phylogenetic relationship to the mostly semicolonial and colonial members of its genus. It is at least as likely, however, that its breeding aggregations result because the enlarged and hollow eucalyptus knots in which it nests come in clusters. Each tree often has as many as a dozen or so knothole cavities suitable for tree martin nesting.

Since two groups of the more typical open-cup builders are reported to nest solitarily (*H. griseopyga* and the two blue swallows of Africa, the *H. atrocaerula* group) (Table II), we can rule out the notion that all members of the genus *Hirundo* breed in groups owing to phylogenetic inertia. The fact that these two species usually nest in sites (e.g., aardvark burrows, vertical holes in the ground) that are likely to be in short supply is more consistent with the nest site limitation hypothesis. Since the blue swallow has recently been observed nesting on human artifacts (Snell 1969), we would predict that it will soon be reported more often as nesting semicolonially.

Our assertion that swallow coloniality is a response to nest and nest site shortages is not meant to imply that previous studies documenting some benefit for individuals living in groups for specific swallow species were mistaken. Indeed, we would be surprised if swallows that found themselves living in groups did not use other group members and the information they might provide to increase their foraging efficiency (e.g., in *Riparia riparia*, Emlen and Demong 1975; or in *Hirundo petrochelidon*, Brown 1985) or reduce their risk of predation (e.g., *Riparia riparia*, Hoogland and Sherman 1976; Brown 1985). We would consider such group effects as secondary adaptations, however, that arose and were maintained after nest site shortages had forced the members of each taxon into breeding groups.

In this light, the temporal and spatial patchiness of the family's aerial insect food resource would be a permissive rather than causative condition of group living. Rather than favoring grouping directly, because of the potentially greater harvesting efficiency of individuals in a central group, the patchy food distribution would *permit* grouping in response to habitat shortage, because food is sufficiently abundant and indefensible to make the exclusion of new settlers at a traditional site too costly (Slobodchikoff and Schultz this volume). Similarly, once nest site shortages forced swallows into a group, it might pay individuals to follow successful group members to better foraging sites or simply join foraging group members whenever they were detected. Any benefits from the increased foraging efficiency that might result, however, might be balanced by increased crowding costs (e.g., greater ectoparasitism or social competition, Brown 1985; M ller 1987a).

We believe that the comparative data on the pattern of swallow coloniality are broadly consistent with the traditional aggregation hypothesis. In this light, the entire family might best be considered facultatively colonial. As Snapp (1973, 1976) first suggested, it is differences in nesting behavior, which are associated with differences in the abundance and dispersion of ideal nest sites, that are finally translated into different modal states of coloniality for each swallow taxon. Thus, swallow colonies or aggregations might best be considered epiphenomena. They result from adaptive decisions to tolerate some crowding in order to deal with shortages of traditional nests and nest sites. Such nests and sites, in turn, are the critical ecological resource limiting swallow reproduction.

V. SUMMARY AND CONCLUSIONS: A PROSPECTIVE PERSPECTIVE ON THE PAST

We believe that the TAG hypothesis for the origin and maintenance of group living in the swallows is a special case of a more general principle that future prospects for survival and reproduction are often (always?) contingent upon the past. History and an appropriate use of historical information can influence the prospects for survival and reproduction profoundly and positively in almost any organism. Adaptive decisions, including those about where, when, and with whom to settle (habitat choice), must be based on "good" information about the environment. Such information, in turn, may come to an individual as part of its genetic inheritance, as a result of direct experience with its environment, or as a cultural heritage (Lorenz 1965; Bonner 1980; Plotkin and Odling-Smee 1981; Pulliam and Dunford 1980).

Since it is a necessary and sufficient condition for adaptive decision making, good information is always going to be a critical resource. When environments vary cryptically so that habitat quality cannot be assessed by casual inspection, information would be a resource in short supply. Organisms that assessed habitat quality by noting their own reproductive success or failure at a site should make better decisions about staying or dispersing than those that fail to take personal experience into account. They would be using information that they had gathered during their own lifetime in making habitat choice decisions.

Individuals ready to settle at a site new to them, in contrast, could not use their own experience as a guide. Those that use the experience of other breeders at a site might do better than those settling apparently suitable sites at random. Rather than basing their decisions on personal experience, however, they would be using the experience of others to inform their settling decisions. The critical information on site quality permitting adaptive habitat choice would become available to site-naive prospectors via simple observation or even cultural transmission (e.g., when prior use of a site leaves artifacts like mud nests).

This habitat choice rule would result in continuous use of high-quality traditional sites. Once settled, an intrinsically suitable site would act as a magnet for additional settlers. Prospectors using the decision rule--follow the successful leader--would settle at the better sites. If such sites were defendable, continuous occupation of traditional territories could result, as it does in some lizards (Stamps 1987). For other species, like the swallows, an absence of overwhelming crowding costs or sufficient benefits of aggressive exclusion, would produce aggregations. This would not be due to any direct advantages of group living, but rather due to the danger and wasted energy associated with settling new sites.

Once an information shortage and the resulting benefits of tradition forced a species into colonies, adaptation to group living would be expected to follow.   Given the appropriate circumstances, then, we would expect animals living in groups to use their neighbors to reduce predation, increase foraging efficiency, or both. In essence, their behavior should evolve in ways that minimize the costs and maximize the benefits of group living even though they have been forced into groups by a shortage of ideal habitat or easily accessed information about ideal habitat.

The pattern of barn swallow semi-coloniality suggested the TAG hypothesis and we tested it as a potential explanation of sociality in the entire swallow family.   The hypothesis was supported sufficiently that we now take it as our working hypothesis for the family. We feel that it can help to explain group living even more widely.   For example, Ashmole (1962) actually presaged the TAG hypothesis when he emphasized the importance of tradition in controlling habitat choice in colonial seabirds. He suggested that the presence of breeding adults might be indicative of safer sites for prospecting juveniles. If true, then settling novel sites should be quite risky for seabirds that breed in groups at traditional sites, as it is in the swallows.

We believe that the traditional-site hypothesis could be relevant and warrants testing in the many species that breed in dense groups when nearby and apparently suitable sites remain empty (e.g., some colonial arthropods, including some spiders, ants, and termites, Wilson 1971, 1975; island or beach breeding reptiles like sea turtles, Baker 1978; other birds including many herons, finches, and aerial insectivores, Lack 1968; many mammals, including ground squirrels, pinnepeds, and many terrestrial primates, Wilson 1975; and the many marine invertebrates that are known to use the presence of conspecifics as a settling cue, Thorson 1950; and general reviews in Orr 1970; Baker 1978; and Shields 1982).

Finally, and on a much more speculative note, we would suggest that the potential wide occurrence of traditional sites and traditional aggregations might be another illustration of a more crucial role of tradition in adaptation.   Just as individuals do better by settling where others have settled before, they might do better to transmit genomes that have been successful in the past rather than blindly exploring novel genome space (Shields 1987).   Similarly, learned or culturally transmitted behaviors that characterize successful individuals or groups of organisms are likely to provide adaptive role models for individuals or groups developing a behavioral repertoire in the absence of any other information (Campbell 1975; Boyd and Richerson 1985).   Santayana's aphorism, "Those who cannot remember the past are condemned to repeat it," emphasizes avoiding the errors of the past. Good advice perhaps, but we would make the related point: "Those who cannot remember the past are condemned to grope blindly." Thus, we emphasize the notion that being traditional as long as it is biased toward mimicking the successes of the past, is likely to stand one in good stead today and in the future.

ACKNOWLEDGMENTS

We appreciate the field assistance of numerous students and volunteers over the years, including S. Balzano, M. Berg, L. Birr, H. Broce, K. Buchanen, S. Cogswell, T. Crossman, E. Defries, T. Ehlers, S. Gaynor, M. Grogan, J. Herter, W. Kappelman, P. Mason, M. Matiz, S. Muller, E. Myers, S. Peters, E. Schering, N. Skolnick, C. Smith, M. Somerville, S. Webb, M. Woodrey and especially M. Ehlers and C. LeBlanc. We also are grateful to L. Rathman and his crew for their assistance and forbearance in the field. During the study M. Hebblethwaite, J. Crook, and S. Wiles-Ehmann received financial support from Sigma Xi. J. Crook received assistance from a Chapman grant from the American Museum of Natural History, a Fuertes award from the Wilson Ornithological Society, and a Herbert and Betty Carnes Award from the American Ornithologists Union. Financial support for W. Shields was provided by the State University of New York, College of Environmental Science and Forestry, the Cranberry Lake Biological Station (R. Brocke, Director), and a grant (BSR 83-19559) from the National Science Foundation.

REFERENCES

Alexander, R.D. 1974. The evolution of social behavior. *Annual Review of Ecology and Systematics* 5:325-383.

Ali, S. and S.D. Ripley. 1972. *Handbook of the birds of India and Pakistan.* Vol. 5. Bombay, India:Oxford University Press.

Allen, R.W. and M.M. Nice. 1952. A study of the breeding biology of the purple martin (*Progne subis*). *American Midland Naturalist* 47:606-645.

Andersson, M. 1980. Nomadism and site tenacity as alternative reproductive tactics in birds. *Journal of Animal Ecology* 49:175-184.

Anonymous. 1978. *Complete book of Australian birds.* Sydney, Australia:Readers Digest Services.

Ashmole, N.P. 1962. The black noddy *Anous tenuirostris* on Ascension Island. Part 1. General biology. *Ibis* 103b:458-473.

Baker, R.R. 1978. *The evolutionary ecology of animal migration.* London:Hodder and Staughton.

Ball, G. 1982. Barn swallow fledgling successfully elicits feeding at a non-parental nest. *Wilson Bulletin* 94:352-363.

Ball, G.F. 1983. Evolutionary and ecological aspects of the sexual division of parental care in barn swallows. Ph.D. dissertation, Rutgers University, Newark, N.J.

Bannerman, D.A. and W.M. Bannerman. 1971. *Handbook of the birds of Cyprus.* Edinburgh:Oliver and Boyd.

Barclay, R.M. In press. The costs and benefits to barn swallows (*Hirundo rustica*) of using old nests. *Auk*.

Barrentine, C.D. 1978. The biology of bridge nesting barn and cliff swallows in central Washington. M.S. thesis, Central Washington University.

Beecher, M.D. and I.M. Beecher. 1979. Sociobiology of bank swallows: Reproductive strategy of the male. *Science* 205:1282-1285.

Benson, C.W. 1982. Migrants in the Afrotropical region south of the equator. *Ostrich* 53:31-49.

Bent, A.C. 1942. *Life histories of North American flycatchers, larks, swallows, and their allies.* New York:Dover Publications.

Blancher, P.J. and R.J. Robertson. 1985. Site consistency in kingbird breeding performance: Implications for site fidelity. *Journal of Animal Ecology* 54:1017-1027.

Bonner, J.T. 1980. *The evolution of culture in animals.* Princeton, N.J.: Princeton University Press.

Boyd, A.W. 1935. Report on the swallow enquiry. 1934. *British Birds* 29:3-21.

Boyd, A.W. 1936. Report on the swallow enquiry. 1935. *British Birds* 30:98-116.

Boyd, R. and P.J. Richerson. 1985. *Culture and the evolutionary process.* Chicago:University of Chicago Press.

Brown, C.R. 1978. Clutch size and reproductive success of adult and subadult purple martins. *Southwestern Naturalist* 23: 597-603.

Brown, C.R. 1979. Territoriality in the purple martin. *Wilson Bulletin* 91:583-591.

Brown, C.R. 1984. Laying eggs in a neighbor's nest: Benefit and cost of colonial nesting in swallows. *Science* 224:518-519.

Brown, C.R. 1985. The costs and benefits of coloniality in the cliff swallow. Ph.D. dissertation, Princeton University, Princeton, N.J.

Brown, C.R. and M.B. Brown. 1986. Ectoparasites as a cost of coloniality in Cliff Swallows (*Hirundo pyrrhonota*). *Ecology* 67:1206-1218.

Brown, J.L. 1964. The evolution of diversity in avian territorial systems. *Wilson Bulletin* 76:160-169.

Brown, J.L. 1969. Territorial behavior and population regulation in birds: A review and re-evaluation. *Wilson Bulletin* 81:293-329.

Brown, J.L. 1975. *The evolution of behavior.* New York:W.W. Norton.

Brown, J.L. and G.H. Orians. 1970. Spacing patterns in mobile animals. *Annual Review of Ecological Systems* 1:239-262.

Bryant, D.M. 1973. The factors influencing the selection of food by the house martin *Delichon urbica* (L.). *Journal of Animal Ecology* 42:539-564.

Bryant, D.M. 1975. Breeding biology of house martins *Delichon urbica* in relation to aerial insect abundance. *Ibis* 117: 180-216.

Bryant, D.M. 1978. Environmental influences on growth and survival of nestling house martins (*Delichon urbica*). *Ibis* 120:271-283.

Bryant, D.M. 1985. "The swallow." pp. 571-572. In: B. Campbell and E. Lack, eds. *A dictionary of birds.* Vermillion, England:Buteo Books.

Campbell, D.T. 1975. On the conflicts between biological and social evolution and between psychology and moral tradition. *American Psychologist* 30:1103-1126.

Caraco, T. 1981a. Energy budgets, risk and foraging preferences in dark-eyed juncos (*Junco hymelais*). *Behavioral Ecology and Sociobiology* 8:213-217.

Caraco, T. 1981b. Risk-sensitivity and foraging groups. *Ecology* 62:527-531.

Chapman, B.R. 1973. The effects of nest ectoparasites on cliff swallow populations. Ph.D. dissertation. Lubbock, Texas:Texas Tech University.

Chapman, L.B. 1955. Studies of a tree swallow colony. *Bird Banding* 26:45-70.

Collias, N.E. and E.C. Collias. 1984. *Nest building and bird behavior*. Princeton, N.J.:Princeton University Press.

Crook, J.H. 1965. The adaptive significance of avian social organisations. *Symposium of the Zoological Society of London* 14:181-218.

Crook, J.R. 1984. Barn swallow (*Hirundo rustica*) social behavior: Nest attendance, aggression, and sexually selected infanticide. Ph.D. dissertation, State University of New York, College of Environmental Science and Forestry, Syracuse, N.Y.

Crook, J.R. and W.M. Shields. 1985. Sexually selected infanticide by adult male barn swallows. *Animal Behaviour* 33:754-761.

Crook, J.R. and W.M. Shields. 1987. Non-parent nest attendance in the barn swallow: Helping or harrassment? *Animal Behaviour*, in press.

Darling, F.F. 1938. *Bird flocks and the breeding cycle*. Cambridge, England:Cambridge University Press.

Delacour, J. 1947. *Birds of Malaysia*. New York:MacMillen Co.

Dement'ev, G.P. 1968. *Birds of the Soviet Union*, Vol. 6. Jerusalem:Israel Program for Scientific Translations.

DeSteven, D. 1978. The influence of age on the breeding biology of the tree swallow (*Iridoprocne bicolor*). *Ibis* 120:516-523.

Dow, H. and S. Fredga. 1983. Breeding and natal dispersal of the goldeneye, *Bucephala clangula*. *Journal of Animal Ecology* 53:679-692.

Dyrcz, A. 1984. Breeding biology of the mangrove swallow *Tachycineta albilinea* and the gray-breasted martin *Progne chalypea* at Barro Colorado Island, Panama. *Ibis* 126:59-66.

Edson, J.M. 1943. A study of the violet-green swallow. *Auk* 60: 396-403.

Emlen, J.T. 1952. Social behavior in nesting cliff swallows. *Condor* 54:177-199.

Emlen, J.T. 1954. Territory, nest building and pair formation in the cliff swallow. *Auk* 71:16-35.

Emlen, S.T. and N.J. Demong. 1975. Adaptive significance of synchronized breeding in a colonial bird: A new hypothesis. *Science* 188:1029-1031.

Emlen, S.T. and L.W. Oring. 1977. Ecology, sexual selection, and the evolution of mating systems. *Science* 197:215-223.

Erskine, A.J. 1979. Man's influence on potential nesting sites and populations of swallows in Canada. *Canadian Field-Naturalist* 93:371-377.

Freer, V.M. 1979. Factors affecting site tenacity in New York bank swallows. *Bird Banding* 50:349-357.

Fretwell, S.D. 1972. *Populations in a seasonal environment.* Princeton, N.J.:Princeton University Press.

Fretwell, S.D. and H.L. Lucas, Jr. 1970. On territorial behavior and other factors influencing habitat distribution in birds. I. Theoretical development. *Acta Biotheoretica* 19: 16-36.

Friedmann, H., L. Griscom and R.T. Moore. 1950. *The birds of Mexico.* Berkeley, California:Cooper Ornithological Club.

Ghiselin, M.T. 1974. *The economy of nature and the evolution of sex.* Berkeley, CA:University of California Press.

Gould, J. 1967. *Birds of Australia.* London:Methuen and Co.

Greenwood, P.J. and P.H. Harvey. 1982. The natal and breeding dispersal of birds. *Annual Review of Ecology and Systematics* 13:1-21.

Hails, C.J. 1984. The breeding biology of the pacific swallow *Hirundo rustica* in Malaysia. *Ibis* 126:198-211.

Hall, B.P. and R.E. Moreau. 1970. *An atlas of speciation in African passerine birds.* London:Trustees of the British Museum.

Hamilton, W.D. 1964. The genetical evolution of social behaviour. I and II. *Journal of Theoretical Biology* 7:1-52.

Hamilton, W.D. 1971. Geometry for the selfish herd. *Journal of Theoretical Biology* 31:295-311.

Harrison, C. 1982. *An atlas of birds in the western palearctic.* Princeton, New Jersey:Princeton University Press.

Haverschmidt, F. 1968. *Birds of Surinam.* Edinburgh:Oliver and Boyd.

Hill, J.R., III. 1982. Reproductive success in the barn swallow (*Hirundo rustica*): Effects of nest reuse and age. M.S. thesis, Pennsylvania State University, State College, Penn.

Holroyd, G.L. 1975. Nest site availability as a factor limiting population size of swallows. *Canadian Field-Naturalist* 89: 60-64.

Hoogland, J.L. and P.W. Sherman. 1976. Advantages and disadvantages of bank swallow (*Riparia riparia*) coloniality. *Ecological Monographs* 46:33-58.

Horn, H.S. 1968. The adaptive significance of colonial nesting in the Brewer's blackbird (*Euphagus cyanocephalus*). *Ecology* 49:682-694.

Hosking, E. and C. Newberry. 1946. *The swallow.* London:Collins.

Jackson, J.A. and P.G. Burchfield. 1975. Nest-site selection of barn swallows in east-central Mississippi. *American Midland Naturalist* 94:503-509.

Johnston, R.F. and J.W. Hardy. 1962. Behavior of the purple martin. *Wilson Bulletin* 74:243-262.

King, B.F. and E.C. Dickinson. 1975. *A field guide to the birds of southeast Asia.* London:Collins.

Koenig, W.D. and F.A. Pitelka. 1981. "Ecological factors and the role of kin selection in the evolution of cooperative breeding in birds." pp. 261-280. In: R.D. Alexander and D.T. Tinkle, eds. *Natural selection and social behavior.* New York:Chiron Press.

Kuerzi, R.G. 1941. Life history studies of the tree swallow. *Proceedings of the Linnean Society of New York* 52-3:1-52.

Kuzniak, S. 1967. Observations on the breeding biology of the swallow (*H. rustica* L.). *Acta Ornithologica* 10:177-211.

Lack, D. 1954. *The natural regulation of animal numbers*. Oxford, England:Oxford Clarendon Press.

Lack, D. 1966. *Population studies of birds*. Oxford, England:Oxford Clarendon Press.

Lack D. 1968. *Ecological adaptations for breeding in birds*. London: Methuen.

Leffelaar, D. and R.J. Robertson. 1984. Do male tree swallows guard their mates? *Behavioral Ecology and Sociobiology* 16:73-79.

Leffalaar, D. and R.J. Robertson. 1985. Nest usurpation and female competition for breeding opportunities by tree swallows. *Wilson Bulletin* 97:221-224.

Lin, N. and C.D. Michener. 1972. Evolution of sociality in insects. *Quarterly Review of Biology* 47:131-159.

Lohoefener, R.R. 1977. Comparative nesting ecology of solitary and colonial nesting barn swallows in west-central Kansas. M.S. thesis, Fort Hays State University, Fort Hays, Kansas.

Lohoefener, R.R. 1980. Comparative breeding biology and ecology of colonial and solitary nesting barn swallows (*Hirundo rustica*) in east-central Mississippi. Ph.D. thesis, Mississippi State University, Jackson, Miss.

Lohrl, H. and H. Gutscher. 1973. Zur Brutoecologie der Rauschswalbe (*Hirundo rustica*) in einem sudwestdeuschen Dorf. *Journal für Ornithologie* 114:399-416.

Lombardo, M.P. 1984. Relations between breeders and nonbreeders in a presocial species, the tree swallow (*Tachycineta bicolor*). Ph.D. dissertation, Rutgers University, New Brunswick, N.J.

Lombardo, M.P. 1985. Mutual restraint in tree swallows: A test of the tit for tat model of reciprocity. *Science* 227:1363-1365.

Lorenz, K. 1965. *Evolution and modification of behavior*. Chicago:University of Chicago Press.

Lunk, W.A. 1962. The rough-winged swallow *Stelgidopteryx ruficollis* (Vieillot). *Publications of the Nuttal Ornithological Club* 4:1-155.

Mackworth-Praed, C.W. and C.H.B. Grant. 1955. *Birds of eastern and northeastern Africa*. London:Longmans, Green and Co.

Mackworth-Praed, C.W. and C.H.B. Grant. 1973. *Birds of west central and west Africa*. London:Longmans.

Mason, E.A. 1936. Parasitism of bird's nests by *Protocalliphora* at Groton, Massachussetts. *Bird Banding* 7:112-121.

Mason, E.A. 1953. Barn swallow life history data based on banding records. *Bird Banding* 24:91-100.

Mayhew, W.W. 1958. The biology of cliff swallows in California. *Condor* 60:7-37.

Mayr, E. and J. Bond. 1943. Notes on the generic classification of the swallows, Hirundinide. *Ibis* 85:334-341.

Mayr, E. and J.C. Greenway, Jr. 1960. *Peter's checklist of birds of the world*. Vol. IX. Cambridge, Mass.:Museum of Comparative Zoology, Harvard.

McLachlan, G.R. and R. Liversidge. 1963. *Robert's birds of South Africa*. Capetown, South Africa:Cape and Transvaal Printers.

Mead, C.J. 1979a. Colony fidelity and interchange in the sand martin. *Bird Study* 26:99-106.

Meyer de Schauensee, R. 1964. *Birds of Columbia*. Narberth, Pennsylvania:Livingston Publishing Co.

Meyer de Schauensee, R. 1984. *The birds of China*. Washington, D.C.: Smithsonian Institution.

Meyer de Schauensee, R. and W.H. Phelps, Jr. 1984. *Birds of Venezuela*. Princeton, New Jersey:Princeton University Press.

Moller, A.P. 1983. Breeding habitat selection in the swallow *Hirundo rustica*. *Bird Study* 30:134-142.

Moller, A.P. 1984. Parental defence of offspring in the barn swallow. *Bird Behaviour* 5:110-117.

Moller, A.P. 1985a. The social behavior of swallows (*Hirundo rustica*) during the breeding season. Ph.D. dissertation, Aarhus University, Aarhus, Denmark.

Moller, A.P. 1985b. Mixed reproductive strategy and mate guarding in a semi-colonial passerine. *Behavioral Ecology and Sociobiology* 17:401-408.

Moller, A.P. 1987a. Advantages and disadvantages of coloniality in the swallow, *Hirundo rustica*. *Animal Behaviour* 35:819-831.

Moller, A.P. 1987b. Intraspecific nest parasitism and anti-parasite behaviour in swallows, *Hirundo rustica*. *Animal Behaviour* 35:247-254.

Moreau, R.E. 1939. Numerical data on African birds' behaviour at the nest: *Hirundo s. smithii* Leach, the wire-tailed swallow. *Proceedings Zoological Society London* 109:109-125.

Moreau, R.E. 1940. The biology of *P. holomelaena* and the nature of the data. *Ibis* 14:234-244.

Moreau, R.E. 1947. Relations between number in brood, feeding-rate and nestling period in nine species of birds Tanganyika territory. *Journal of Animal Ecology* 16:205-209.

Moreau, R.E. and W.M. Moreau. 1939. Observations of swallows and house martins at the nest. *British Birds* 33:146-151.

Moss, W.W. and J.H. Camin. 1970. Nest parasitism, productivity and clutch size in purple martins. *Science* 168:1000-1003.

Muldal, A., H.L. Gibbs and R.J. Robertson. 1985. Preferred nest spacing of an obligate cavity-nesting bird, the tree swallow. *Condor* 87:356-363.

Murray, B.G., Jr. 1979. *Population dynamics: Alternative models*. New York:Academic Press.

Murray, B.G., Jr. 1984. A demographic theory on the evolution of mating systems as exemplified by birds. *Evolutionary Biology* 18:71-140.

Murray, J.J. 1962. Barn swallow nesting in the mouth of a cave. *Auk* 79:117.

Myers, G.R. 1977. The evolution of helping behavior in barn swallows (*Hirundo rustica*). M.S. thesis, Kent State University, Kent, Ohio.

Myers, G.R. and D.W. Waller. 1977. Helpers at the nest in barn swallows. *Condor* 59:311-316.

Orians, G.H. 1969. On the evolution of mating systems in birds and mammals. *American Naturalist* 103:589-603.

Orr, R.T. 1970. *Animals in migration.* London:MacMillan.

Park, P. 1981a. Results from a nesting study of welcome swallows in southern Tasmania. *Corella* 5:85-90.

Park, P. 1981b. A colour-banding study of welcome swallows breeding in southern Tasmania. *Corella* 5:37-41.

Petersen, A.J. 1955. The breeding cycle in the bank swallow. *Wilson Bulletin* 67:235-286.

Pforr, M. and A. Limbrunner. 1982. *The breeding birds of Europe.* London:Croom Helms.

Pizzey, G. 1980. *A field guide to the birds of Australia.* Princeton, New Jersey:Princeton University Press.

Plotkin, H.C. and F.J. Odling-Smee. 1981. A multiple-level model of evolution and its implications for sociobiology. *Behavioral and Brain Sciences* 4:225-268.

Pulliam, H.R. and C. Dunford. 1980. *Programmed to learn: An essay on the evolution of culture.* New York:Columbia University Press.

Pulliam, H.R. and T. Caraco. 1984. "Living in groups: Is there an optimal group size?" pp. 122-147. In: J.R. Krebs and N.B. Davies, eds. *Behavioural ecology. An evolutionary approach.* Oxford:Blackwell Scientific Publications.

Redmond, R.L. and D.A. Jenni. 1982. Natal philopatry and breeding area fidelity of long-billed curlews (*Numenius americanus*): Patterns and evolutionary consequences. *Behavioural Ecology and Sociobiology* 10:277-279.

Ridgely, R.S. 1981. *A guide to the birds of Panama.* Princeton, New Jersey:Princeton University Press.

Rowley, I. 1983. "Remating in birds." pp. 331-360. In: P. Bateson, ed. *Mate choice.* London:Cambridge University Press.

Samuel, D.E. 1971a. Field methods for determining the sex of barn swallows (*Hirundo rustica*). *Ohio Journal of Science* 71:125-128.

Samuel, D.E. 1971b. The breeding biology of barn and cliff swallows in West Virginia. *Wilson Bulletin* 83:284-301.

Sheppard, C.D. 1977. Breeding in the tree swallow, *Iridoprocne bicolor*, and its implications for the evolution of coloniality. Ph.D. dissertation, Cornell University, Ithaca, New York.

Shields, W.M. 1982. *Philopatry, inbreeding, and the evolution of sex.* Albany:State University of New York Press.

Shields, W.M. 1984a. Barn swallow nobbing: Self-defence, collateral kin defence, group defence, or parental care? *Animal Behaviour* 32:132-148.

Shields, W.M. 1984b. Factors affecting site fidelity in Adirondack barn swallows *Hirundo rustica*. *Auk* 101:780-789.

Shields, W.M. 1987a. "Dispersal and mating systems: Investigating their causal connections." pp. 3-24. In: B.D. Chepko-Sade and Z.T. Halpin, eds. *Patterns of dispersal among mammals and their effect on the genetic structure of populations.* Chicago:University of Chicago Press.

Shields, W.M. 1987b. Sex and adaptation. Ch. 15. In: R.E. Michod and B.R. Levin, eds. *The evolution of sex: An examination of current ideas.* Sunderland, Mass.:Sinauer Associates.

Shields, W.M. and J.R. Crook. 1987. Barn swallow coloniality: A net cost for group living in the Adirondacks? *Ecology,* in press.

Sikes, P.J. and K.A. Arnold. 1984. Movement and mortality estimates of cliff swallows in Texas. *Wilson Bulletin* 96:419-425.

Skutch, A.F. 1952. Life history of the blue and white swallow. *Auk* 69:342-406.

Skutch, A.F. 1960. Life histories of Central American birds II. Berkeley, California:Cooper Ornithological Society.

Skutch, A.F. 1981. *New studies of tropical American birds.* Cambridge, Mass.:Publications Nutall Ornithological Club, No. 19.

Slobodchikoff, C.N. 1984. "Resources and the evolution of social behavior." pp. 227-251. In: P.W. Price, C.N. Slobodchikoff and W.S. Gaud, eds. *A new ecology: Novel approaches to interactive systems.* New York:John Wiley and Sons.

Smith, J.M. and H.B. Graves. 1978. Some factors influencing mobbing behavior in the barn swallows (*Hirundo rustica*). *Behavioral Biology* 23:355-372.

Snapp, B.D. 1973. The occurrence of colonial breeding in the barn swallow (*Hirundo rustica*) and its adaptive significance. Ph.D. dissertation, Cornell University, Ithaca, New York.

Snapp, B.D. 1976. Colonial breeding in the barn swallow (*Hirundo rustica*) and its adaptive significance. *Condor* 78:471-480.

Snell, M.L. 1969. Notes on the breeding of the blue swallow. *Ostrich* 40:65-74.

Speich, S.M., H.L. Jones and E.M. Benedict. 1985. Review of the natural nesting of the barn swallow in North America. *American Midland Naturalist* 115:248-254.

St. Peters, D. 1973. "The swallows." pp. 182-191. In: B. Grzimek, ed. *Animal life encyclopedia.* Vol. 9. Birds III. New York:Van Nostrand Reinhold.

Stamps, J.A. 1987. Conspecifics as cues to territory quality: A preference of juvenile lizards (*Anolis aeneus*) for previously used territories. *American Naturalist* 129:629-642.

Stoner, D. 1936. Studies on the bank swallow, *Riparia riparia riparia* (Linneaus). *Roosevelt Wildlife Bulletin* 4:126-233.

Stutchbury, B.J. and R.J. Robertson. 1985. Floating populations of female tree swallows. *Auk* 102:651-654.

Thompson, A.L. 1964. "Swallow." pp. 790-793. In: A.L. Thompson, ed. *A new dictionary of birds.* London:Nelson.

Thorson, G. 1950. Reproductive and larval ecology of marine bottom invertebrates. *Biological Reviews* 25:1-45.

Turner, A.K. 1984. Nesting and feeding habits of brown-chested martins in relation to weather conditions. *Condor* 86:30-35.

Vansteenwegen, C. 1982. Longevite et structure des nid d'Hirondelles des cheminee (*Hirundo rustica*). *Aves* 19:247-251.

Vietinghoff-Riesch, A.F. 1955. *Die Rauchscwalbe*. Berlin, West Germany:Duncker and Humblot.

von Haartman, L. 1957. Adaptation in hole-nesting birds. *Evolution* 11:294-347.

Waltz, E.C. 1983. On tolerating followers in information centers, with comments on testing the adaptive significance of coloniality. *Colonial Waterbird* 6:31-36.

Ward, P. and A. Zahavi. 1973. The importance of certain assemblages of birds as 'information centres' for food finding. *Ibis* 115:517-534.

Waser, P.M. and W.T. Jones. 1983. Natal philopatry among solitary mammals. *Quarterly Review of Biology* 58:355-390.

Wetmore, A., R.F. Pasquier and S.L. Olson. 1984. *The birds of the Republic of Panama*. Washington, D.C.:Smithsonian Institution Press.

Whistler, H. and N.B. Kinnear. 1949. *Popular handbook of Indian birds*. London:Gurney and Jackson.

White, C.M.N. 1947. The pearl-breasted swallow in northern Rhodesia. *Ibis* 89:359-360.

Whitham, T.G. 1980. The theory of habitat selection examined and extended using *Pemphigus* aphids. *American Naturalist* 115:449-466.

Wilkinson, G.S. and G.M. English-Loeb. 1982. Predation and coloniality in cliff swallows (*Petrochelidon pyrrhonota*). *Auk* 99:459-467.

Williams, G.C. 1966. *Adaptation and natural selection*. Princeton, N.J.: Princeton University Press.

Wilson, E.O. 1971. *Insect societies*. Cambridge, Mass.:Harvard University Press.

Wilson, E.O. 1975. *Sociobiology*. Cambridge, Mass.:Harvard University Press.

Withers, P.C. 1977. Energetic aspects of reproduction by the cliff swallow. *Auk*:94:718-725.

Wittenberger, J.F. and J.L. Hunt, Jr. 1985. "The adaptive significance of coloniality in birds." pp. 1-75. In: D.S. Farner, J.R. King and K.C. Parkes, ed. *Avian biology*. vol. III. New York:Academic Press.

Wolinski, R.A. 1981. Pair-formation, nest-site selection and territory in the barn swallow (*Hirundo rustica*). M.S. thesis, Eastern Michigan University.

Wynne-Edwards, V.C. 1962. *Animal dispersion in relation to social behaviour*. Edinburgh:Oliver and Boyd.

Yoshimura, J. and W.M. Shields. 1987. Probabilistic optimization of phenotype distributions: A general solution for the effects of uncertainty on natural selection? *Evolutionary Ecology* 1, in press.

Chapter 10

TERRITORY QUALITY: KEY DETERMINANT OF FITNESS IN THE GROUP-LIVING GREEN WOODHOOPOE

J. David Ligon and Sandra H. Ligon

Department of Biology
University of New Mexico 87131

I. INTRODUCTION

Cooperative breeding in birds is now widely recognized as an outcome of certain environmental, demographic, and social features (e.g.,
Koenig and Pitelka 1981; Ligon 1981, 1983a; Emlen 1982; Rowley 1983;
Wolfenden and Fitzpatrick 1984; Zack and Ligon 1985a,b; Stacey and
Ligon 1987). With regard to the first of these, limitation of some resource, restricted either in space or time, is thought to promote the evolution of cooperative breeding Emlen (1982). For many species a key
demographic character, a low adult mortality, has been identified as a
factor promoting cooperative breeding, specifically when annual reproduction of young birds exceeds the mortality of breeders (Ricklefs 1975).
In an environment where nearly all territories are continuously occupied,
this creates a queue of individuals waiting for a breeding opportunity in
the form of a territorial vacancy (Wiley and Rabenald 1984). When such
a vacancy appears, competition for it often is intense; in several species,
unisexual "teams" compete for the space (Ligon and Ligon 1978, 1983;
Koenig 1981). Finally, for some species, cooperation *per se* is of great
importance to each individual and this may have led to the prolongation
of family ties and the evolution of mutualistic interactions, including
cooperative breeding (Ligon 1983a,b; Stacey and Ligon 1987).

In some important respects, the green woodhoopoes (*Phoeniculus
purpureus*) that we studied near Lake Naivasha in Kenya (Ligon and
Ligon 1978, 1982, 1983) appear to conform to the typical pattern of
avian cooperative breeding, if such a thing really exists. On our primary
study site, Morendat Farm, the woodhoopoes exhibit: (1) stability in
number of territories and thus breeding individuals, since only one pair
breeds in each territory, regardless of group size; (2) team competition
for breeding vacancies; and (3) dependence on a critical limited resource, the roost cavity. In other respects, however, these birds do not
"follow the script" for avian cooperative breeding. Specifically, (1) annual mortality is about twice that of some other well-studied cooperative breeders; (2) reproductive effort per year is extensive, with up to
three broods produced in a six-month period if environmental conditions
are suitable (cf. Woolfenden and Fitzpatrick 1984, p. 270-272); (3) turnover of family lineages is high; (4) turnover of territories appears to be
high (i.e. territories come into and go out of existence with some frequency); (5) both sexes are highly philopatric, with more females than
males inheriting breeding status on their natal territories (cf. Greenwood
1980); and (6) merger of unrelated birds of the same sex to form new
social units occurs frequently in females and occasionally in males.

We believe that virtually all of the most interesting life history attributes of green woodhoopoes, both those that conform to the general
pattern of avian cooperative breeding and those that do not, are based in
large part on one key requirement, the roost cavity, and that this dependence is largely responsible for the great interterritory variation
observed in natality and mortality (see below).

A. Overview of the Woodhoopoe Project and
   Structure of the Study Population

We began our study of green woodhoopoes in mid-1975 and ended it in mid-1984. Table I provides an outline of the periods of field work. We ultimately marked a total of 386 woodhoopoes in 35 flocks, most of which (269) were banded on or adjacent to our primary study site, Morendat Farm. By January 1982, at the end of the main body of our project (Table I), 93 percent of the green woodhoopoes on Morendat Farm were of known parentage. This kind of coverage was possible only because mortality was high (i.e., most adult birds present at the onset of the study and thus of unknown parentage soon disappeared and were replaced), and because most woodhoopoes of both sexes are extremely sedentary.

The woodhoopoe groups that we marked and studied are part of a semi-isolated population of birds that occupy the yellow-barked acacia (*Acacia xanthophloea*) woodland ringing Lake Naivasha in the Rift Valley of Kenya. Based on the suitable habitat available around the lake, and knowledge of territory sizes, we estimate that in 1975 about 105 social units made up the Lake Naivasha population of green woodhoopoes. Since that time the number of birds undoubtedly has declined coincident with the additional destruction by man of acacia woodland. Thus the 35 groups that we marked represent a sizable fraction of this population of woodhoopoes.

TABLE I. Periods of field work conducted on green woodhoopoes during the decade 1975-1984.

Field Work	Investigators
July 1975 - June 1976	J.D. Ligon, S.H. Ligon
January 1977	JDL, SHL
June - August 1977	JDL, SHL
January 1978	JDL, SHL
June - August 1978	JDL, SHL
January 1979	JDL
June - December 1979	JDL, SHL, S. Zack
January - December 1980	S. Zack, D.C. Cole
June - August 1981	JDL, SHL
August - December 1981	S. Zack
January 1982	JDL, S. Zack
June - July 1984	JDL, J.C. Bednarz

II. THE ROOST CAVITY: A KEY CRITICAL RESOURCE

Green woodhoopoes in our study area *always* roost either in a tree cavity or, lacking an available cavity, under loose dead bark of acacias. Most of the birds that we banded were captured at roost holes. Although we have followed hundreds of birds to roost sites, we have never recorded a woodhoopoe spending the night in the open. One aspect of this dependence on cavities is the frequent occurrence of several birds in one hole. Often, because of pronounced sexual dimorphism in size, a cavity holds only flock members of one sex. For example, we have recorded as many as eight females roosting together. This unisexual "dormitory" pattern of roosting has an important effect on demography and dispersal: When a predator captures woodhoopoes at the roost, it often removes most or all flock members of one sex, thus opening the territory to immigration by birds of the decimated sex (Ligon and Ligon 1982).

A. Roost Cavities and High Mortality

Although woodhoopoes, without exception, use cavities or other cover for roosting, a large cost apparently is associated with this behavior. All of our evidence suggests that most predation on these birds occurs at the roost holes. The woodhoopoes exhibit a higher rate of mortality than most other well-studied cooperative breeders--about 40 percent per year for males and 30 percent per year for females (Ligon 1981; Table II).

In most avian species, immatures survive far less well than do adults. This usually is as true of group-living species as it is of more typical birds (e.g., yearling Florida scrub jays, *Aphelocoma c. coerulesences*, 34% per year, adults 82% [Woolfenden and Fitzpatrick 1984]; first-year acorn woodpeckers in California 57%, adult breeders 77% [Koenig et al. 1986]). In contrast, in the woodhoopoes rates of disappearance of birds of different age and/or social status are almost identical within each sex. We believe that this provides additional evidence for the importance of nocturnal, roost-site predation. That is, increased age and experience provide no benefit if predation occurs primarily while the birds are roosting. For example, since males of all ages frequently roost together, all are vulnerable when in the roost cavity. We have occasionally recorded the overnight disappearance of one or two birds, at the same time that their roost partners have lost their tails. In such cases we assume that the predator was unable to reach all birds in the cavity and pulled the tail out of the fortunate survivor that was just out of reach. (When woodhoopoes in cavities are threatened, they point the tail toward the entrance, elevate their rump feathers and produce a strong odor via a substance released from the oil gland [Ligon and Ligon 1978]).

Thus the cavity-roosting habit appears to be risky. Our evidence suggests that major nocturnal predators are driver ants, Tribe Dorylini, and large-spotted genets (*Genetta tigrina*). Birds occupying cavities in weak wood, or that are shallow and/or have large entrance holes, are

TABLE II.   Mean percent minimum annual[1] survival in 22-27 flocks of individually marked green woodhoopoes.

	Males			Females		
	Breeders	Helpers	Imma-tures[2]	Breeders	Helpers[3]	Imma-tures[2]
% Survival	62	60	61	70	67	73
Range	(50-82)	(30-81)	(50-79)	(57-86)	(53-78)	(60-78)

[1]Based on six years, January 1976 - January 1982.
[2]Birds from about 1 to 6 months of age at the beginning of their first calendar year.   The survivorship data are based on the first full year (January-January) of life, thus nestlings and many individuals in juvenal plumage are not included.
[3]A few female helpers undoubtedly dispersed and were not relocated.

vulnerable to genets and possibly to wild and feral cats (*Felis libyca* and *F. domesticus*). All cavities are susceptible to invasion by raiding driver ants.

B. Dependence on Cavities Reflects
a Physiological Character

It seems paradoxical that green woodhoopoes should inevitably roost in cavities where they apparently are so susceptible to capture. Earlier, we suggested that perhaps these birds are unable to remain endothermic in the absence of insulation afforded by the roost cavities (Ligon and Ligon 1978). However, we have only recently been able to test this suggestion. We measured metabolic rates and body temperatures (T_B) of three green woodhoopoes and discovered that the birds did indeed appear to be incapable of maintaining normal T_B's at the relatively moderate ambient temperature (T_A) of 19°C, and that their behavioral responses to such nocturnal T_A's were inappropriate; i.e., the birds increased their activity as T_A declined, which increased the rate of heat loss (Ligon et al. submitted). At 2400 h and at a T_A of 19°C, T_B's of the three woodhoopoes dropped from a daytime level of 40-42°C to 33-36°C.
In short, it appears that the critical limiting resource for green woodhoopoes is the roost cavity and, moreover, that this dependence on cavities is imposed on the birds by their physiology. Thus, contemporary green woodhoopoes apparently must avoid the low (7-10°C) nocturnal temperatures characteristic of the study site by roosting in tree cavities. The severe interspecific competition for cavities (honeybees,

rodents, other birds) means that safe cavities vary in number both spatially and temporally and thus usually are scarce as well as critically important to the birds. Because of the frequent takeover of cavities by other species, a key element of territory quality is number of potential back-up or secondary roost sites.

C. Philopatry by Both Sexes

In many kinds of birds, including most cooperative breeders that have been studied, a sexual difference exists in dispersal patterns and distances, with females being the "dispersing sex" (e.g., Greenwood 1980). Green woodhoopoes do not follow this rule. Although a few females are known to have moved several territories, whereas we have recorded no male as having done so, most females, like males, breed either in or close to their natal territory (Table III). In fact, more females than males breed in their natal territories. This pattern appears to be unique among the cooperatively breeding birds studied to date.

We believe that the extreme philopatry shown by both male and female woodhoopoes is directly related to their dependence on roost cavities. Most woodhoopoes of each sex move no farther than to an adjacent territory. Woodhoopoes occupy very large territories and flocks commonly intrude into neighboring territories when the territory owners are elsewhere. A conspicuous activity of an intruding flock is exploration of the owners' roost cavities and other tree holes in the territory. As a result of this kind of behavior, an individual woodhoopoe presumably learns the location and quality of cavities in surrounding territories. This information, we believe, is critical to a bird's decision to move or not move into a nearby territory when the opportunity arises (i.e., when a bird or birds of its sex die, creating a breeding vacancy). As will be considered below, the quality and quantity of available and potential roost sites within a given territory probably is the key factor determining the overall lifespan and reproductive success of the birds occupying it.

TABLE III. Dispersal by sex from natal to breeding territory.

	Bred in Natal Territory	Moved[1] (territory no.)				
		1	2	3	6	13[2]
Females	18	14	1	2	2	1
Males	9	20	4			

[1]Because many birds move as twosomes, the younger, beta bird may or may not actually have attained breeding status and reproduced (Ligon and Ligon 1983).
[2]Estimated.

D.Inbreeding

An inevitable outcome of extreme philopatry by both sexes is the presence of neighborhoods of territories that contain close relatives. For example, in 1981, 0.5 genetic relationships among *breeding* birds were common in our study population (Fig. 1). This means, of course, that many young birds produced in this neighborhood were related at the first cousin (r = 0.125) level. The relatedness of young birds in neighboring flocks, together with the conservative pattern of dispersal, suggests that relatives from different flocks should sometimes pair and breed. (In contrast, all of our evidence suggests that close relatives in the same flock, such as parents and offspring or siblings, never mate [Ligon 1987]).

To investigate matings between relatives, JDL next visited the study site in mid-1984 (Table I), by which time a few of the birds hatched in 1980 and 1981 should have attained breeding status. We found that during the two and one-half year interval, January 1982-June 1984, survival of breeders had been relatively high, with only two 1980-hatched woodhoopoes and no 1981-hatched birds having attained breeding status. In fact, most birds of the 1981 age cohort had disappeared.

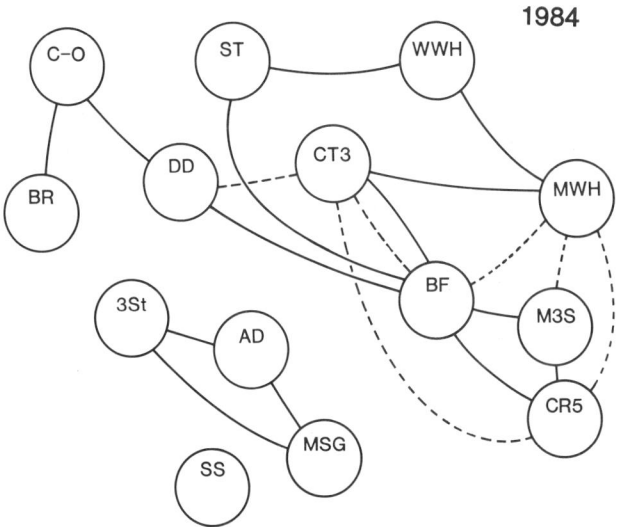

Fig. 1.   Genetic ties among **breeding** *woodhoopoes on Morendat Farm in 1981. Solid lines connect territories containing breeders related at the 0.5 level, dashed lines where breeders are related at the 0.25 level. The "p" indicates probable relationships. In these latter cases one breeder was hatched prior to the onset of the study. For example, the breeding female in the CR5 territory came from the BF territory. Her probable mother in the BF territory was the certain mother of female breeders in the CT3 and M3S territories.*

However, relatedness among breeders in neighboring territories again was high and in two flocks the breeding pair was closely related. In the C-O flock, the relationship of the breeders was nephew-aunt (r = 0.25), while in the BF territory it was somewhat less (0.188). In the first case, the male simply moved "next door" into his father's natal territory, whereas in the second case the movement pattern was more circuitous and thus more interesting (Fig. 2). Male B/WR was hatched in the CT3 territory in early 1980; in early 1981 he and his father deserted Female BR (B/WR's mother) in the CT3 territory and moved into the ST territory, one of the best territories on the study area (see below). Male B/WR was still a non-breeding helper in the ST territory in January 1982. By June 1984, he was a successful breeder in the BF territory, and was mated to Female G/WR (a half-aunt plus second cousin, r = 0.188).

Given the conservative, short-distance movement by both males and females, the fact that close relatives occasionally form pair bonds is not surprising. What is surprising is the fact that we have recorded it so infrequently. Our limited data do not suggest that relatives more distantly related than parents and offspring, or siblings, avoid mating with each other. Rather, the infrequent occurrence of such matings appears to be due to the high annual rate of mortality, together with the generally

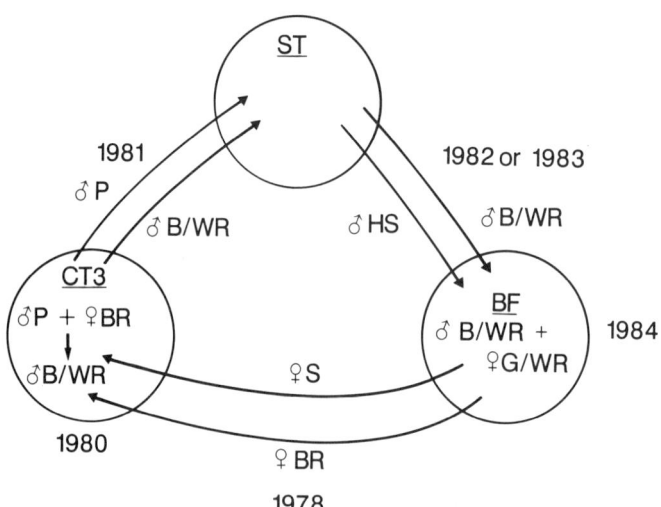

Fig. 2. Schematic illustration of the events leading to the pairing of close relatives, male B/WR and female G/WR. In 1978, female BR emigrated with a sibling (S) to the CT3 territory. In 1980, female BR and her mate, male P, produced one offspring, male B/WR. In 1981, male P and his son (B/WR) emigrated to the superior ST territory. In 1982 or 1983, B/WR moved with a male half sibling (HS) to the BF territory, and there mated with female G/WR, his mother BR's half-sister and probable cousin.

long period of time between hatching and first breeding (Table IV). Most close relatives in different flocks do not have the opportunity to breed with each other simply because mortality typically intervenes. When this is not the case, close relatives (e.g., $r = 0.25$) may form pair bonds and breed, as described above.

E. Evidence Suggesting Inbreeding Depression

Although a consideration of inbreeding at the population level is outside the scope of this chapter, certain empirical data suggest that inbreeding depression may exist in the green woodhoopoes that we studied. One result of inbreeding in both wild and domestic birds is reduced hatchability (Romanoff 1972; van Noordwijk and Scharloo 1981). In the woodhoopoes the percentage of unhatched eggs is high and strikingly similar regardless of clutch size or year. One or more eggs did not hatch in 82 percent (49/60) of the nests for which we know both original clutch size and number of eggs hatched. Of the 32 unhatched eggs that we examined from 22 nests, more than half contained obvious embryos that had died at various stages of development up to the point of hatching (Table V); thus many of the unexamined eggs that did not hatch probably also were fertile. Separation of the data by year (Table VI) reveals a similar pattern: About one-third of all eggs laid do not hatch. We know of no other species for which hatching failure, apart from predation, is so consistently high.

In part, we interpret the philopatry and hatching failure in a way similar to Bengtsson (1978): The decision to disperse and avoid inbreeding, or to be philopatric and risk inbreeding depression, should be dependent on the relative costs of inbreeding versus dispersal. For the woodhoopoes we suggest that the costs of long-distance dispersal are so great that the risk of inbreeding below the level of parent and offspring or siblings is accepted. Moreover, although there may be a cost of inbreeding in the form of reduced hatchability of eggs, it is not apparent that this cost is particularly important. Starvation of nestlings is almost nonexistent (Table VI), possibly as a result of the reduction in brood size brought about by the presumed inbreeding, and a single pair of birds may rear up to three broods during the potential breeding period June–December (Ligon and Ligon 1978). As a result of the high survivorship of

TABLE IV. Age at first breeding.[1]

	Males	Females
Mean	3.26	3.93
N	19	15
Range[1]	1.5–5.0	1.5–7.0

[1] In years.

nestlings and the multi-brooded reproductive pattern of woodhoopoes, loss of offspring via inbreeding depression may exert little effect on the fitness of breeders. That is, other factors in a sense compensate for the mortality of embryos. This seems to parallel the situation in another naturally inbred population where inbred individuals or pairs composed of relatives suffer no reduction in reproductive success, as measured by number of recruited adults per egg (van Noordwijk and Scharloo 1981).

To summarize this section, we suspect that hatching failure is a manifestation of inbreeding depression resulting from the extreme philopatry exhibited by green woodhoopoes, in addition to the small, semi-isolated population at Lake Naivasha, and that the philopatry, in turn, is driven by the critical dependence of these birds on safe roost cavities.

TABLE V. Clutch size and unhatched eggs in green woodhoopoe nests[1] in 1975, 1978-81.

| | | | | | Number Not Hatched | | |
Clutch Size	No. Nests	No. Eggs	No. Hatched	Percent Hatched	With Embryo	Without Embryo[2]	Contents Unknown
2	6	12	8	67	1	2	1
3	29	87	54	62	9	3	21
4	24	96	61	64	7	10	18
5	1	5	4	80			1
Totals	60	200	127	(64)	17	15	41

[1]Only nests where normal hatching took place.
[2]No embryo discernible in eggs incubated full term.

TABLE VI. Hatching and fledging success of green woodhoopoe eggs.[1]

Year	No. Nests	No. Eggs	No. Egg Hatched	Percent Hatched	No. Young Fledged	Percent Fledged
1975	6	20	13	65	12	92
1978	4	14	9	64	9	100
1979	6	21	13	62	12	92
1980	7	22	15	68	14	93
1981	11	38	25	66	24	96
Totals	34	115	75	65	71	95

[1]Only nests where number of eggs laid, number hatched, and number fledged known.

III.   THE DEMOGRAPHIC PATTERNS OF GREEN WOODHOOPOES
        AND THE EVOLUTION OF RECIPROCITY

Although we have presented our view of the relationship between high mortality and the existence of reciprocity in the woodhoopoes (Ligon 1983; Ligon and Ligon 1983), here we make a few points not previously emphasized.

Green woodhoopoes exist in a social environment composed of groups of cooperating individuals. The outcome of most intergroup competitive interactions is dependent on relative sizes of the competing groups. Larger groups tend to dominate smaller ones, and very small groups (2-4) often cannot successfully defend territorial space if neighboring groups are numerically superior. Thus alliances are a critical component of an individual woodhoopoe's existence. This is most clearly seen in the group emigration of two or more same-sex individuals to a breeding vacancy in a nearby territory (Ligon and Ligon 1978, 1982, 1983; see also Koenig 1981). Team emigration greatly increases the chances that the members will be able to acquire and retain possession of the new territory in the face of competition and aggression by neighboring groups. Thus present-day woodhoopoes occupy a social environment where cooperative alliances are critically important. Typically, unisexual groups of allies are relatives.

However, the very high mortality of these birds not infrequently leads to a situation where an individual loses its same-sex relatives, or in the case of females, a bird might emigrate alone to a new area. Such single birds are at an overwhelming competitive disadvantage, and it appears that allying with an unrelated individual is a sound and commonly employed strategy (Table VII). In short, formation of teams by unrelated birds, particularly females, is not a rare event. However, we have recorded such alliances only among females acquiring low-quality territories. The more productive territories tend to be passed from parent to offspring (see below). Thus, formations of teams composed of unrelated birds clearly is one kind of strategy for procurement of breeding space and status. Ligon and Ligon (1983) describe other kinds of personal benefits derived from the development of alliances by unrelated, same-sex individuals.

TABLE VII. Composition of unisexual "teams" of two or more birds immigrating to a territory, 1975-1981.

	Kin	Non-kin	?
Male	14	4	2
Female	8	14	6

Summary: 18 of 40 were composed of non-relatives.

IV. TERRITORY QUALITY: TEMPORAL STABILITY
AND PERSISTENCE OF LINEAGES

If suitable roost holes are both critical and scarce, availability of high-quality cavities should be the single most important feature of territory quality for green woodhoopoes. In this section we look at temporal stability of territories, reproductive success in the different territories, and the persistence of family lineages. If number of occupied territories remains about the same over time, this might suggest habitat saturation, at least locally. However, constancy in number of territories occupied does not in itself indicate stability in flock lineages. It has been suggested that territories, especially superior ones, are likely to be passed on from generation to generation within family lineages, and that "reserving" breeding sites for descendants is one important aspect of territorial behavior in cooperatively breeding birds (Brown 1974). Here we describe a means of assessing territory quality over time, based on the fates of birds occupying specific territories. It appears that variation in territory quality is largely responsible for the great differences seen in lifetime fitness among green woodhoopoes that have attained breeding status.

A. Stability of Territory Numbers Over Time

During the 10 years of this study we found that four of the 13 original established territories on Morendat Farm disappeared (Table VIII). During this same interval, five new territories were initiated, one of which lasted only a few months. Thus no net change in number of territories occurred over the 10 years, despite the great fluctuation in the number of woodhoopoes occupying the territories considered (Table VIII).

B. Persistence of Family Lineages

Because territories are restricted in number, one possible aspect of territoriality is that such behavior serves to reserve a future breeding site for offspring of the current breeders; i.e., some young birds that remain in their natal territories will inherit breeding status there. Green woodhoopoes of either sex may breed within their natal territories. Thus in this species, inheritance of breeding space and status does occur.

However, at first glance our data do not appear to support strongly the territory inheritance concept (Fig. 3). In 10 of the 14 territories present in 1975, all group members and all of their descendants disappeared in 10 years or less. In other words, in 71 percent of the territories, inheritance of breeding status either did not occur at all, or did so for only one generation before complete turnover of the genetic lineage took place.

TABLE VIII. Temporal stability of territories, 1975-1984.

	'75	'76	'77	'78	'79	'80	'81	'82	'83	'84
Territory										
CO	--
LH	------------------------------x[1]
ST	--
DD	--
WWH	--
BF	--
KG	-----------------------x
H5	----------------------------x
AD	--
SS	--
CR	----------------------------x
MSG	--
CR5	--
3ST	---
BRF	---
CT3	---
WV	---------------------------------
ST_2	------x

[1]Indicates territory ceased to exist. Maximum number of woodhoopoes: (102; 1/78). Minimum number of woodhoopoes: (51; 6/81). Number at last census: (69; 6/84)

C. Territory Quality

Meaningful measures of territory quality are difficult to obtain. Although one can measure various vegetational attributes of a territory, just what the measured features mean to the animals occupying them often is in doubt. Similarly, assessments of food quality and quantity, especially for insectivorous species, may be unreliable. Therfore, we have not emphasized these proximate measures, but instead focus on the ultimate effects of territory quality: natality (N) and mortality (M). Some of our colleagues have cautioned us about circularity in this approach. We make three replies to this concern. First, an element of circularity can be productive if it helps to advance understanding of a system or concept. The development of optimal foraging theory is an example of a productive approach initially based on circularity. Here, we do make the reasonable assumption that the most productive territories, N/M > 1 (those in which the number of woodhoopoes produced is greater than the number dying) are the best ones, from the perspectives of the breeding birds occupying them.

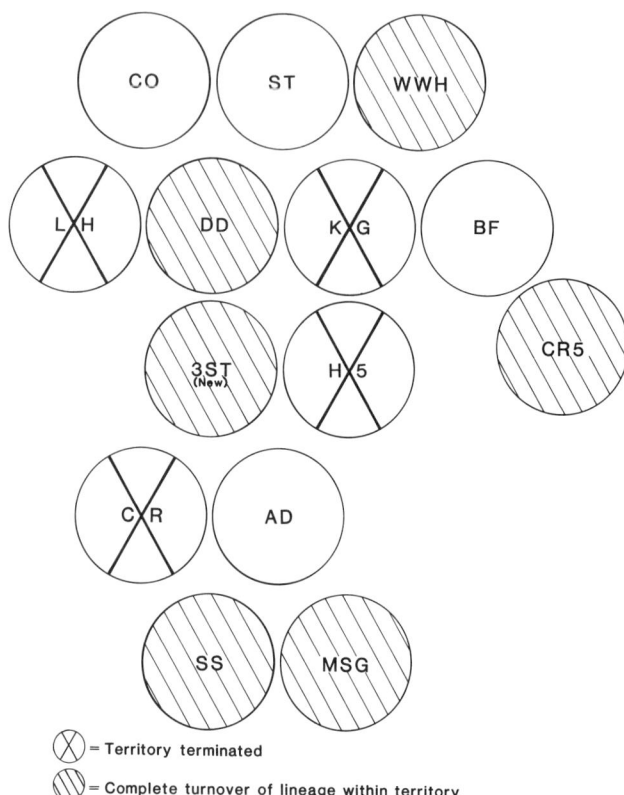

$\bigotimes$ = Territory terminated

$\bigoplus$ = Complete turnover of lineage within territory

Fig. 3. Diagrammatic illustration of the green woodhoopoe territories present on and adjacent to Morendat Farm in 1975. Territories x'ed out ceased to exist during the ensuing decade, while hatched territories had a complete turnover of genetic lineage.

Second, although we assume that territories with an N/M ratio of more than one are more productive than the others, this tells us nothing about either relative natality or mortality across all territories. The ratio is used simply to divide the territories into two categories: net exporters (N/M > 1) or net importers (N/M < 1) of woodhoopoes.

Third, to provide an independent measure of territory quality, we compared various vegetational characteristics of 11 territories measured in 1975-76 (Fig. 4) by use of a modified point-quarter method (Cottam and Curtis 1956). Although boundaries have changed with changes in flock sizes over the following decade, the core areas of the surviving territories have remained more or less constant. Three of these territories have N/M ratios greater than one. We compared tree sizes as measured by DBH measurements and found that tree sizes in these three territories were almost significantly larger (0.05 < P < .1, Mann-Whitney U) than in the other eight territories (Table IX). We believe that tree size is a very important aspect of territory quality for green woodhoopoes, for two reasons. First, large trees provide better foraging substrate and more food than do smaller ones (our unpub. data).   Second, and even more critical, is the relationship between tree size and cavities: bigger trees have a greater likelihood of holding cavities suitable either for roosting or nesting by the woodhoopoes. In some cases loss of a few or even only one key cavity tree can lead to the disappearance of an entire social unit (Ligon and Ligon 1978).   Thus a causal relationship may exist between tree sizes on a given territory and rates of natality, and especially, mortality.

In addition, we conducted a stepwise multiple regression on the 1976 data, with flock size as the dependent variable.   Territory size and number of trees per territory accounted for 78 percent of the variance among flocks. Simple correlations of flock size versus number of trees

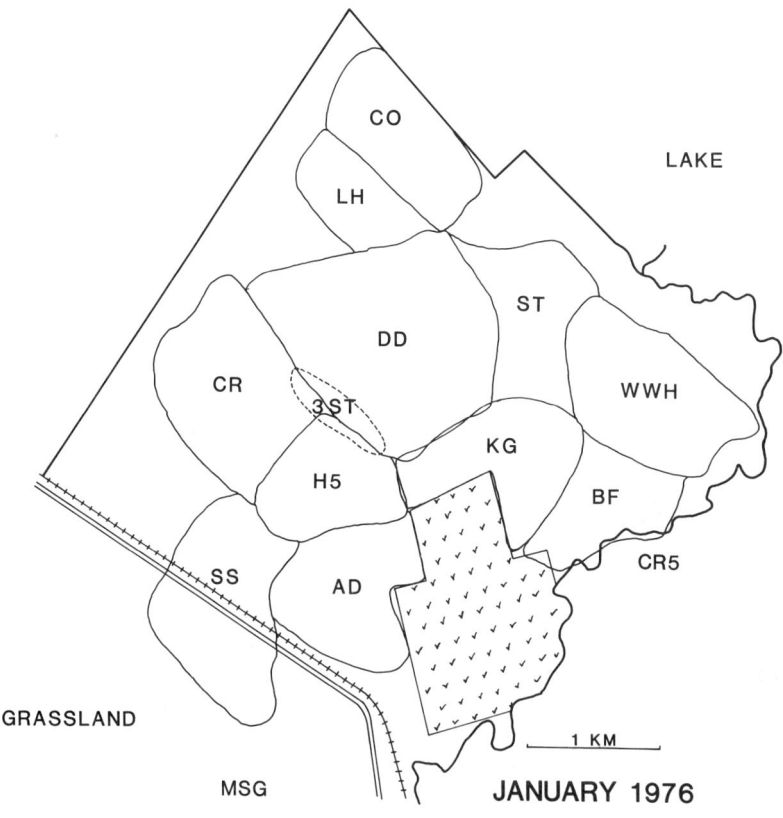

Fig. 4. The green woodhoopoe territories on Morendat Farm as they existed in late 1975 and early 1976. See Table IX.

TABLE IX. The natality/mortality ratio and tree size in territories of green woodhoopoes.

Flock	Mean Mortality/Mean Natality[1]	Mean Tree Size[2]
ST	2.70	0.528 (120)[3]
BF	2.24	0.742 (120)
AD	1.97	0.577 (120)
CO	0.60	0.426 (164)
LH	0.25	0.371 (120)
DD	0.86	0.517 (144)
WWH	0.66	0.540 (124)
KG	0.50	0.685 (120)
H5	0.60	0.516 (160)
SS	0.76	0.498 (120)
CR	0.57	0.562 (140)

[1] See Figure 5.
[2] DBH in meters.
[3] Number of trees measured.

and territory area also were significant (flock size-number of trees: $r = 0.865$, $P < 0.01$; flock size-territory area: $r = 0.813$, $P < 0.01$). These findings suggest that tree number and tree size are important aspects of territory quality for green woodhoopoes. They also emphasize that the relationship between a social unit of woodhoopoes and its territory is a dynamic one, with the number of birds affecting territory quality as measured by size and number of trees, as well as vice versa (Ligon and Ligon 1978).

We measured natality as the number of juvenile birds surviving to the end of the year in the territory in which they were hatched. Mortality is based on marked birds lost in each year. Because long-distance emigration occurs in mature female helpers, unaccounted-for disappearances of such birds are not included in our tabulations. This information alone does not tell us specifically how one territory is superior to another, but since the number and quality of cavities are critical both to reproduction and survival, we believe that these data reflect in large part the cavity situation in a given territory, relative to other territories.

Only five of 16 territories had a N/M ratio of greater than one (Fig. 5). These five (High Quality = HQ) had both significantly higher natality and significantly lower mortality than did the 11 Low Quality (LQ) territories (Mann-Whitney U's: $P < 0.0025$ and $P < 0.005$, respectively). In all five of the HQ territories, offspring of one or both of the original breeders has inherited breeding status. In fact, in 1984 a *daughter* of at least one of the original breeders held reproductive status in four of these five

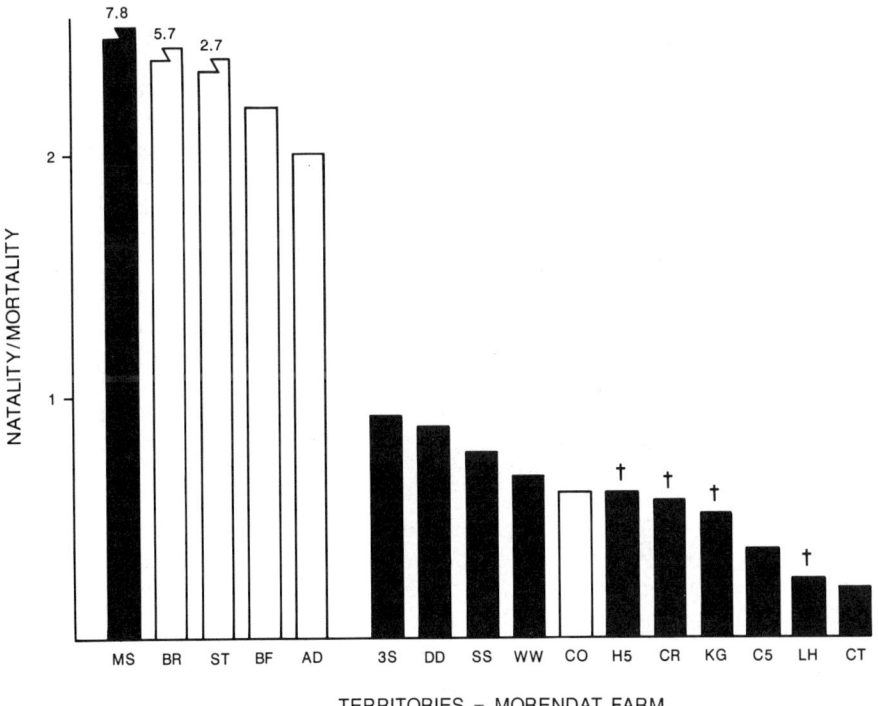

NATALITY/MORTALITY

TERRITORIES – MORENDAT FARM

Fig. 5. The ratio of natality to mortality in territories of green woodhoopoes on Morendat Farm. The value for each territory represents an average based on from 5 to 9 years of data. Shaded bars indicate territories where turnover of genetic lineage occurred at least once. Crosses indicate territories that ceased to exist.

territories. (In nearly all other cooperatively breeding species, territories are inherited largely or exclusively by males.) In the fifth, the MSG territory, a son inherited breeding status and bred successfully in late 1979 and 1980. However, after losing his mate late in 1980, this male moved with four sons to the then-maleless AD territory. Desertion of an HQ territory by several birds, leaving it temporarily unoccupied, was an event unique in our experience.

Figure 5 provides one line of evidence for differential quality of territories. However, it provides little direct information about the relative fitness of the individual breeders occupying the different territories.

D. Reproductive Success and Long-term Survival Related to Territory Quality

To look more closely at the fates of individual breeders, we tabulated the number of new breeders per year for each territory and found that the five HQ territories had significantly less turnover of breeders (Mann-Whitney U, P < 0.001) than did the remaining 11 territories; i.e., individuals persisted longer as breeders in the HQ territories. We also compared number of young woodhoopoes produced per individual breeder in the two categories and found that each breeding bird in an HQ territory produced on average significantly more offspring over its lifetime than did breeders in the LQ territories (Mann-Whitney U, P < 0.001).

To contrast further the differences in potential fitness of breeders in the HQ territories with those in the LQ territories, we tallied the number of known emigrants to another territory, the number of those that attained breeding status, and the number that actually bred successfully, producing at least one surviving offspring (Table X). As with the other comparisons, breeders in HQ territories produced more total emigrants, more emigrants that became breeders, and more that bred successfully. Table X makes one additional, important point: An emigrant from a LQ territory can, and sometimes does, attain breeding status in a HQ territory (e.g., CO to BRF, SS to AD, CT3 to BF, see Fig.

TABLE X.   Relative fitness parameters of breeders occupying specific territories.

Territory	No. Known Emigrants	No. Emigrants Becoming Breeders	No. Emigrants Breeding Successfully
MSG	13	10	7
BRF	4	1	1
ST	4	3	3
BF	8	6	4
AD	5	4	4
CO	1	1	1 (in BRF)
LH	2	1	0
DD	8	6	3 (in LH,KG,H5)
WWH	3	2	1
KG	4	1	1
H5	3	1	1
SS	2	1	1 (in AD)
CR	0	-	-
CR5	0	-	-
3ST	0	-	-
CT3	1	1	1 (in BF)

2). What this means is that a chance exists that a breeder in a LQ territory will produce an offspring that ultimately breeds in a HQ territory; i.e., LQ territories can produce big winners. Thus, from a genetic perspective, it can pay a non-breeder from a HQ territory to move into a LQ one. Since most territories are not very productive, the potential for this kind of genetic "reward" to birds in the LQ territories may help to keep the system going, for the costs of moving into a LQ territory appear to be great. As an example of the costs of immigrating into a LQ territory, the original breeders in the DD territory produced emigrants that bred successfully. However, they did so in LQ territories that ceased to exist during the course of the study (LH, KG, H5; see Fig. 3). Only one descendant of the original DD breeders (a great-granddaughter) was known to be alive in 1984.

We close this section with a cautionary message. Our data indicate that neither flock nor territory size at a single point in time may be a sound indicator of the relative fitness of the breeders or the lineages occupying a particular territory. Figure 6 shows maximum group sizes during the first year of the study, 1975-76, and the number of known living descendants of the original breeders a decade later, in mid-1984. Note the lack of success, as measured by number of descendants, shown by the three largest groups present in 1975. Possibly of greatest interest and significance is the striking dichotomy among groups in number of descendants after 10 years (Table XI). The original breeders of 1975 either have a good many descendants (13-18), or none (7 pairs with 0, 1 with 1), with no intermediate values. All four of the lineages occupying HQ territories have been very successful, as have two individuals hatched in LQ territories. Both of these latter birds (1 male, 1 female) gained success as a result of obtaining breeding status in an HQ territory. The success of the two birds emigrating from LQ to HQ territories suggests that territory quality, rather than genetic lineage, is indeed the primary factor determining reproductive success and survival of offspring over time. This suggests in turn that HQ territories in particular should be passed on from parent to offspring and, as previously mentioned, this has been the case.

V. SUMMARY AND CONCLUSIONS

Many critical life history features of green woodhoopoes (*Phoeniculus purpureus*) appear to be related to a particular trait. These birds apparently lack the physiological means to contend with even moderately low ambient temperatures. To avoid the nocturnal temperatures (7-10°C) characteristic of our study site at Lake Naivasha in Kenya, the woodhoopoes roost in cavities in trees and several birds typically roost together. Cavities are limited in number and thus are a limited resource.

The necessity of cavity roosting appears to be responsible for the comparatively high rate of predation on woodhoopoes (40% mortality per

TABLE XI. Group size and descendants a decade later.

Territory	Maximum Group Size 1975-1976	Number Breeding Emigrants	Number Known Living Descendants of Original Breeding Pair as of 1984
MSG	6	10	14
ST	8	3	13
BF	7	6	13
AD	4	4	18
CO	5	1 (to BRF)	12
LH	4	1	0
DD (+)[1]	16	6	1
WWH	6	2	0
KG +	11	1	0
H5	4	1	0
SS	5	1 (to AD)	14
CR +	10	0	0
CR5	5	0	0
3ST	5	0	0

[1]Complete turnover of lineage within the territory.
[2]Disappearance of the territory.

year for males, 30% for females). Our evidence indicates that large-spotted genets (*Genetta tigrina*) and driver ants (Tribe Dorylini) are the principal predators of roosting birds. Because of predation pressure and competition with other species for cavities (e.g., honey bees, acacia rats, dormice, other kinds of birds), cavity quality (i.e., size of entrance, size of cavity, strength of wood surrounding the cavity) and number of potential roost holes probably are the major factors determining reproductive success and survival of birds occupying a particular territory.

In addition to its effect on mortality rate, the dependence of green woodhoopoes on holes in trees for roosting, together with the scarcity of high-quality cavities, has affected dispersal patterns and this, in turn, has affected the birds' reproductive biology in several ways, possibly including the evolution of cooperative breeding. Here we list the major factors influenced by the physiological intolerance of woodhoopoes to low nighttime temperatures and point out how or where each factor affects other traits:

1. Intolerance of low temperatures has led to cavity roosting.
2. Obligatory use of roost cavities results in high levels of nocturnal predation. Cavities differ in quality with regard to invasion by predators.

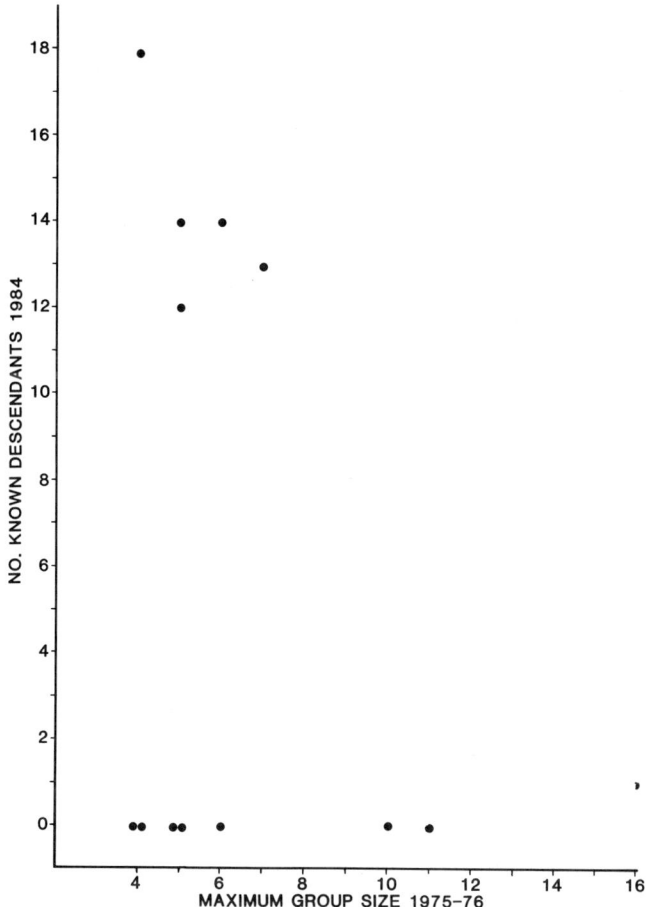

Fig. 6.   The relationship between maximum group size in 1975-76 and the number of known living descendants a decade later, in mid-1984. The three largest groups (10, 11 and 16) were represented by only one known descendant in 1984.   In contrast, five initially small social units had from 12 to 18 known living descendants after 10 years.

3.   The critical dependence on cavities has led to extremely philopatric dispersal.   Most woodhoopoes of both sexes that attain
breeding status do so either in their natal territory or in an adjacent one.

4.   The extreme philopatry by both sexes, plus the semi-isolated
nature of the Lake Naivasha population of green woodhoopoes,
appears to have led to inbreeding depression, manifested in the
form of embryo mortality.   About one-third of all eggs laid do

not hatch. Most unhatched eggs that we examined held recognizable embryos.

5. Possibly as a result of the pre-hatch mortality and the consequent reduction in brood size, starvation of nestlings is very rare. This, together with the fact that a pair of woodhoopoes may produce up to three broods per year, suggests that the overall cost of the putative inbreeding depression is not great.

6. The scarcity of safe roost sites probably has favored retention of offspring within the parental territory. Since a correlation does not exist between number of helpers and number of young produced (Ligon and Ligon, in press), this may be viewed in part as a form of extended parental care.

7. Retention of young birds in their natal territories has led to various forms of cooperation (e.g., group defense of the territory), including cooperative breeding.

8. Group living in this population of woodhoopoes has placed a premium on alliances. This has led to the regular occurrence of unisexual "teams" of related or unrelated individuals. Teamwork is most apparent in the context of territory acquisition and defense.

All of the above factors associated with cavity roosting relate to territory stability and ultimately to territory quality. We have found that over a 10-year period (1975-1984) the number of territories on our primary study area has remained constant. However, some territories disappeared during this interval while others were created; thus numbers alone do not accurately reflect stability. Lineages of woodhoopoes occupying these territories have proved to be much less stable, with 10 of 14 original lineages disappearing during the decade.

By comparing mortality and natality among territories over several years, we were able to divide them into net exporters or net importers of woodhoopoes (Natality/Mortality > or < 1). Individual breeders in "High Quality" territories (N/M > 1) lived longer and produced more offspring on average than did birds from "Low Quality" (NM/ < 1) territories. An independent measure of territory quality based on territory size, number of trees and size of trees supports the notion that those territories with an N/M ratio of more than one are superior in those resources most critical to green woodhoopoes.

The High Quality territories, and thus the breeding birds occupying them, were most productive in terms of emigrants (i.e., a N/M ratio of more than one really does reflect export of woodhoopoes). Over the period of study the five HQ territories produced an average of 6.8 emigrants, and of these an average of 4.8 became breeders. In contrast, the LQ territories produced an average of 2.2 emigrants, with only an average of 1.3 of these attaining breeding status.

Where a bird became a breeder was the critical factor in its own reproductive success. Three birds hatched in LQ territories bred in HQ territories. Two of these, one of each sex, were extremely successful, leaving 12 and 14 living descendants as of the end of the study, in mid-

1984. In contrast, the original breeders in eight of the other nine LQ territories had no known descendants, while the other had only one, a great-granddaughter of the original breeders.

The picture that has emerged from this analysis is that territory quality, in terms of characteristics of the trees--especially cavities--is the primary determinant of reproductive success and survivorship of green woodhoopoes. Birds originating in inferior territories can be highly successful if they obtain breeding status in a superior territory. Although most territories have either ceased to exist or have seen complete turnover of the original lineage, the HQ territories have been passed on from parent to offspring. In four of the five territories designated as HQ, daughters of the original breeders held breeding status at the end of this project. In contrast, a descendant held breeding status in only one of 11 LQ territories. We conclude that the fitness of an individual bird, as measured by its lifetime reproductive success and the success of its offspring in attaining breeding status and breeding successfully, is to a great extent determined by the quality of the territory that it occupies. Although HQ territories tend to be retained within the family lineage, birds from inferior territories on occasion have the opportunity to move into HQ territories. When this happens, such individuals can have reproductive success as great as those hatched in the HQ territories and later breeding there.

ACKNOWLEDGMENTS

Our field studies of green woodhoopoes in Kenya were made possible via the support of R.E. Leakey and G.R. Cunningham-van Someren of the National Museums of Kenya. E.K. Ruchiami, of the Office of the President of Kenya, kindly granted research permits. S. Zack, D.C. Schmitt and J.C. Bednarz assisted with the field work and W. and R. Hillyar and R. and B. Terry provided logistical support. This project was generously supported by the National Science Foundation, the National Geographic Society, the American Museum of Natural History, the National Fish and Wildlife Laboratory, and the University of New Mexico. We thank all of these individuals and institutions.

REFERENCES

Bengtsson, B.O. 1978. Avoiding inbreeding: At what cost? *Journal of Theoretical Biology* 73:439-444.
Brown, J.L. 1974. Alternate routes to sociality in jays--with a theory for the evolution of altruism and communal breeding. *American Zoologist* 14:63-80.
Cottam, G. and J.T. Curtis. 1956. The use of distance measures in phytosociological sampling. *Ecology* 37:451-460.
Emlen, S.T. 1982. The evolution of helping. I. An ecological constraints model. *American Naturalist* 119:29-39.

Greenwood, P.J. 1980. Mating systems, philopatry and dispersal in birds and mammals. *Animal Behaviour* 28:1140-1162.

Koenig, W.D. 1981. Space competition in the acorn woodpecker: Power struggles in a cooperative breeder. *Animal Behaviour* 29:396-409.

Koenig, W.D., R.L. Mumme and F.A. Pitelka. 1987. *Population ecology of the cooperatively breeding acorn woodpecker*. Princeton, New Jersey:Princeton University Press.

Koenig, W.D. and F.A. Pitelka. 1979. Relatedness and inbreeding avoidance: Counterploys in the communally breeding acorn woodpecker. *Science* 206:1103-1105.

Koenig, W.D. and F.A. Pitelka. 1981. "Ecological factors and kin selection in the evolution of cooperative breeding in birds." pp. 261-280. In: R.D. Alexander and D.W. Tinkle, eds. *Natural selection and social behavior*. New York:Chiron.

Ligon, J.D. 1981. Demographic patterns and communal breeding in the green woodhoopoe (*Phoeniculus purpureus*). pp. 231-243. In: R.D. Alexander and D.W. Tinkle, eds. *Natural selection and social behavior*. New York:Chiron.

Ligon, J.D. 1983a. Cooperation and reciprocity in avian social systems. *American Naturalist* 121:366-384.

Ligon, J.D. 1983b. "Commentary on cooperative breeding strategies among birds." pp. 120-127. In: A.H. Brush and G.A. Clark, Jr., eds. *Perspectives in ornithology*. New York:Cambridge University Press.

Ligon, J.D. 1987. The question of parent-offspring conflict in cooperatively breeding birds. Proceedings of the XIX International Ornithology Congress. Ottawa, Canada. In press.

Ligon, J.D., C. Carey and S.H. Ligon. 1988. Cavity roosting and cooperative breeding in the green woodhoopoe reflect a physiological trait. *Auk* 105:in press.

Ligon, J.D. and S.H. Ligon. 1978. The communal social system of the green woodhoopoe in Kenya. *Living Bird* 17:159-197.

Ligon, J.D. and S.H. Ligon. 1982. The cooperative breeding behavior of the green woodhoopoe. *Scientific American* 247:126-134.

Ligon, J.D. and S.H. Ligon. 1983. Reciprocity in the green woodhoopoe (*Phoeniculus purpureus*). *Animal Behaviour* 31:480-489.

Ligon, J.D. and S.H. Ligon. In press. "The cooperative breeding system of the green woodhoopoe in Kenya." In: P.B. Stacey and W.D. Koenig, eds. *Cooperative breeding in birds: Long-term studies of behavior and ecology*.

Ricklefs, R.E. 1975. The evolution of cooperative breeding in birds. *Ibis* 117:531-534.

Romanoff, A.L. 1972. *Pathogenesis of the avian embryo*. New York:John Wiley and Sons.

Rowley, I. 1983. "Commentary on cooperative breeding strategies among birds." pp. 127-133. In: A.H. Brush and G.A. Clark, Jr., eds. *Perspectives in ornithology*. New York:Cambridge University Press.

Stacey, P.B. and J.D. Ligon. 1987. Territory quality and dispersal options in the acorn woodpecker, and a challenge to the habitat saturation model of cooperative breeding. *American Naturalist*. In press.

Van Noordwijk, A.J. and W. Scharloo. 1981. Inbreeding in an island population of great tits. *Evolution* 35:674-688.

Wiley, R.H. and K.N. Rabenold. 1984. The evolution of cooperative breeding by delayed reciprocity and queuing for favorable social positions. *Evolution* 38:6009-621.

Wilson, E.O. 1975. *Sociobiology. The new synthesis.* Cambridge, MA: Belknap Press of Harvard University Press.

Woolfenden, G. and J.W. Fitzpatrick. 1984. *The Florida scrub jay. Demography of a cooperative-breeding bird.* Princeton, NJ:Princeton University Press.

Zack, S. and J.D. Ligon. 1985a. Cooperative breeding in *Lanius* shrikes. I. Habitat and demography of two sympatric species. *Auk* 102:754-765.

Zack, S. and J.D. Ligon. 1985b. Cooperative breeding in *Lanius* shrikes. II. Maintenance of group-living in a nonsaturated habitat. *Auk* 102:766-773.

Chapter 11

RESOURCE AND CLIMATIC VARIABILITY: INFLUENCES ON
SOCIALITY OF TWO SOUTHWESTERN CORVIDS

John M. Marzluff
Russell P. Balda

Department of Biological Sciences
Northern Arizona University, Flagstaff AZ 86011

I.   INTRODUCTION

The social organization of a species can be viewed in part as an
adaptation to prevailing environmental conditions. This view leads to
three predictions concerning variability in social organization. (1) Organisms living in different environments facing different selective pressures
may exhibit different social organizations. This is certainly not a new

prediction. Crook (1965) observed that members of the same families of birds often differ in social organization depending on the distribution of food, nest sites, and predators in their environments. (2) Identical environmental conditions may give rise to different social organizations. Identical selective pressures may be adapted to in a variety of ways. Pandas solve the need for an opposable thumb differently from primates (Gould 1980). Similarly, we may expect different species to solve similar problems posed by their environment with different social organizations. Just as phylogeny may constrain morphological adaptation so may phylogeny constrain behavioral adaptation. One needs only to spend a few hours in a given environment to realize that all species inhabiting that environment do not have identical social structures. (3) Individuals of the same species, even those in close spatial proximity, may face different environmental conditions and therefore may exhibit different social organizations. Environmental conditions also can change dramatically from year to year which may favor species with plastic social structures because they would be able to track continually shifting selective pressures. Variability in social structure within a species has been documented in few instances (Stacey and Bock 1978; Trail 1980; Bekoff and Wells 1981; Brown et al. 1983; Brown and Brown 1985).

Acorn woodpeckers (*Melanerpes formicivorus*) are an excellent example of a species with a plastic social structure that varies with prevailing environmental conditions. In southern Arizona, acorn productivity varies spatially and temporally. Acorn woodpeckers that live in productive habitats form permanent social groups, typical of this species in other parts of its range (Stacey and Bock 1978). However, neighboring woodpeckers in less productive habitat cannot store enough acorns to survive over the winter; instead of forming social groups they live on territories as less social pairs that migrate to presumably more productive areas each winter (Stacey and Bock 1978). Moreover, in some years acorn productivity is high throughout the area and pairs forego migration and remain on their territories. Juveniles may be tolerated on these territories so that the more typical social organization of this species is approximated (Stacey and Bock 1978). Even in typically productive habitats of Arizona, woodpecker populations are low after acorn crop failures which may result in a lower incidence of helping at the nest because suitable habitat is not saturated with breeding woodpeckers (Trail 1980).

Many environmental factors have been demonstrated or postulated to affect social structure. Density and sex ratio of the species in question, the distribution and abundance of food, and the abundance of predators are the primary candidates influencing social structure (Brown 1974, 1976; Stacey and Bock 1978; Reyer 1980; Bekoff and Wells 1981; Waser 1981). Climatic conditions may influence survivorship and fecundity of predators and their prey which in turn may affect social organization as indicated above. Harsh weather is especially likely to influence subordinates because they may be physically smaller or they may have reduced access to limited resources (Lack 1954). Dominance is commonly age and/or sex dependent (Gauthreaux 1978). Harsh weather may therefore influence mate and/or helper availability which could affect social structure. We postulate that climatic variability is a major reason why many factors affecting social structure are also variable. Variability in environmental factors may thus result in variable social organization. If this is true, species living in variable climates should exhibit more social plasticity than species in more constant climates.

Here we focus on the social plasticity of a population of pinyon jays (*Gymnorhinus cyanocephalus*). Our study population inhabits the upper elevational limits of this species' distribution. This montane environment is characterized by extremely variable, harsh, and unpredictable climate. The abundance of piñon pine (*Pinus edulis*) seeds, the favored food of pinyon jays, is also extremely variable both spatially and temporally. We document the relative influence of variability in food abundance and climate on flock size, survivorship, sex ratio, and breeding biology of this population.

We also broadly outline how a limited resource can influence the development of two different social systems in two species of birds in the family Corvidae. Both scrub jays (*Aphelocoma coerulescens*) and pinyon jays inhabit the woodlands of the southwestern United States, where seeds of the piñon pine are an important, but spatio-temporally variable, resource. These two corvids, however, have grossly different social organizations (Brown 1974). Pinyon jays specialize on these seeds to a greater extent than scrub jays which may explain differences in these species' social systems.

II. RESOURCE AND CLIMATIC VARIABILITY IN THE SOUTHWESTERN UNITED STATES

Two important habitats in the Southwest are the ponderosa pine (*Pinus ponderosa*) forests, inhabiting an elevation band from 2200 to 2800 m, and the piñon-juniper (*Pinus edulis-Juniperus* spp.) woodland, found at slightly lower elevations. Ponderosa pines grow on dry sites in open, park-like stands. Seeds of this pine are small and winged for wind dispersal; however, pinyon jays harvest and cache these seeds before they are shed from cones (Balda and Bateman 1972). The piñon-juniper woodland is the driest plant community in North America dominated by trees. Because of the relatively severe aridity, plant reproduction, especially seed germination and seedling establishment, is difficult. Piñon pines and junipers have responded to the limited number of "safe" germination sites in two ways: (1) junipers produce large numbers of relatively small fleshy berries that are eaten by birds and the seed broadcast over the landscape by defecating birds; and (2) piñon pines produce relatively large, high-energy seeds that are harvested, eaten, or carried off and stored by birds and mammals.

Weather factors are extremely variable in the montane Southwest. Winter temperature in the ponderosa pine forest averaged above freezing during four winters from 1972 to 1986 and well below freezing in others years (Fig. 1). Piñon-juniper woodlands are warmer on average in winter, but equally variable (Fig. 1). Spring snowfall is potentially very

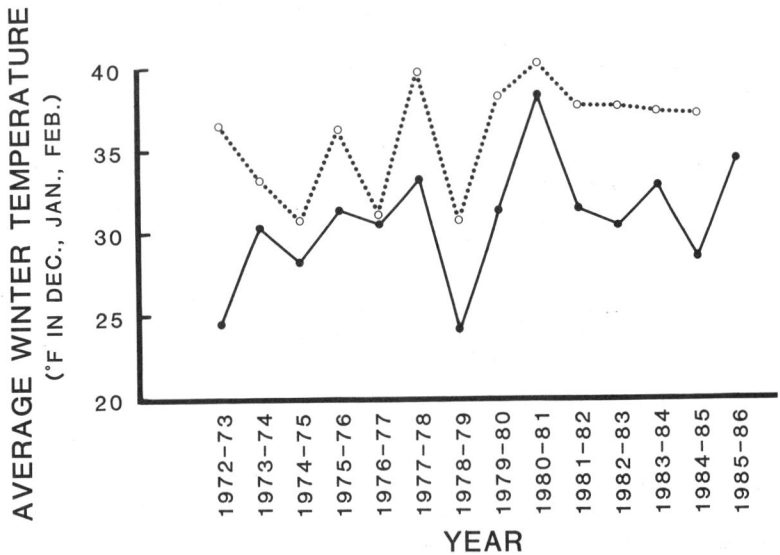

Fig. 1. Annual variation in average winter temperature in South-western pinyon pine (open circles, dotted line) and ponderosa pine (closed circles, solid line) forests. Data was obtained from National Oceanic and Atmospheric Administration local climatological data sheets for Flagstaff, AZ (ponderosa pine) and Winslow, AZ (piñon pine).

important to jays, because it is at this time that nesting is initiated. Ponderosa pine forests have extreme annual variation in this factor, in some years receiving nearly 100 in of snow, but in others receiving vir-tually none (Fig. 2). Piñon-juniper woodlands rarely have substantial spring snowfall (Fig. 2).

Piñon pine seeds are an irregular and unpredictable, but occasionally abundant, food source (Barger and Ffolliott 1972; Ligon 1978). Irregular and/or unpredictable cone production may result from climatic variabil-ity during the three growing seasons cones take to mature (Little 1938a,b). Unpredictability and synchrony in seed production may have evolved in response to seed predation by vertebrate and invertebrate animals (Vander Wall and Balda 1977; Ligon 1978; Smith and Balda 1980). An irregular boom and bust production of seeds was evident on our study area (Fig. 3). From 1972 to 1985 there were seven years when practical-ly no pine seeds were harvested in northern Arizona by Native Ameri-cans (Fig. 3). During such bust years virtually all seeds may be eaten and stored by jays for later use (Ligon 1978). This is in contrast to four bumper crops in the years 1974, 1976, 1980, and 1983 when up to 130,000 lbs of seeds were harvested (Fig. 3). Large, synchronized crops, such as these, allow seed-caching birds to hide large numbers of seeds in subterranean caches. Several pinyon jays lived over the entire time span illustrated in Fig. 3. Scrub jays have similar lifespans (Woolfenden and

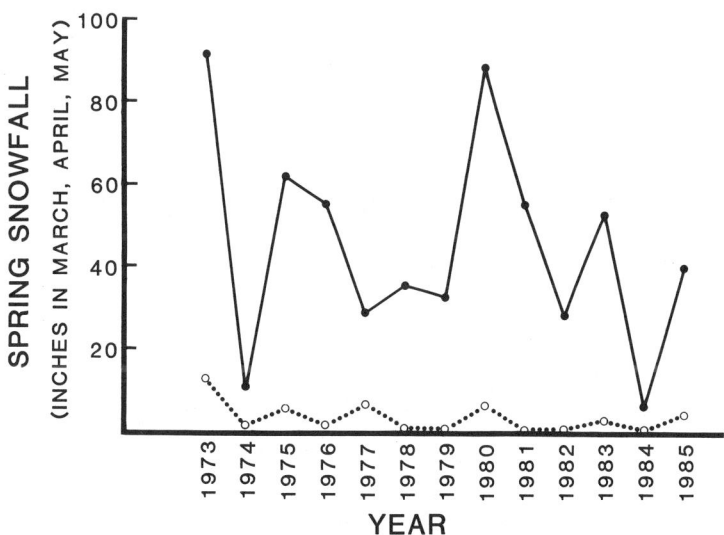

Fig. 2. Annual variation in total spring snowfall in Southwestern piñon pine (open circles, dotted line) and ponderosa pine (closed circles, solid line) forests. Source of data as in Fig. 1.

Fitzpatrick 1984; C. Collins pers. comm.); thus, some individuals are likely to experience both boom and bust crops several times during their lifetime.

III. RESOURCE VARIABILITY AND THE EVOLUTION OF SOCIALITY IN TWO SOUTHWESTERN CORVIDS

Both pinyon and scrub jays inhabit the extremely variable (Figs. 1-3) piñon-juniper woodland, yet they exhibit grossly different social systems. They are social and asocial endpoints, respectively, on the colonial route to sociality proposed by Brown (1974). Scrub jay pairs may defend classic, year-round all-purpose territories (Bent 1946; Hardy 1961; Brown 1974). In contrast, pinyon jays live in integrated, social flocks throughout the year, nest in colonies, occasionally have helpers at the nest, and only defend an area immediately around the nest (Balda and Bateman 1971, 1972; Balda and Balda 1978; Ligon 1978).

A key difference between scrub and pinyon jays that may have influenced the evolution of their different social systems is their reliance on piñon pine seeds. In Arizona, scrub jays eat pine seeds when they are available on or near their territory (Bent 1946); however, they are only slightly morphologically and behaviorally specialized to exploit this

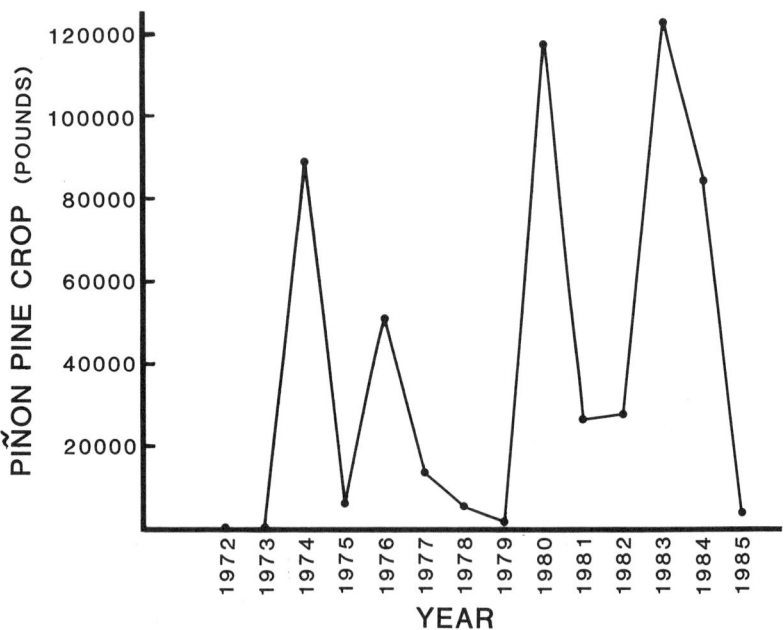

Fig. 3. Annual variation in pounds of piñon pine seeds collected in northern Arizona by Native Americans.

resource (Vander Wall and Balda 1981). Pine seeds are a major component in the diet of adult and nestling pinyon jays throughout the year (Bateman and Balda 1973; Ligon 1978). Pinyon jays are specialized morphologically and behaviorally to efficiently exploit pine seeds (Balda and Bateman 1971, 1972; Ligon and Martin 1974; Ligon 1978; Vander Wall and Balda 1981). The importance of pine seeds to pinyon jays is further indicated by their eruptive movements out of areas experiencing seed crop failures (Westcott 1964; Vander Wall and Balda 1981; Bock 1982).

We have outlined a scenario for the evolution of sociality in scrub and pinyon jays that is based on their differential specialization on the unpredictably variable piñon pine crop (Fig. 4). Scrub jays have relatively poor long-distance flight capabilities (Vander Wall and Balda 1981) which may confine them to small home ranges. Their generalized diet may have evolved as a strategy to survive extreme fluctuations in resources that occur over small regions of piñon woodland (Fig. 3). A small home range containing a mixture of usable food types may be economically defendable; hence, territoriality should evolve (Brown 1969). Provided dispersal costs are less than benefits young should disperse from the natal territory leaving pairs to jointly defend all-purpose territories (Brown 1974; Woolfenden and Fitzpatrick 1984). Pinyon jays are strong fliers (Vander Wall and Balda 1981) and are able to move long distances in search of locally abundant pine crops. Large areas, parts of which

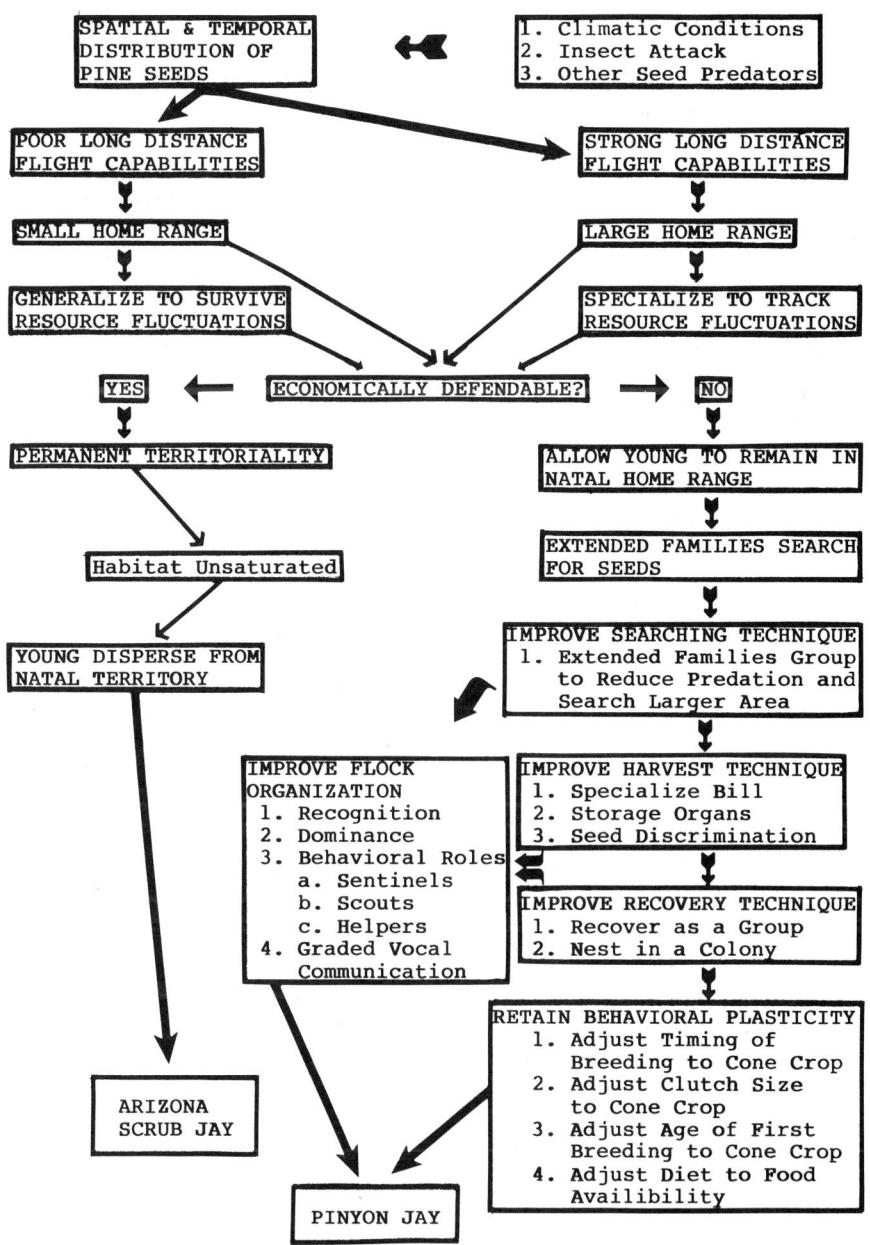

Fig. 4.   Scenario for evolution of different social systems in two corvids inhabiting the same environment.   See text for explanation.

occasionally produce seeds, may not be economically defendable, and hence territoriality is not expected to evolve (Brown 1969). Indeed, neighboring flocks do not exclude each other from their home ranges (Balda and Bateman 1971).

The ability to track a spatially and temporally variable food source over a large area may be improved by group foraging (Horn 1968; Ward and Zahavi 1973; Cody 1974). Extended families may form the initial flock, but family groups should join together to more efficiently search large areas with reduced risk of predation. As flock size increases and morphological specialization develops, flock organization should improve so the harvest and recovery of pine seeds can proceed in a coordinated manner. Reliance on localized caches of seeds (Balda and Bateman 1971, 1972; Ligon 1978; Vander Wall and Balda 1981) may favor group recovery to reduce the risk of predation and reduce the risk of cheating jays stealing unattended caches. Males continue to forage as a group throughout the breeding season while females incubate eggs and brood young (Balda and Bateman 1971; Ligon 1978). In addition to using clumped seed caches at this time, unpredictably clumped insects are searched for which may favor the evolution of colonial nesting (Horn 1968; Ligon 1978). We view resource unpredictability as an important evolutionary pressure favoring the flocking and colonial nesting habits of pinyon jays.

Pinyon jays, although specialized to track a fluctuating resource, appear to have retained behavioral flexibility enabling them to both capitalize on abundant pine crops and survive poor crops. Reproductive success may be maximized following large cone crops because jays nest sooner after such crops ripen (Balda and Bateman 1972; Ligon 1971, 1974, 1978) and lay larger clutches after large crops (Balda and Bateman 1972; this report). The consequences of poor crops may be lessened by reducing reproductive effort (delaying breeding and laying smaller clutches), using alternative food sources, or wandering. Pinyon jays thus appear to have evolved to take advantage of boom piñon crop years. Scrub jays, on the other hand, appear adapted to survive bust years, although they may capitalize on boom crops by breeding early and increasing reproductive output (Ligon, pers. comm.)

Pinyon jays and scrub jays respond more similarly to other selective pressures in the piñon-juniper woodland. Both respond to variable climate by being opportunistic foragers and by storing food. Both conceal nests, post sentinels, and mob potential predators to reduce the risk of predation. Flocking and colonial nesting in pinyon jays also may thwart predators. Major differences in sociality exhibited by these two corvids appear related to their different responses to the selective pressure of temporally and spatially patchy food.

IV.   ECOLOGICAL CORRELATES OF FLOCK SIZE, SURVIVORSHIP, AND BREEDING BIOLOGY IN A FLOCK OF PINYON JAYS

A.   Methods

This section deals exclusively with a flock of individually color-banded pinyon jays that inhabits the ponderosa pine forest in and around Flagstaff, AZ. This population of jays is unique in two respects: (1) its diet is supplemented at feeders in Flagstaff, and (2) Flagstaff is at the upper altitude of this species' elevational distribution. Despite these peculiarities, jays in this flock still rely heavily on piñon and ponderosa pine seeds and the flock has a similar age and sex structure as an unprovisioned flock (Marzluff and Balda, in press a).

Beginning in 1972, activities of the Flagstaff Town Flock have been closely monitored at local feeding stations. Every day that the jays visited the feeders the color combination of all birds seen was recorded. From these "roll call counts" we calculated annual survivorship of each age class for year X as the percent of birds surviving from 31 Dec. of year X - 1, to 31 Dec. of year X. Reliance on feeding stations was quantified as the number of days during October and November that the flock visited local feeders. Total flock size was calculated after the breeding season, on 1 August. Flock sex ratio and number of wanderers was also calculated at the end of each December. Wanderers are apparently dispersing juveniles and yearlings that are seen in the fall and spring of each year. These birds are assumed to be from other flocks because they are unbanded and sometimes have morphologies distinct from Town Flock birds.

Each year known breeding areas of the Town Flock were searched for current nesting activity. Onset of breeding, breeding success at several stages, and the age composition of pairs were determined at this time. For this paper, we use clutch size, number of fledglings, and the number of juveniles each pair produces as an indication of per-pair breeding success. We will also discuss trends in the percent of nests found that had helpers and that had yearling breeders.

The piñon crop each fall, and weather factors for the winter, spring, and summer of year X and X - 1 were related to characteristics of the flock in year X. Seeds were collected by Native Americans and sold to a local feed store. Chet Anderson, owner of the store, kindly provided us with his purchasing receipts which we converted to pounds of seeds by assuming an average purchase price of $1.20/lb. Weather data was obtained from N.O.A.A. local climatological data sheets for Flagstaff, AZ. Weather variables are defined and abbreviated in tables as follows: Winter = Dec., Jan., Feb., during which time we recorded total snowfall (W SNOW) and average monthly temperature (W TEMP). Spring = March, April, May, during which time we recorded total snowfall (SP SNOW), total snowfall from 15 February to 15 March (BREED SNOW), total precipitation (SP PPT), and average monthly temperature (SP TEMP). Summer = June, July, Aug., during which time we recorded total precipitation (SU PPT) and average monthly temperature (SU TEMP). Weather variables measured in year X - 1 have the same abbreviations, except they are preceded by a P (e.g., PSU PPT = previous summer precipitation).

Weather and pine crop measures were correlated with measures of flock size, age composition, survivorship, and breeding biology. We report Pearson correlation coefficients as r and partial correlation coefficients as r_{pr}. Partial correlations were used to investigate the influence of weather while holding pine seed levels constant and vice versa. We present numerous correlations between related variables, therefore a few may appear significant by chance alone. Skepticism is justified, but as noted by Brown et al. (1982), we justify this approach because we present our results as testable hypotheses, not as statements of cause and effect. We are most interested in exploring the relative influence of each independent variable; therefore, we compare two-tailed probabilities even if they are not significant at the 5 percent level. We employ some hypothesis testing at the end of this section and use $P < 0.05$ as our significance level.

B.   Annual Variation in Flock Size and Composition

Flock size and age composition of the flock varied annually (Fig. 5). Flock size averaged 177 jays, but has varied from 129 in 1985 to 292 in 1978. Flock size appears to be a reflection of great variability in the

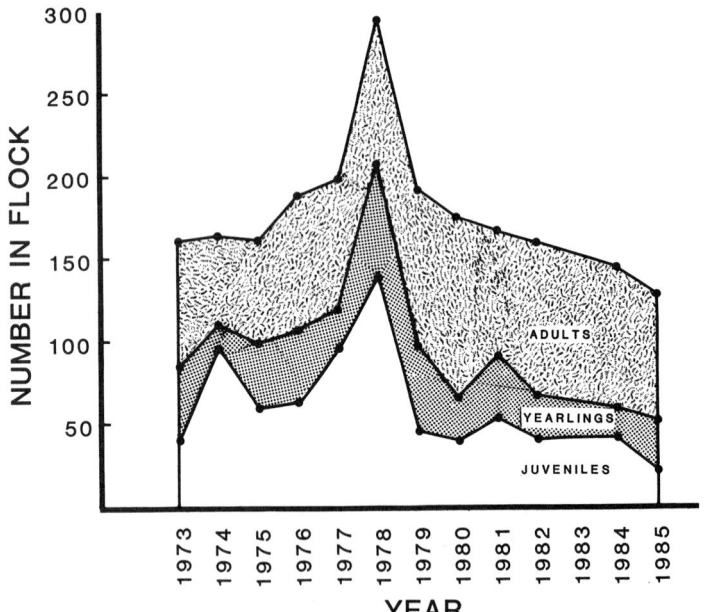

Fig. 5. Annual variation in age composition prior to dispersal within a population of pinyon jays. Data was unavailable for 1983 because jays rarely visited feeding stations.

number of juveniles (SD = 31.93) present in the flock before dispersal. In 1978, 138 juveniles were banded from the Town Flock compared to only 22 in 1985. In contrast, the number of adults in the flock remained fairly close to the mean of 84, ranging only from 63 to 110 (SD = 14.6). Yearlings showed intermediate variability around a mean of 35 (SD = 15.8).

Variability in flock size was most strongly correlated with factors endogenous to the flock (Table I). This is to be expected because flock size is the sum of these factors (juveniles, yearlings, adults, immigrants). The relative importance of these components did differ: juveniles and yearlings were closely related to flock size (Table I), but adults and immigrants had only minor influences (r = 0.17, P = 0.30; r = 0.24, P = 0.26, respectively).

Flock size was influenced to a lesser degree by food and weather conditions. Flock size was larger during warm winters (Table I) and this correlation was substantially stronger when the masking effect of previous pine cone crop was removed (r_{pr} = 0.43, P = 0.09). This correlation appears to be primarily a reflection of increased numbers of juveniles following warm winters (Table I). Flock size was smaller the year after a good pine crop (Table I) and this effect was strengthened when winter temperature was held constant (r_{pr} = -0.55, P = 0.04). This correlation with flock size appeared to be mediated by the number of juveniles, yearlings, and adults in the flock. These age classes were less abundant the year after good pine crops; however, adults were more common the winter immediately following large fall pine crops (Table I). Larger flock size following mild winters is intuitive, but an inverse relationship with pine crop is not. As we will show later, however, survivorship of adults is low following large pine crops and breeding is earlier in the spring. In Flagstaff, breeding at this time is a risky strategy that is not likely to be successful because of the extremely variable and harsh spring snowstorms.

Juveniles were more numerous in the flock following warm winters and in years after poor cone crops, but neither relationship was very strong (Table I). Other factors related to breeding success, such as number of predators or availability of other food may influence the number of juveniles to a greater extent than the food and weather variables that we investigated.

The number of adults in the flock was influenced equally by weather and food factors. More adults were present following large cone crops and during years when winter snow was heavy (Table I). More adults following snowy winters is counterintuitive; perhaps insects, an important food item (Ligon 1978), are more numerous following wet winters.

Yearlings in the flock and immigrants showed an interesting relationship to weather and pine seeds (Table I). Yearlings were primarily influenced by weather factors, being less abundant following wet springs and summers. They were also slightly less abundant after heavy piñon crops. Immigrants showed the opposite relationship, being more common after good piñon crops (r_{pr} = 0.53, P = 0.11) and wet springs (r_{pr} = 0.61, P = 0.05). These opposing correlations may result because some of the

Table I. Factors most strongly correlated with flock size and its components. Pearson correlations are given. P values with asterisks indicate that the partial correlation between those variables, holding other strongly correlated factors constant, was significant at P < 0.05. Abbreviations of weather factors are given in methods.

Variable	Piñon Crop	Previous Crop	Weather Factors		Endogenous Factors	
Flock Size	r = -0.20 N = 12 P = 0.27	-0.42 0.09*	W TEMP 0.19 0.28		JUVS[1] 0.86 <0.01*	YEAR[2] 0.71 0.01*
# Juveniles	r = -0.08 N = 12 P = 0.41	-0.28 0.19	W TEMP 0.28 0.19			
# Yearlings	r = -0.36 N = 13 P = 0.11	-0.31 0.16	SP PPT -0.63 0.01*	SU PPT -0.51 0.04		
# Adults	r = 0.50 N = 13 P = 0.04*	-0.29 0.17	W SNOW 0.53 0.03*			
# Immigrants	r = 0.06 N = 9 P = 0.44	0.08 0.28	SP PPT 0.46 0.11*		Yearling Sex Ratio 0.75 0.01*	

[1]Number of juveniles
[2]Number of yearlings

decline in yearlings reflects emigration, which is thus balanced by immigration. Wet years following pine crops are apparently good years for dispersal of yearlings from and into flocks. Entrance into a flock by dispersers, however, was also strongly correlated with the sex ratio of yearlings in the flock (Table I). In years when the sex ratio of yearlings was biased in favor of males, more wanderers immigrated into our study flock. Virtually all immigrants, and emigrants, were females.

In summary, flock size was primarily related to the number of juveniles and yearlings (Fig. 6). Immigrants and adults had lesser positive relationships. Warm winters were associated with more juveniles and poor cone crops were associated with more juveniles and more yearlings, hence flock size was larger following warm winters and poor cone crops. Wet springs and good pine crops were apparently favorable for dispersal by female juveniles and yearlings because immigration was higher and the number of yearlings was lower after such conditions. Immigration of wanderers appeared closely related to flock sex ratio; immigration was high into a strongly male-biased flock.

C. Annual Variation in Survivorship

Our survivorship calculations for juveniles and yearlings include disappearances from the flock both from death and from emigration. The majority of emigrants do not establish themselves as breeders in neighboring flocks; thus, we assume disappearance reflects death before and during dispersal. Moreover, if successful dispersal is fairly constant between years our comparisons of annual survivorship, based on disappearance, are valid. Because we have no evidence that adults disperse, survivorship of adults is probably not confounded by dispersal.

Annual survivorship was variable, but clearly age dependent; adults survived better than yearlings and yearlings survived better than juve-

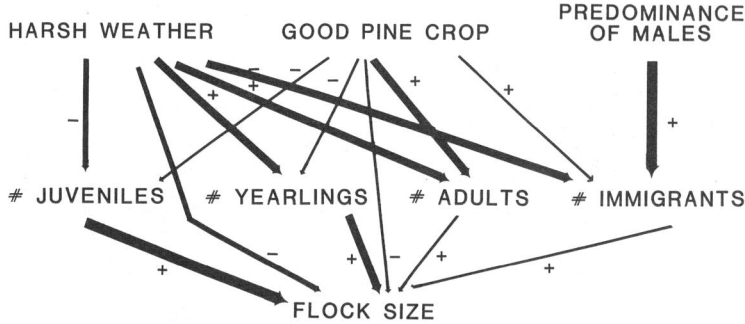

Fig. 6. Summary of the influences of weather, pine crop, and sex ratio on flock size. Width of arrows indicates relative influence based on P-values from correlation analyses. Negative influences are indicated by (-); positive influences by (+). Harsh weather is defined as cold, snowy winters and springs or hot, dry summers.

niles (Fig. 7). Averaging over the years, 74 percent of adults, 63 percent of yearlings, and 41 percent of juveniles survived each year. Yearling survivorship showed the most year-to-year variability (SD = 19.8), followed by juvenile survivorship (SD = 13.3) and adult survivorship (SD = 12.0).

Survivorship of all age classes was more strongly correlated with weather factors than with variations in the piñon crop (Table II). Juveniles survived better following large pine crops and when springs were warm and wet. The influence of pine crop on juvenile survivorship was not apparent until the overriding influence of weather was removed by partial correlation (r_{pr} = 0.58, P = 0.05). Yearlings survived better over wet springs, even if it was also cold and snowy. Yearlings that, as juveniles, were raised during good pine crops also survived better than those raised during poor crops (note relatively strong correlation between yearling survivorship and previous cone crop). Adults, like yearlings, survived wet weather better than dry, but showed an interesting curvilinear relationship to pine crop (Fig. 8). They survived best during intermediate cone crops. Lack of high quality food during crop failures intuitively should reduce survivorship, but low survivorship during bumper cone crops is more difficult to explain. Perhaps increased activity during harvest, and exposure in unfamiliar habitat to predators

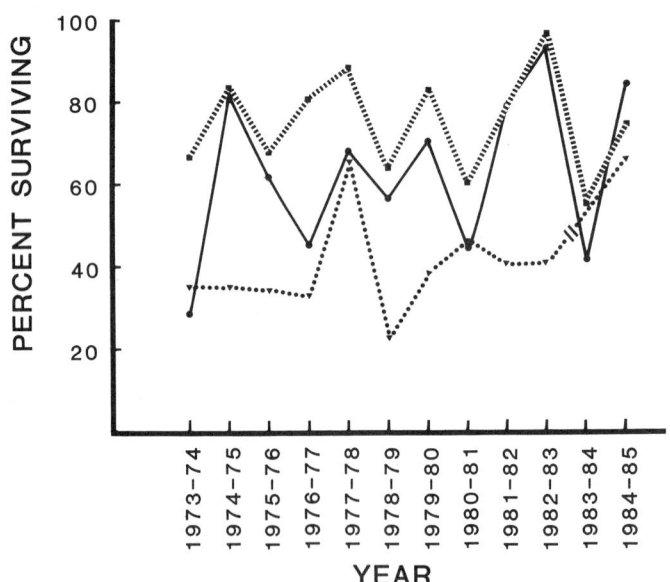

Fig. 7. Annual variation in survivorship of adults (solid squares, hatched line), yearlings (solid circles, solid line), and juveniles (solid triangles, dotted line). Data was unavailable for 1983-84 because jays rarely visited feeding stations.

Table II.   Factors most strongly correlated with age-specific survivorship.   See legend in Table I.

Variable	Piñon Crop	Previous Crop	Weather Factors		
			SP TEMP	PSUM PPT	PSP PPT
Juvenile Lx	r = 0.31	0.48	0.47	0.84	-0.59
	N = 11				
	P = 0.17*	0.07	0.07*	0.01*	0.03*
			SP TEMP	SP PPT	PSP SNOW
Yearling Lx	r = -0.11	0.46	-0.57	0.76	-0.79
	N = 12				
	P = 0.37	0.06	0.03*	<0.01*	<0.01*
			SP PPT	SP TEMP	SP SNOW
Adult Lx	r = -0.20	-0.04	0.55	-0.54	0.43
	N = 9				
	P = 0.30	0.46	0.03*	0.04*	0.08*

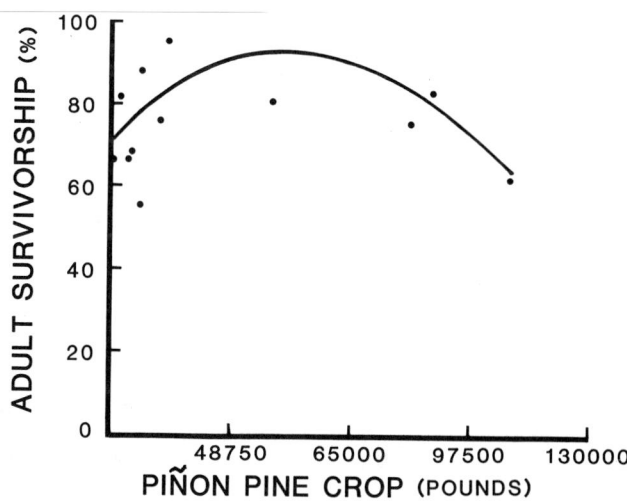

Fig. 8.   Adult survivorship as a function of size of the piñon pine crop.   The curvilinear relationship: Lx = 70.67 + 0.0006 (piñon) – 0.000000006 (piñon)^2 provided the best fit the these data (R = 0.78, $F_{2,9}$ = 6.79, P < 0.05).

while harvesting accounts for low survivorship. Two observations support this. First, adult mortality is heaviest in the fall of the year (Balda and Marzluff, in prep.). Second, heavy and predictable reliance on feeding stations, which may be similar to reliance on localized piñon crops was also strongly associated with reduced adult survivorship (Fig. 9). Reliance on feeders may indicate more than predictable exposure to predators. Jays use feeders most following poor pine crops and dry summers, which suggests that lack of preferred foods (piñon seeds and insects) may drive birds to use less preferred foods offered at feeders. Lack of preferred foods coupled with predictable exposure to predators may result in the strong negative correlation between survivorship and dependence on feeders.

To summarize relative influences on survivorship: weather has the primary influence on survivorship in this population. However, what we felt *a priori* was harsh weather (cold wet springs and winters) actually was associated with high survivorship of yearlings and adults (Fig. 10). This indicates moisture shortages, not low temperature may be of prime importance to this population, perhaps because moisture influences insect and seed production. Reliance on locally abundant food sources like bumper piñon crops and feeders has a cost in terms of lower survivorship. Juveniles survived better, however, during years of bumper pine crops than during crop failures and this influence appeared to be carried over into their higher survivorship as yearlings.

D. Sex Ratio of the Flock and its Influence on Options of Yearlings

Males survive better than females (Balda and Marzluff, in prep.) and this, coupled with yearly variability in survivorship (Fig. 7), results in annual variability of the flock sex ratio (Fig. 11). The sex ratio of adults was always biased toward males, but it varied twofold from 1.3 males per female in 1983-84 to 2.7 males per female in 1976-77. The sex ratio of yearlings was extremely variable, ranging from equality in 1980-81 to a females bias of one male to three females in 1985-86 and a male bias of 5.5 males per female in 1975-76. On average, 64 percent of adults and 61 percent of yearlings in the flock were males. There was a slight inverse relationship between the adult and yearling sex ratio seen in all years except 1977-1980.

The tendency for adult and yearling sex ratios to vary inversely with each other was evident from correlation analyses (Table III). In addition, when the flock was small, male bias among adults was stronger. Perhaps because females are lower on the dominance hierarchy (Balda and Balda 1978) they have higher mortality when conditions are unfavorable and flock size decreases. The importance of spring temperature appears to be a spurious result of its correlation with flock size and the sex ratio of the other age class because controlling these factors increased the P-value for spring temperature to a minimum of $P = 0.29$, whereas the influence of sex ratio and flock size were unchanged by controlling spring temperature. Piñon crop was unrelated to sex ratio.

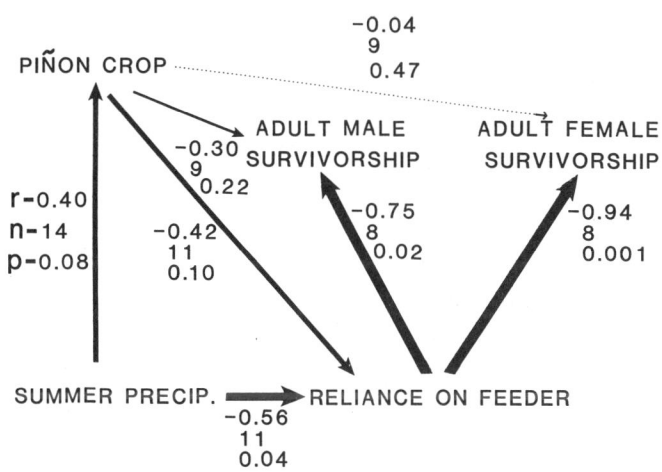

Fig. 9. *Correlations between adult survivorship and food availability. Width of arrows indicates relative strength of correlations. Pearsons' r, N, and two-tailed P values are given for each correlation.*

A potential result of variability in sex ratios is variability in the breeding options of yearlings (Brown 1974; Emlen 1978; Reyer 1980). Yearlings in the Town Flock exercise three options during the breeding season: (1) join a nonbreeding, nonhelping subflock (Balda and Bateman 1971), (2) help their parents at the nest (Balda and Bateman 1971), or (3) breed (Ligon 1971; Marzluff and Balda, in press b.). Whether yearlings breed or not and whether they help or not varies annually (Fig. 12). On average, 11.1 percent of nests each year had a helper; however, this ranged from 2 to 47 percent of nests. Forty-seven percent may be an overestimate because only 15 nests were observed in 1977, seven of which had helpers. Males rarely bred as yearlings. On average only 3.4

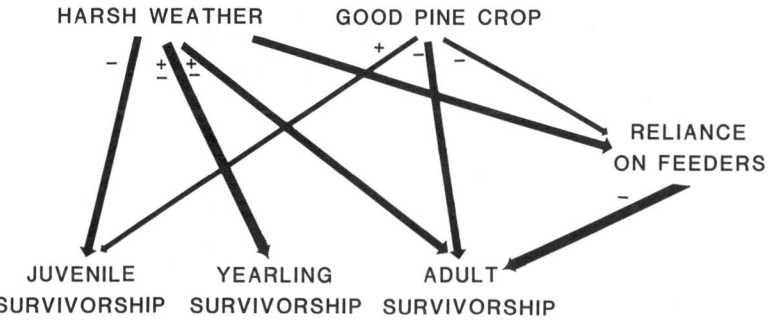

Fig. 10. *Summary of correlations relating weather, pine crop, and reliance on feeders to age-specific survivorship. See legend of Fig. 6.*

Table III. Factors most strongly correlated with sex ratio and options exercised by yearlings. See legend in Table I.

Variable	Piñon Crop	Previous Crop	Weather Factors		Endogenous Factors	
Yearling Sex Ratio	r = -0.11	0.17	SP TEMP		AD SEX[1]	FLOCK[3]
	N = 12		-0.54		-0.34	-0.42
	P = 0.37	0.30	0.04		0.15	0.09
Adult Sex Ratio	r = 0.08	0.03	SP TEMP		YR SEX[2]	
	N = 12		0.42		-0.34	
	P = 0.41	0.46	0.09		0.15	
% Helpers	r = -0.04	-0.39	None[4]		AD SEX	FLOCK
	N = 10				0.66	0.65
	P = 0.46	0.13			0.03*	0.03*
% Male Yr Breeders	r = -0.54	-0.23	SP TEMP	SP SNOW		
	N = 10		0.55	-0.48		
	P = 0.05*	0.26	0.05*	0.08*		
% Female Yr Breeders	r = 0.39	0.37	PSU TEMP		YR SEX	
	N = 10		-0.56		0.53	
	P = 0.13	0.15	0.05*		0.07*	

[1]Sex ratio of adults. [2]Sex ratio of yearlings. [3]Flock size. [4]No weather factors even moderately correlated with percent helpers.

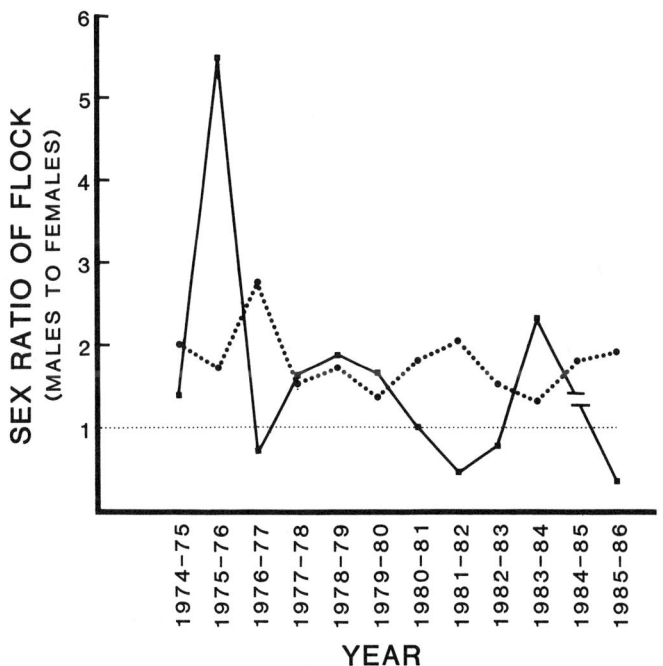

Fig. 11. Annual variation in sex ratio of yearlings (solid squares, solid line), and adults (solid circles, dotted line). Small dotted line indicates a 50:50 sex ratio; points above represent a surplus of males, points below a surplus of females.

percent of nests were occupied by male yearlings, but in some years up to 10 percent of nests were attended by male yearlings. Females breed as yearlings more commonly (Marzluff and Balda, in press b.); an average of 8.8 percent of nests had yearling females (usually mated with two-year-old males).

Sex ratio was closely correlated with the occurrence of helping at the nest and breeding by female yearlings (Table III). When adult females were in short supply, more yearling males helped their parents (r_{pr} = 0.75, P = 0.03), and when yearling males were common more females bred as yearlings (r_{pr} = 0.63, P = 0.05). Both of these trends are hypothesized to result from a shortage of potential female mates in the flock. The less dominant yearling males (Balda and Balda 1978) presumably cannot compete for this limited resource with older, more dominant males and hence helping is an alternative strategy that may provide inclusive fitness benefits and yield experience to the helper (Brown 1974). A higher percentage of nests had female yearling breeders when there were many yearling males. Just as important to this option, however, may be the conditions of yearling females because more bred following

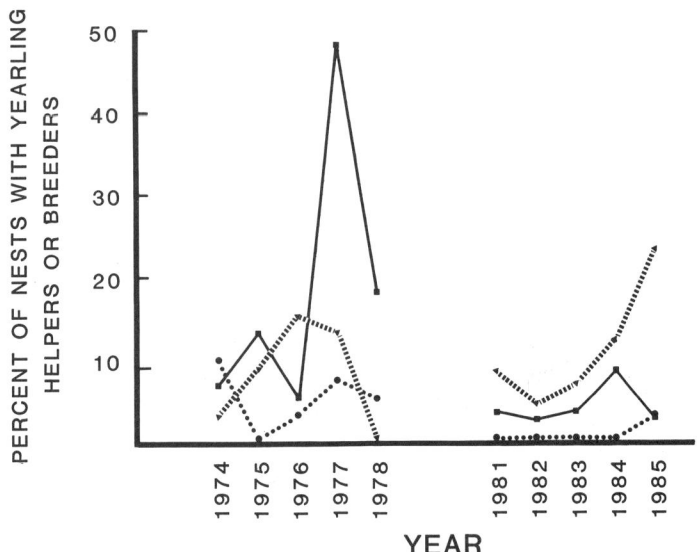

YEAR

Fig. 12. Annual occurrence of yearlings as breeders or helpers.
Percent of nests with helpers (solid squares, solid line), male yearling
breeders (solid circles, dotted line, and female yearling breeders (solid
triangles, hatched line) each year is graphed. No data is presented for
1979 and 1980 because few nests were found in those years. Sample
sizes of nests beginning with 1974 and ending with 1985 are: 31, 23,
40, 15, 22, 26, 52, 50, 43, 39.

good pine crops (r_{pr} = 0.63, P = 0.06) and if they were raised during mild
summers (r_{pr} = -0.66, P = 0.03) (Table III).

E.   Onset of Breeding and Breeding Success

Previous work by Balda and Bateman (1971, 1972) and Ligon (1971,
1974, 1978) has centered on the influence of pine seeds on the timing of
breeding and breeding success in pinyon jays. These workers found earlier breeding and larger clutch sizes associated with large piñon crops, and
weather conditions during breeding were of lesser importance. We also
found this to be generally the case. However, weather was consistently
better correlated with the timing and success of breeding than was piñon
crop (Table IV).

Variability in onset of breeding was great in this population. Piñon
crop and snowfall during the onset of breeding (15 February-15 March)
both appear to be important correlates (Fig. 13). The earliest breeding
was on 19 February in 1977; in contrast, in 1985 first breeding attempts
did not occur until 13 April (52 days later). Approximately equal snow
fell in 1974 (10.7 in), 1977 (10.2 in), and 1982 (12.1 in). During these

Table IV.   Factors most strongly correlated with onset of breeding and breeding success.   See legend in Table I.

Variable	Piñon Crop	Previous Crop	Weather Factors
Onset of Breeding	r = 0.02	0.31	BREED SNOW 0.26
	N = 10		
	P = 0.48	0.20	0.23
% 5-egg Clutches	r = 0.43	-0.30	BREED SNOW -0.73
	N = 7		
	P = 0.17	0.26	0.05*
% 3-egg Clutches	r = -0.48	0.09	BREED SNOW 0.70
	N = 7		
	P = 0.14	0.42	0.06*
Average Fledglings	r = -0.46	0.20	PSU TEMP 0.57
	N = 12		
	P = 0.07*	0.27	0.03*
Average Juveniles	r = -0.19	0.47	PWINT TEMP 0.67
	N = 12		
	P = 0.27	0.06	0.01*

years cone crops differed and breeding was earlier following larger cone crops (Fig. 13).   Heaviest snowfall was recorded in 1982 (35.4 in), yet onset of breeding was equal to that following snowfall one-third as heavy (10.7, 17.8, and 8.7 in), perhaps because of the large piñon crop in 1981 (Fig. 13).   Breeding was very early in 1984 following little snow (1.5 in) and a large cone crop.   Partial correlation analysis indicated neither snowfall nor pine crop was significantly related to onset of breeding, but snowfall was more closely correlated to onset (r_{pr} = 0.27, P = 0.22) than was size of the pine crop (r_{pr} = -0.05, P = 0.44, contra Balda and Bateman 1972).

Clutch size in our population was most commonly four, but in some years over 40 percent of all nests contained three or fewer eggs and in other years over 30 percent contained five eggs (Fig. 14).   Large clutches were more common and small clutches less common during years of abundant piñon seeds (Table IV, Fig. 14), especially when

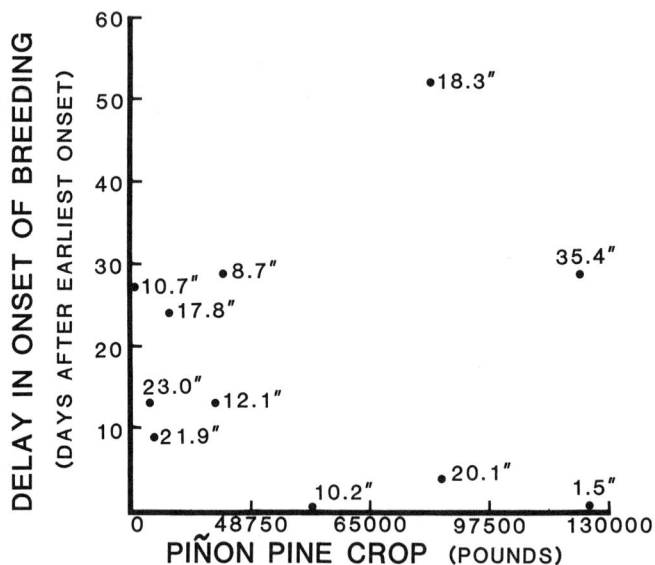

Fig. 13. *Onset of breeding as a function of pine crop and snowfall. Earliest breeding is represented as a 0 delay in onset of breeding. Inches of snowfall from 15 February–15 March are given adjacent to data points.*

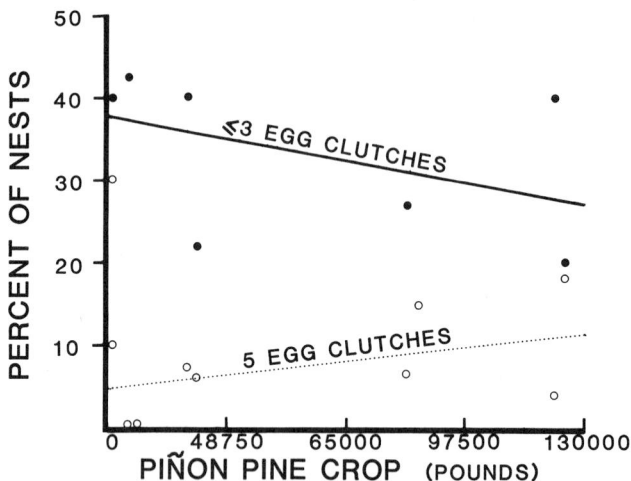

Fig. 14. *Frequency of small (≤3 eggs) and large (5 eggs) clutches as a function of piñon pine crop. Each point represents one year. Least-squares regression lines are plotted. Percent ≤3 eggs = 0.38 – 0.00000088 (piñon), R^2 = 23%, $F_{1,5}$ = 1.51. Percent 5 eggs = .05 + 0.00000046 (piñon), R^2 = 18%, $F_{1,5}$ = 1.13.*

the influence of snowfall was held constant ($\leq$ 3 eggs r_{pr} = -0.76, P = 0.07; 5 eggs r_{pr} = 0.85, P = 0.04) (Table IV), perhaps because females began incubating before clutches were complete (Balda and Bateman 1972) and did not continue to lay when conditions turned cold and snowy.

Warm summers and warm winters the year prior to breeding were positively correlated with average per-pair production of fledglings and yearlings (Table IV). Piñon crop was negatively correlated wth per-pair success, especially the production of fledglings (Table IV). This seems counter to expectation as high-quality food would be available for developing young; however, jays bred earlier following large pine crops (Fig. 14), and spring snow after eggs were laid commonly reduced per-pair success substantially during these years.

In summary, breeding biology was strongly influenced by piñon crop, weather, and flock sex ratio (Fig. 15). Jays bred slightly earlier, laid larger clutches, but produced fewer juveniles following large pine crops. Harsh weather, particularly spring snowfall, which is common at this upper elevational limit for pinyon jays (Fig. 2), often masks the potential influence of the pine crop because breeding was delayed, clutches were smaller, and breeding was less successful during snowy springs. More female yearlings bred following mild spring weather, good cone crops, and when the yearling sex ratio was biased in favor of males. Helping by sons of the breeders was more common when the adult sex ratio was biased toward males. Helping was not strongly correlated with pine crop or climatic factors.

F.   Did Town Flock Jays Read the Literature?

The results we have presented thus far were based on two-tailed probabilities to ascertain the relative importance of weather, food, and endogenous factors on the social structure of pinyon jays. Now we will

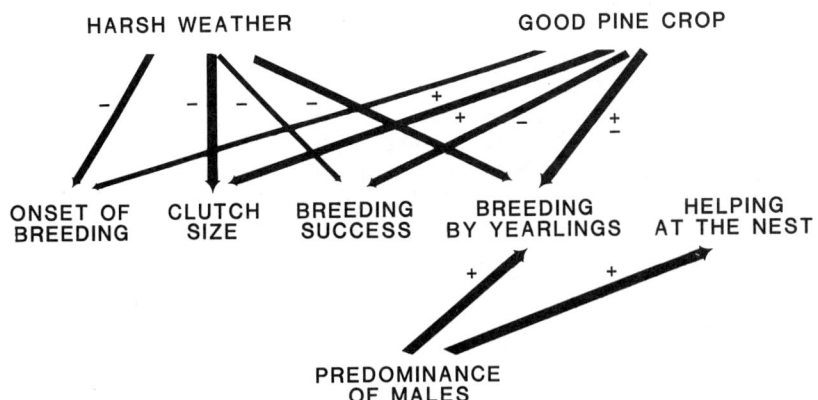

Fig. 15. Summary of correlations relating weather, pine crop, and sex ratio to breeding biology. See legend of Fig. 6.

test predictions from the literature with one-tailed probabilities to determine if this flock of jays responds to fluctuations in weather and food in manners similar to other flocks.

Piñon pine seeds are a very important food source for pinyon jays; therefore, following large pine crops breeding success, survivorship, and flock size should all be higher. Breeding was shown to be earlier in the seasons following bumper pine crops (Ligon 1971, 1974, 1978; Balda and Bateman 1972). We observed a similar, but nonsignificant, trend over nine seasons (r_{pr} = 0.27, P = 0.11). Clutch size should be larger when food is more abundant (Lack 1947; Balda and Bateman 1972). When we held snowfall constant, clutches were significantly larger after bumper pine crops (≤3 eggs r_{pr} = -0.76, P = 0.04; 5 eggs r_{pr} = 0.72, P = 0.05). Survivorship of less dominant members of the flock (young and females; Balda and Balda 1978) should increase with increasing food supply (Lack 1954), but we only observed this to be true for juveniles (r_{pr} = 0.58, P = 0.025). Because sex-specific survivorship was not correlated with pine crop, predictions of fewer male helpers, fewer female yearling breeders, and fewer immigrants in years following abundant food (Brown 1974; Reyer 1980) were not supported. Ligon (1978) predicted that flock size may increase the year following large pine crops because of additive effects of breeding success and survivorship. We observed the opposite trend (r_{pr} = -0.55, P = 0.04), possibly because reliance on a localized food source was associated with lower survivorship and early breeding attempts in our area commonly fail because of unpredictably harsh spring snows.

Our predictions about the influence of harsh weather relate primarily to its potential effect on foods other than pine seeds, primarily insects. Cold snowy winters presumably require more energy to maintain thermal homeostasis and hence if food is in short supply, juveniles and females should experience lower survival (Emlen 1940; Lack 1954) and enter the breeding season in poor shape, thus breeding later (Balda and Bateman 1972) and with reduced success (Lack 1947, 1954, 1968). These effects should translate into skewed sex ratios and smaller flock size. Winter weather, harsh as it can be in the ponderosa pine forest (Fig. 1), had few of these predicted effects. Large flock size was correlated with warm winters (r_{pr} = 0.43, P = 0.04), but breeding and survivorship were unaffected. Reliance on feeders may account for some of these results.

Spring snowfall is another form of harsh weather that appeared to be influential on this population. Spring snow may reduce food availability and increase energetic costs of breeding. In years of heavy spring snow, breeding should be delayed (Lack 1954, 1968), success should be lower (Lack 1968; Balda and Bateman 1971, 1972), survivorship of young and females should be lower (Lack 1954; Balda and Bateman 1972), and sex/age ratios should become more skewed toward older males (Morton and Sherman 1978; North and Morgan 1979). In years with cold, snowy springs, jays bred later (r_{pr} = 0.27, P = 0.11), laid fewer eggs (r_{pr} = 0.85, P = 0.02), and juveniles suffered disproportionately more mortality than older birds (r_{pr} = 0.86, P = 0.001). Mortality was not sex related; therefore, sex ratio and yearling options were unaffected.

Spring and summer drought can be harsh in the Southwest and this may influence birds by reducing insect availability. This should result in lower breeding success (Ligon 1978), but we observed the opposite effect. Breeding was earlier and more successful during years with dry springs, presumably because moisture at this time is usually in the form of detrimental snowfall. Drought had other important influences on this flock; dispersal was reduced (Table I), yearling survivorship was lower (Table II, Fig. 9), and reliance on feeders (and their associated bird baths) was higher which correlated with lower survivorship of adults (Fig. 9).

V.   SUMMARY AND CONCLUSIONS

Coping with stressful environments or variable resources may lead to a variety of adaptations, making it difficult to predict social structure solely on the basis of the distribution of resources. For example, we show that scrub jays and pinyon jays in Arizona inhabit an environment characterized by extreme variability in climate and variability in availability of an important food source, the seeds of piñon pine, yet they exhibit grossly different social organizations. Scrub jay pairs appear to defend territories. In contrast, pinyon jays live in a highly organized permanent flock throughout the year, nest in colonies, and occasionally have helpers at the nest. We explain these differences in sociality as different solutions to the problem of spatio-temporal variability in food resources. Scrub jays are poor long-distance fliers and are unable to efficiently search large areas for locally abundant cone crops. To survive poor cone crops they remain generalists, using whatever resources are available on their home ranges. Small home ranges that supply a variety of foods are economically defendable, thus territoriality evolved. Young disperse from natal territories because the costs of settling in a new territory are low due to a surrounding habitat unsaturated with scrub jays. Pinyon jays, in contrast, have strong flight capabilities and have become piñon pine seed specialists. They are able to fly long distances in search of areas with abundant seeds. Flocking could be adaptive because of increased search efficiency and reduced predation. Locally abundant but widely scattered crops are not economically defendable, so territoriality did not evolve. Reliance on patches of cached seeds and spatio-temporally variable insects during the breeding season may have favored colonial nesting. The breeding biology of pinyon jays appears to have evolved to take maximum advantage of bumper pine crops. They breed earlier, and lay more eggs when pine seeds are abundant. These jays have retained behavioral flexibility that allows them to survive poor cone crops.

We investigated the response of one flock of pinyon jays to resource variability over the last 14 years. We used partial correlation analyses to investigate the relative influence of climate, piñon crop, and sex ratio on flock size, individual survivorship, and breeding biology in this flock.

The number of juveniles prior to dispersal varied greatly between years and was closely correlated with annual fluctuations in flock size. Juveniles were more abundant, and thus flock size larger, following warm winters. Flock size was smaller following abundant pine crops. This counterintuitive result may be peculiar to our study flock because it inhabits the upper elevational reaches of the species' distribution. Early breeding following abundant seed crops is usually not successful at this elevation because of unpredictably harsh spring snow storms. Adult survivorship also was low following good pine crops, perhaps because of increased activity during harvest and continued exposure, in unfamiliar habitat, to predators. Lower juvenile recruitment and lower adult survivorship may explain our peculiar finding of reduced flock size after abundant seed crops.

Immigration of yearlings into our study flock was correlated with large fall pine crops and wet springs. More importantly, however, immigrants (which are primarily females) entered our flock more often when the sex ratio of adults was biased in favor of males. Shortage of available mates may thus influence immigration.

Annual survivorship was variable, but clearly age dependent; adults survived better than yearlings and yearlings survived better than juveniles. Our estimates of juvenile and yearling survivorship are conservative because some disappearances represent unobserved, successful emigration. Weather had the primary influence on survivorship; however, what we felt *a priori* was harsh weather (cold wet springs and winters) was actually associated with high survivorship of yearlings and adults. Moisture, not temperature, may thus be limiting to this population. Reliance on locally abundant food sources such as bumper piñon crops and feeders was correlated with lower survivorship. Juvenile survivorship was higher during years of abundant pine seed production.

Sex ratio varied annually but was not correlated with changes in resources or climate. Helping at the nest by sons of breeders was closely correlated to the abundance of yearling females (potential mates) in the flock. When females were in short supply, helping was more common.

Variability in timing of breeding and breeding success was correlated with resource abundance and snowfall prior to breeding. Weather was consistently better correlated with these factors than was piñon crop. Breeding was delayed and clutches smaller following snowy springs and poor pine crops. Production of independent young was inversely related to pine crop in the previous fall, perhaps because early breeding attempts are rarely successful at the elevation this flock inhabits.

Climatic conditions, abundance of food, and abundance of mates were important correlates of social structure, survivorship, and breeding biology in this flock of jays. Weather was of primary importance, which was expected because this flock lives at the upper elevational limit of the species' distribution. When weather effects were held constant, piñon crop was correlated with flock dynamics in many predicted ways, even though this population receives supplemental food from feeders. Sex ratio was typically biased toward males and was a primary factor influencing options taken by yearlings in the flock.

Resource variability was correlated with variability in social structure between species and with variability within a society. Between species, different social structures appear to result from different degrees of specialization on the variable resource. Increased specialization by pinyon jays on the variable resource of pinyon pine seeds is apparently important to the evolution of their flocking and colonial nesting behaviors. This specialist has retained flexibility. Within a population of jays, onset of breeding and breeding success were related to resource availability thus enabling a rapid response to changing conditions.

ACKNOWLEDGMENTS

We are indebted to A.R.C.S., the Frank M. Chapman Memorial Fund, Sigma Xi, the Wilson Ornithological Society, Joseph and Elizabeth Marzluff, and the sponsored research department of Northern Arizona University for financial support. Much of the data summarized herein was collected by Gene Foster, Jane Balda, Diana Gabaldon, Pat McArthur, Dan Cannon, and Larry Clark. Long-term monitoring of this population would have been impossible without the aid of these researchers and without the aid of Katherine Bartlett and Bill and Judy Burding, who allowed us to watch jays from their homes. Chet Anderson gave freely of his time and energy to provide us with yearly measures of pine seed abundance. David Ligon reviewed this manuscript and has shared his ideas about jays with us over the past 15 years. Lastly, Colleen Marzluff provided financial, social, and technical support necessary for the completion of this project.

REFERENCES

Balda, R.P. and J. Balda. 1978. The care of young piñon jays and their integration into the flock. *Journal für* Ornithologie 119:146-171.
Balda, R.P. and G.C. Bateman. 1971. Flocking and annual cycle of the Piñon Jay, *Gymnorhinus cyanocephalus*. Condor 73:287-302.
Balda, R.P. and G.C. Bateman. 1972. The breeding biology of the piñon jay. *Living Bird* 11:5-42.
Barger, R.L. and P.F. Ffolliott. 1972. Physical characteristics and utilization of major woodland tree species in Arizona. *United States Department of Agriculture Research Paper* RM-83.
Bateman, G.C. and R.P. Balda. 1973. Growth, development, and food habits of young Piñon Jays. *Auk* 90:39-61.
Bekoff, M. and M.C. Wells. 1981. Behavioural budgeting by wild coyotes: The influences of food resources on social organization. *Animal Behaviour* 29:794-801.
Bent, A.C. 1946. Life histories of North American jays, crows and titmice. *United States National Museum Bulletin* 191. Part II.
Bock, C.E. 1982. Synchronous fluctuations in Christmas bird counts of Common Redpolls and Piñon Jays. *Auk* 99:383-384.

Brown, J.L. 1969. Territorial behavior and population regulation in birds. *Wilson Bulletin* 81:293-329.

Brown, J.L. 1974. Alternate routes to sociality in jays--with a theory for the evolution of altruism and communal breeding. *American Zoologist* 14:63-80.

Brown, J.L. 1976. *The evolution of behavior.* New York:W.W. Norton and Company, Inc.

Brown, J.L. and E.R. Brown. 1985. Ecological correlates of group size in a communally breeding jay. *Condor* 87:309-315.

Brown, J.L., E.R. Brown and S.D. Brown. 1982. Morphological variation in a population of Grey-crowned babblers: Correlations with variables affecting social behavior. *Behavioral Ecology and Sociobiology* 10:281-287.

Brown, J.L., D.D. Dow, E.R. Brown and S.D. Brown. 1983. Socio-ecology of the Grey-crowned Babbler: Population structure, unit size and vegetation correlates. *Behavioral Ecology and Sociobiology* 13:115-124.

Cody, M.L. 1974. Optimisation in ecology. *Science* 183:1156-1164.

Crook, J.H. 1965. The adaptive significances of avian social organizations. *Symposium of the Zoological Society of London* 14:181-218.

Emlen, J.T. 1940. Sex and age ratios in survival of the California Quail. *Journal of Wildlife Management* 4:92-99.

Emlen, S.T. 1978. "The evolution of cooperative breeding in birds." pp. 245-281. In: J.R. Krebs and N.B. Davies, eds. *Behavioural ecology an evolutionary approach.* MA:Sinauer Associates, Inc.

Gauthreaux, S.A. Jr. 1978. "The ecological significance of behavioural dominance." pp. 17-54. In: P.P.G. Bateson and P.H. Klopfer, eds. *Perspectives in ethology*, vol. 3. London:Plenum Press.

Gould, S.J. 1980. *The Panda's thumb.* New York:W.W. Norton and Company, Inc.

Hardy, J.W. 1961. Studies in behavior and phylogeny of certain New World jays (Garrulinae). *University of Kansas Science Bulletin* 42.

Henderson, J. 1920. Migrations of the pinyon jay in Colorado. *Condor* 22:36.

Horn, H.S. 1968. The adaptive significance of colonial nesting in the Brewer's Blackbird (*Euphagus cyanocephalus*). *Ecology* 48:682-694.

Lack, D. 1947. The significance of clutch size. *Ibis* 89:302-352.

Lack, D. 1954. *The natural regulation of animal numbers.* Oxford:Clarendon Press.

Lack, D. 1968. *Ecological adaptations for breeding in birds.* London: Methuen and Co., Ltd.

Ligon, J.D. 1971. Late summer-autumnal breeding of the Piñon Jay in New Mexico. *Condor* 73:147-153.

Ligon, J.D. 1974. Green cones of the piñon pine stimulate late summer breeding in the piñon jay. *Nature* 250:80-82.

Ligon, J.D. 1978. Reproductive interdependence in piñon jays and piñon pines. *Ecological Monographs* 48:111-126.

Ligon, J.D. and D.J. Martin. 1974. Piñon seed assessment by the piñon jay, *Gymnorhinus cyanocephalus. Animal Behaviour* 22:421-429.

Little, E.L. Jr. 1938a. The earliest stages of pinyon cones. *United States Forest and Range Experimental Station Research Note* 46.

Little, E.L. Jr. 1938b. Stages of growth of pinyons in 1938. *United States Forest and Range Experimental Station Research Note* 50.

Marzluff, J.M. and R.P. Balda. In press a. Pairing patterns and fitness in a free-ranging population of Pinyon Jays: What do they reveal about mate choice? *Condor.*

Marzluff, J.M. and R.P. Balda. In press b. Pair bonds in a Pinyon Jay society: Advantages of, and constraints forcing, mate fidelity. *Auk.*

Morton, M.L. and P.W. Sherman. 1978. Effects of a spring snowstorm on behavior, reproduction, and survival of Belding's ground squirrels. *Canadian Journal of Zoology* 56:2578-2590.

North, P.M. and B.J.T. Morgan. 1979. Modeling heron survival using weather data. *Biometrics* 35:667-681.

Reyer, H.-U. 1980. Flexible helper structure as an ecological adaptation in the pied kingfisher (*Ceryle rudis rudis* L.). *Behavioral Ecology and Sociobiology* 6:219-227.

Smith, C.C. and R.P. Balda. 1980. Competition among insects, birds and mammals for conifer seeds. *American Zoologist* 19:1065-1083.

Stacey, P.B. and C.E. Bock. 1978. Social plasticity in the acorn woodpecker. *Science* 202:1298-1300.

Trail, P.W. 1980. Ecological correlates of social organization in a communally breeding bird, the Acorn Woodpecker, *Melanerpes formicivorus. Behavioral Ecology and Sociobiology* 7:83-92.

Vander Wall, S.B. and R.P. Balda. 1977. Coadaptations of the Clark's nutcracker and the piñon pine for efficient seed harvest and dispersal. *Ecological Monographs* 47:89-111.

Vander Wall, S.B. and R.P. Balda. 1981. Ecology and evolution of food-storage behavior in conifer-seed-caching corvids. *Zeitschrift für Tierpsychologie* 56:217-242.

Ward, P. and A. Zahavi. 1973. The importance of certain assemblages of birds as 'information-centres' for food-finding. *Ibis* 115:517-534.

Waser, P.M. 1981. Sociality or territorial defense? The influence of resource renewal. *Behavioral Ecology and Sociobiology* 8:231-237.

Westcott, P.W. 1964. Invasion of Clark nutcrackers and piñon jays into southeastern Arizona. *Condor* 66:441.

Woolfenden, G.E. and J.W. Fitzpatrick. 1984. The Florida Scrub Jay demography of a cooperative-breeding bird. *Monographs in Population Biology* 20: N.J.:Princeton University Press.

Chapter 12

FOOD SHARING IN THE RAVEN, *CORVUS CORAX*

Bernd Heinrich

Department of Zoology
University of Vermont, Burlington 05405

285

I. INTRODUCTION

According to the logic of the theory of evolution, only that behavior evolves which promotes gene-copying success. Helping behavior, where one animal gives up time, energy or even direct reproductive potential to aid another, has understandably attracted detailed attention. It is often apparent that a group of cooperating individuals could derive benefits that are available to independent individuals, provided none of the individuals cheat. However, the challenging task at hand is to determine how helping behavior can evolve in terms of individual fitness.

Helping behavior can commonly be explained by kin selection. It is becoming evident, however, that not all helping is restricted to kin. While relatedness may be a bonus that facilitates the evolution of helping behavior, it is often not the ultimate reason for it. Ecological necessity may sometimes be the driving force, as stressed in the examinations of cooperative breeding in birds (Emlen and Vehrencamp 1985), termites (Thorne 1985) and bees (Michener 1985).

Helpers are often making the best of a poor situation, and they are expectedly in a very competitive situation with their benefactors. Helpers-at-the-nest (Woolfenden and Fitzpatrick 1984) may temporarily forego reproduction because of ecological necessity, and recruitment to food may occur as an anti-predator defense (Elgar 1986). I here explore the case for potential voluntary food sharing of the common raven, *Corvus corax*, that is unlikely to be explicable either in terms of kin selection or of predator defense.

The raven, *C. corax*, is the most widely distributed passerine bird of the world, occurring in Europe, Asia and in North America from the high arctic to Nicaragua (Goodwin 1976). Although most of New England was outside the range of this bird in relatively recent times (Samuels 1872; Furbush 1927; Bull 1974) it has reappeared (Langhlin and Kibbe 1985) in the last 20 years along with the coyote (Hilton 1978). The raven inhabits tundra, forests, deserts and a great variety of other habitat, and is an opportunistic feeder (Nelson 1934; Temple 1974: Stiehl 1978). Ravens take advantage of food provided at refuse dumps (Dorn 1972; Hauri 1958; Stiehl 1978; Brown 1974); however, in the absence of food associated with man, ravens feed on carrion (Hauri 1958; Mylne 1961; Harlow et al. 1975; Dorn 1972) and are well-known commensals of large carnivores during the winter (Holzworth 1930; Mech 1970; Harrington 1978; Allen 1979).

Animals killed by team-hunting canids are a rich, highly concentrated source of food, although one that is likely to be highly patchy in time and space. In addition, even when carcasses are available in the northern forests in winter they are apt to be hidden from view, because of the terrain, vegetation and/or snow. Food discovery is therefore very likely a chancy event that takes considerable time and effort.

Various birds, including hummingbirds (Stiles 1975; Stiles and Wolf 1970; Wolf, Stiles and Hainsworth 1976), sunbirds (Gill and Wolf 1975) and honeycreepers (Carpenter and McMillan 1976) vigorously defend clumped valuable resources, as do stingless bees (Johnson and Hubbell

1974). Blue jays, *Cyanocitta cristata*, who are largely vegetarian (Bent 1964) but who will feed at a carcass when one becomes available to them, vigorously chase each other until (at over 50 baits) rarely more than one pair remain feeding on the bait at any one time (pers. obs.). Similarly, even small birds such as red-breasted nuthatches (*Sitta canadensis*) vigorously defend whole cow or sheep carcasses against conspecifics and black-capped chickadees (*Parus atricapillus*) (pers. obs.). Based on these observations of other relatively small birds, it might be expected that a large and powerful bird like the raven would vigorously define a valuable resource it was lucky enough to find, or at least keep it a secret (Krebs and Davies 1981). However, I have routinely observed a half dozen and up to 40 ravens at fresh animal carcasses in Maine.

The aim of this ongoing but yet preliminary study was to explore the ravens' apparent sharing by asking three questions: are the aggregations of birds found at food bonanzas the result of recruitment?; 2) how is recruitment accomplished?; and 3) how might the food sharing behavior have evolved? I cannot yet provide definitive answers. My primary aim here is to provide a speculative overview resulting from over 700 hours of fieldwork over three falls and two winters.

II.   METHODS

Unless otherwise indicated, the observations were made during the autumns and winters between November 1984 and November 1986 near bait (whole or parts of carcasses of cattle, deer, moose, sheep, goats, slaughterhouse offal, as well as small game animals including cottontail rabbits, snowshoe hare, racoons, beavers and squirrels) deposited on the ground in view of either of two cabins deep in the forests at the edge of Mt. Blue State Park, in Franklin County, Maine. One cabin was at the edge of a 1 ha partially overgrown field, commanding a view of some 3 km over a valley against mountains and ridges. From this cabin bait could be placed as far as 85 m in a 270° semicircle and still be visible (in the winter) through a number of windows set in different directions. The second cabin from which observations were made was separated from the first by 1/2 km of woodland, in a 600 m^2 clearing also surrounded by forest. Bait here could be placed up to 60 m distant. Except for isolated instances where I needed to watch the birds for details of behavior at close range, I maintained a large (>60 m) distance between the bait and the blind. The viewing window of either cabin was usually used with a one-way mirror, and the other windows were covered with newspapers, blankets or tarpaper.

Another set of more informal observations was conducted (also for the same two winters), at a "permanent" feeding station in the forest within visual range of my home in Richmond, Vermont.

In the wild the birds presumably need to move from one temporary feeding site to the next as these become available. Except for experiments indicated, I provided bait at the Maine site only once every two weeks, and unless indicated otherwise, I removed the food after I left;

Bernd Heinrich

the ravens generally did not feed for more than 2-3 days at a time at the site before the food was either eaten or removed, and there were 11-12 days when no food was available. This procedure prevented the accumulation of birds at the site; it usually required 1-3 days after I returned before a raven was seen again.

The ravens did not feed if they detected a human in the vicinity. Throughout the study I therefore entered or left the cabins only under cover of darkness before any birds had arrived or after all had left.

Both Maine and Vermont study areas contained coyotes (*Canis latrans* var.). The Maine site also had black bear (*Ursus americanus*). Both predators could (and did) consume unattended baits left for the birds. Sprinkling the bait and/or the surrounding area with human urine appeared to be an effective deterrent to them, but it did not visibly affect the ravens' behavior.

Voice readings in the field were made with a model P-200 E.P.M. parabolic microphone and recorded on cassette tape with a super ANRS portable stereo casette deck (KD-1636 Mark II). Playback in the field used an Electro-Voice Inc. PA12F. Playback volume was 62-66 db at 75 m (as determined with a Type 1565-A General Radio Co. Portable Sound-level meter), roughly equivalent or greater than the birds' own vocal output. Sonographs from the cassette tapes were prepared with a Kay electronics Digital Sono-Graph 7800 using an analysis filter of 150 Hz.

III.   RESULTS

A.   Raven Populations

Ravens are relatively conspicuous birds both because of their large size and their loud voice. One can hear them from a distance of 2 km, and they can also be seen from that far when they are in flight. Nevertheless, I observed on the average only one bird for each 320 km driving along U.S. Rt. 2 through Vermont, New Hampshire and Maine. In 225 hrs of watching and listening for ravens in areas other than known feeding sites in Maine, I observed one or a pair of ravens for every 5.6 hrs of search.

Ravens were characteristically patchily available at any one time and/or place. Throughout February and early March in 1985 at least 13 ravens fed on a cow carcass 3 km from my home in Richmond, Vermont. For a few days after the cow had been entirely consumed except for bones and hide (by these birds, and by coyotes), I observed ravens singly and in pairs within several kilometers of the area. However, by March 23 most of the birds had left; I provided two calf carcasses and only two ravens (and one turkey vulture, *Cathartes aura*), visited the carcass over 18 days. I conclude that most of the previously observed ravens had moved on. Similarly, on 16 January 1986, I observed approximately 50 ravens at the Bethel, Maine, landfill. But none or only 2-4 were present on at least 15 other dates that I checked. In contrast, at a relatively "permanent" winter feeding station (for bald eagles) along the Atlantic coast at Lamoine, Maine, the number of ravens varied from under 50 to over 500 (M. McCullough, pers. comm.) throughout the winter.

Throughout the 1985-1986 winter I provided meat continuously (from 2 to over 150 kg) at the Vermont site. On 10 December, 46 days after the bait was first put out, it was visited for the first time by two ravens; presumably the same pair who subsequently came almost daily to feed throughout the whole winter and until late March. Nevertheless, on 30 December a third raven was briefly in the vicinity (the pair chased it off), and on 27 November a loose group of five flew high overhead in a southerly direction. Five birds also stopped by for about an hour to inspect a calf on 13 April, but they did not feed (possibly because the meat was decaying). These and other observations indicate that the Vermont site contained a sparse population of nesting resident birds; and larger numbers, when present, were likely due to transients.

At my study site in Maine larger numbers of ravens were reliably present only at specific times. During the first winter (1984-85) 15-20 ravens reliably came to baits, at least until 12 March. Similarly, from September through December 1985, up to 40 ravens routinely showed up at large baits. However, from January to late March of the same (second) winter, only two birds came to food bonanzas, even to meat piles of over 600 kg. However, between 25 and 30 March over 40 ravens were again present, and several carcasses (including the adult cow) that had been left out since the end of January were than consumed. After large numbers of ravens (>50) stopped coming to the Bethel, Maine landfill, 50 km to the southwest in January, they also stopped coming at my study site. Nevertheless, pairs of ravens continued to come to baits at both sites throughout both winters.

In the first winter I observed 15-20 ravens at a temporary (<1 wk) communal nocturnal roost 10 km from my Maine study site, reflecting the numbers of birds who showed up at large baits at that time. These data are entirely consistent with the conclusions of Hauri (1956) in Europe, that some ravens (the juveniles and subadults) form nomadic swarms, while others (breeders on their territories) stay year-round.

In late March 1986 I examined five nests in Vermont, New Hampshire and Maine, and the numbers of young per nest at that time were 2, 2, 2, 3 and 5.

B.   Active Recruitment?

After one raven discovered a bait, birds appeared thereafter every morning, and they often stayed for several hours. My counts of birds utilizing baits represent the maximum number I could see at one time; these are necessarily underestimates.

I attempted to determine whether or not ravens arriving at a bait came singly or in groups. However, birds originally arriving as a group can then leave and return singly, as during successive caching trips. To minimize the counting effort imposed by local movements, I monitored the ravens arriving early in the morning, and those that arrived together after a known (small) number were present.

The large numbers of birds ultimately observed at a bait could result from chance aggregation of individuals, by unsuccessful birds using inadvertent clues given by successful birds, or by recruitment where successful birds advertise. I attempted to answer whether or not recruitment occurred by monitoring numbers through time and by comparing these data with expected increases in numbers of birds under different behavioral scenarios.

Ravens were rarely seen in the absence of bait. As indicated, in the absence of a known bait, I saw one or a pair of birds on the average for every 5.6 hrs of watching. Most of these birds were in level flight just above tree-top level at a distance of up to 2 km away from me. (I observed high-flying soaring birds only near the roost and sometimes in the vicinity of the bait after the bait was exhausted). I only saw birds change their flight path, suggesting they had seen a bait, when they flew within approximately 200 m of the bait. I estimate that, if one bird passes through a 4 km swath near the bait every 5.6 hrs, then on the average to randomly accumulate 20 birds should thus require 1,120 hrs of daylight, or 140 days given 8 hrs of daylight per day. This estimate clearly conflicts with my observations: up to 40 birds accumulated at a carcass that had been available for only two days.

The birds did not forage in flocks in my study area. During the two winters of this study I tabulated 60 raven sightings when no bait was known to be present. Forty-two of these sightings were of single birds (although I could have missed a second bird), 18 were of pairs, and one was a group of five (in Vermont). Similarly, 17 of 24 discoverers of baits were made by single individuals, while seven were made by pairs (Fig. 1). Since both the proportions of flying single birds and pairs at large were similar to those discovering the baits, it can be concluded that the birds were foraging not in flocks, but as single birds or pairs. Therefore, the large number of birds eventually observed at any one bait was not due to the chance arrival of flocks, nor to random independent discoveries. Rather, it seems likely that some kind of social interaction accounts for the gatherings.

The rapid build-up of numbers at a bait could conceivably be the result of social facilitation where birds present provide the main target for other birds who are searching. Indeed it would be very surprising if ravens did *not* avail themselves of such information in their food search. However, given the ravens' low density, this mechanism by itself could not account for the almost explosive accumulation of birds. In general, accumulation of birds, after it started to occur, was complete in 1-2 days (Fig. 2). After a bait had been discovered it was almost invariably visited on successive days in the early morning. The eventual increase in numbers was most pronounced overnight (Fig. 2). New birds arriving near dawn were not likely to have found the carcass by searching that day, and it is more probable that they followed previously successful foragers, possibly from a common roost. The successful foragers apparently provided information that elicited following behavior.

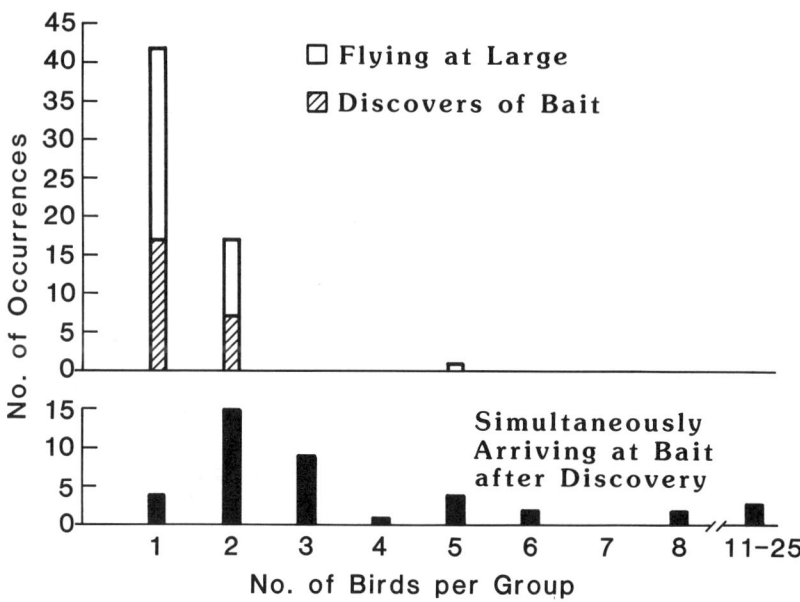

Fig. 1. Numbers of ravens flying together at large discovering baits
(top, hidden bar graph), and arriving at baits after bait discovery.

Fig. 2 demonstrates a typical sequence of bird arrivals at a large
food bonanza, a sheep. The first day after a raven discovered the sheep,
it made several appearances, flying by and vocalizing briefly but not
landing on the bait. On the second morning five ravens came simultaneously and perched nearby in the forest, but they did not feed. On the
third morning seven birds arrived and perched in the trees surrounding
the bait. Within an hour the birds started to feed, and after that the
numbers of birds increased to 20 throughout the day. Overnight on the
dawn of the fourth day the numbers doubled to 40. Many of the newly
arriving birds had white fecal marks on their backs, suggesting that they
had come from a dense roosting aggregation. In other instances (Fig 3A
& 3B), the major increase in numbers occurred throughout the day. In
still another series (Fig. 3C) several birds came, but no further birds
came to this bait with little meat.

   The most dramatic example of apparent recruitment occurred one
late afternoon at a partially eaten deer carcass that I had originally located (by listening to the ravens) and then moved at least 1 km from the
previous feeding site to the field by my cabin. But no raven came near
for the two days that I continually watched the carcass, even though I
frequently heard ravens within several kilometers of the site. I then
broadcast raven calls recorded previously at another food bonanza, and
within 20 sec a raven flew directly towards the sound. I shut off the
tape recorder, and the bird then perched in a tree examining the deer

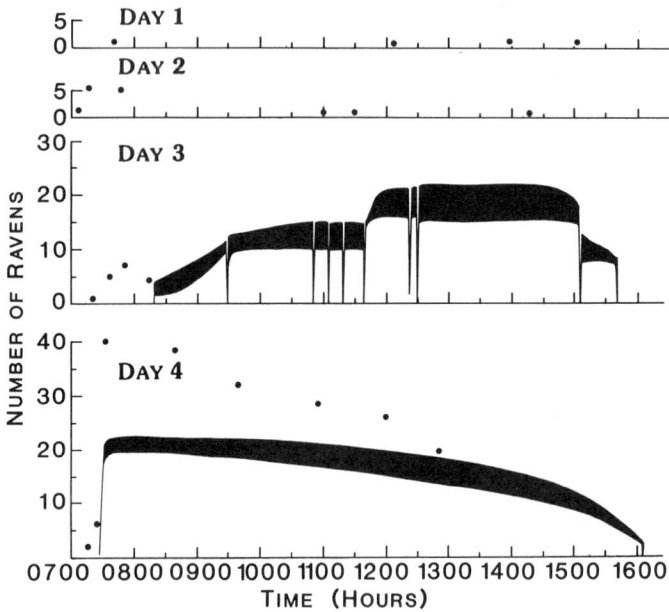

Fig. 2. *The number of ravens seen near an opened sheep carcass the day after it had been discovered (Day 1), when feeding began (Day 2), and the following (two) days until it was consumed. Filled circles indicate birds who were not feeding. Filled areas indicate the range in raven numbers feeding at the carcass at any one time (exact numbers varied on a minute-by-minute basis). Vertical gaps show when all birds left, and again quickly returned.*

carcass for 25 min. No other birds came near, and this raven was relatively quiet, ruffling its feathers, snapping its bill, and making barely audible gurgling noises. It then flew off silently in the direction of the previous feeding site, and within one to two minutes streams of birds (29 in all) were coming back from the same direction the discoverer had departed (Fig. 3A). None of them descended to feed at that time, but approximately 40 birds came on the next dawn and consumed the remains of the carcass.

Rapid accumulation of birds (and rapid consumption of carcasses) also occurs at other sites. In early January, 1986, in Yellowstone National Park, USA, Gillian Bowser (pers. comm.) observed one coyote starting to feed at the carcass of an elk bull (approximate weight = 540 kg) while five magpies and ten ravens were watching. At 1225 hours half of the carcass was consumed, with one coyote, ten magpies and over 25 ravens present at any one time. On the next day the carcass was completely bare, and the ravens were gone.

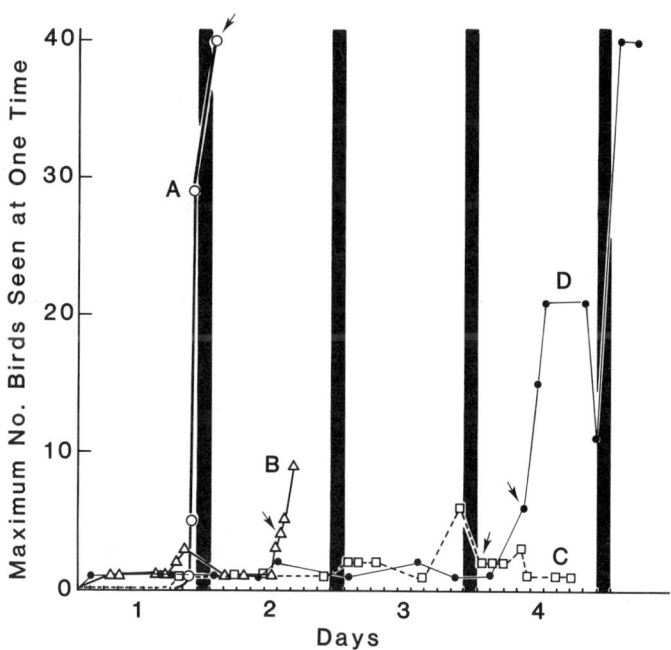

Fig. 3. Maximum number of ravens seen at any one time as a function of time after bait discovery (Day 1). (Nights are indicated by thick vertical bards.) Arrow indicates when feeding first began. A = remains of a deer carcass. B = a Holstein calf (full build-up of birds was not recorded because the observations were terminated on Day 2). C = 1-2 kg meat on a deer skin. D = a sheep.

In Maine, ravens did not leave the carcass at the end of the day in groups. Instead, there was usually a gradual dwindling of numbers (Fig. 2), corresponding often with decreases in the amount of food and increase in agonistic interactions. American crows (not always present), blue jays, woodpeckers, chickadees and nuthatches moved in to pick at the scraps after all of the ravens had left.

C. Recruitment to Playback

Depending on whether or not it is advantageous to permit conspecifics to share a bait, the birds could evolve to make themselves more or less conspicuous at a bait through their vocal or flight behavior. Indeed, if they are conspicuous enough, then local enhancement effects would be functionally similar to local recruitment.

The ravens were highly conspicuous near a bait, at least to humans. Routinely, researchers studying northern predators use ravens to find kills, because they almost always see the birds long before finding the carcasses, especially on snow from an airplane (pers. comm., F. Harrington, H. Hilton, L.D. Mech, and K. Morris). Similarly, it was easy for me on the ground to locate a carcass that had already been discovered by ravens. The primary clues I used were the birds' vocalizations near the bait, although birds in flight were noticeable as well.

If the birds respond to each other's calls and behavior at the bait as well as a human can, then those in the neighborhood for several square kilometers of a bait could hardly fail to be alerted, and in a minute or two they could probably determine the source and/or reason for the calls. I attempted to test the birds' response to a variety of their calls by playback experiments. No response was observed in birds already at the bait and no further birds arrived. However, in the absence of a bait, or when the bait was obliterated (after a snowstorm, or a day following removal) birds that were in the vicinity responded to playbacks of their calls that I routinely heard near baits. In 13 out of 16 trials, birds came directly toward the sound within 20 sec. No birds were visible for at least 10 min prior to the playback. No birds approached in 22 trials of playing the same calls in the forest where no birds were known to be present. Therefore, the calling (at approximately 65 db at 75 m) presumably acts to recruit only over several kilometers. The response does not require individual voice recognition; calls recorded in the Maine study area elicited vigorous response from ravens 300 km east, in Vermont, and 260 km further northwest near Passamaquoddy Bay, Maine.

D. Agonistic Interactions at the Bait

The ravens did not always share their prize peacefully. I observed the birds' interactions from the beginning of feeding on a frozen sheep until it was consumed two days later. Although at least 40 ravens came on the second day after feeding began, only about 20 appeared to have room to feed at the same time. One bird sometimes attempted to grasp a piece of frozen meat that another had hammered out, or it attempted to peck at the same spot where another one was already working. The challenged bird reacted by pecking at the beak of the intruder, or by pulling on its tail or wing feathers. Squabbles sometimes escalated as birds jumped on top of one another or grasped each others' food. In most instances, however, the agonistic interactions lasted no longer than a second or two as one of the birds jumped to the periphery, and then both resumed feeding almost immediately. When many birds were closely crowded around the carcass the evicted bird walked around the perimeter of the feeding birds, and then attempted to gain access at another spot.

In 32 instances, aggressive birds had erect "ear" tufts. In addition, the aggressive birds often also displayed their gular "beard." In 27 birds displaying submissive postures the feathers of the top and sides of the head were fluffed out. No submissives had sleeked feathers at the time of encounter. Almost invariably the birds pecking at apparently "choice" spots on the carcass (where meat was still available) had the feather posture of dominant birds, while the birds on the periphery of the carcass had the fluffed-out heads of submissive birds. Those giving the trill

call and the yell calls (another conspicuous vocalization characteristically heard at baits) showed conspicuous beard and erect "ears."

Agonistic interactions were a direct function of the density of feeding birds (Fig. 4).   With five or fewer birds at a sheep carcass, there
were rarely squabbles.   With ten birds there were approximately 3
squabbles per minute, and with 20 birds present agonistic interactions
increased to nearly ten per minute.

I attempted to discover whether individuals that had been ejected
from a large feeding circle were specifically discriminated against, or
whether they were random individuals that got in the way.   If the displaced birds were a specific subset of the group, then they should have
been ejected consistently, while other birds should have been consistently accepted.   In 53 out of 57 instances the challenged bird went back into
the feeding circle to resume feeding without being challenged within
15 sec.   Similarly, in 55 out of 60 arrivals back to the carcass (primarily
after caching trips) the arriving bird was also not challenged.   Thus
agonistic interactions cannot be accounted for by fights between different groups for possession of the bait.

I did·not see a single bird either injured or permanently excluded
from the bait when large numbers of birds were feeding.   Usually as soon
as the intruder moved a few centimeters away, it was again ignored.
This behavior contrasts sharply with blue jays, where more than two
birds were seldom on one carcass (although in Vermont I saw six blue
jays in the vicinity of one permanent feeder).   A third bird to arrive was

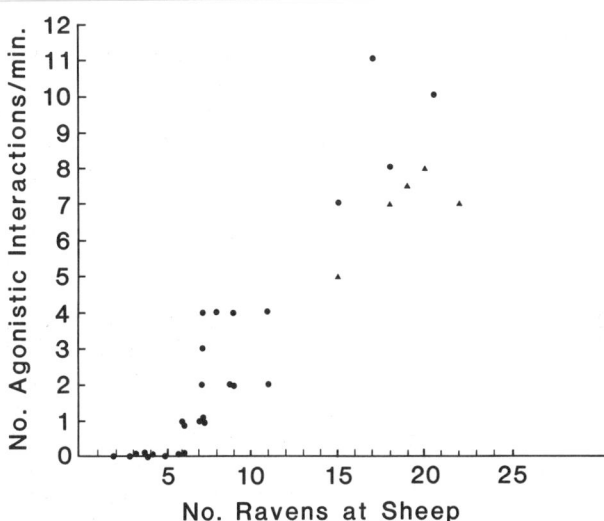

Fig. 4.   *Number of agonistic interactions as a function of number of
ravens feeding at or surrounding a frozen sheep carcass.   Triangles: first
day of feeding.   Closed circles: second day of feeding when most of the
meat was consumed.*

usually vigorously chased, although a second pair sometimes remained at the periphery and later took its turn to feed. Ravens in contexts other than at a bait can be highly aggressive (Gwinner 1964), and there is at least one reported instance of a raven being nearly hammered to death by others in the field (Craighhead 1979, p. 143). I conclude that the agonistic interactions I observed resulted from local interference (occurring primarily after the carcasses were nearly consumed) rather than from attempts to drive all others away from it.

E. Caution

Although the ravens allowed deer and coyotes near them (within several meters) while feeding (pers. obs.), they were highly wary of humans or of "strange" stimuli. The dropping of a spoon in the cabin, a jet passing overhead, or the crack of a tree stem in the woods due to frost would send them to flight. Blue jays and crows, in contrast, paid no obvious attention to these stimuli.

The wariness of the ravens in the bait vicinity was shown also to a high degree at the bait itself. In general the birds flew one or more passes over any dead animal or meat that they found. They came back for additional examinations at several hour intervals lasting up to three days, and before attempting to feed they landed near the bait, walked up to it, pecked it, and jumped up and flew away. On successive approaches the birds eventually only jumped back without flying away, and then they pecked and picked at the bait while making short jumps on and near it. The jabbing-jumping behavior sometimes continued like a dance for several minutes before feeding began.

The time between discovery of the bait and the beginning of feeding appeared to be related to the kind of carrion. Animals native to the area (one moose, one deer, one beaver, two rabbits) as well as two brown Taugenberg goats that looked remarkably like unantlered deer were all pecked within 7 hrs ($\bar{x}$ = 3.5 hrs) of discovery. On the other hand, animals probably unfamiliar to them (two Holstein calves, two Holstein cows, two white sheep) were not pecked until 25 to 74 hrs after discovery ($\bar{x}$ = 49.8 hrs). I conclude bait caution was related to the ravens' perception of "strangeness" of the bait.

The caution, however, was directed not only toward specific kinds of potential prey, but also to specific individuals of that kind. When two Holstein calves were provided simultaneously, feeding by nine ravens eventually commenced on one carcass. None of the nine birds ate the second carcass for two days, and when they landed on the ground behind it they walked a loop around it in order to reach the nearer calf that was being eaten. Similarly, a group of approximately 40 ravens fed on a sheep carcass near my blind until they consumed it, while none fed on another sheep simultaneously present in a field 1/2 km away. The birds flew routinely over this carcass, but seemed oblivious to it. The birds aggregated in time as well. Near the beginning of feeding when one bird flew down to (or away from) the bait, others, if present, followed within seconds, and the birds sometimes advanced toward the bait together (Fig. 5). Clearly the birds were using cues from others' behavior.

I twice observed the presence of goshawks (*Accipiter gentilis*), and three times the presence of red-tailed hawks (*Buteo jamaicensis*) near active feeding sites, but the presence of neither hawk appeared to cause alarm in any of the birds. They ignored the red-tailed hawk and chased one of the goshawks after it approached several of them in flight.

F.   Caching

Ravens cached food shortly after beginning to feed at a large carcass, and they often continued caching for as long as food that was easily torn off was available. There was little or no caching at solidly frozen carcasses.

Typically the birds deposited the food on the ground at least several hundred meters from the carcass and covered it with a leaf or two or other handy debris. Usually birds with meat or fat to cache flew far out of my sight but I located caches in fresh snow at distances up to approximately 1 km from the bait. As far a I could determine, no two caching trips were made to the same spot. In the winter birds stuck food into the snow and covered it with a few sideways shovelling motions of the bill.

After a bait was consumed the birds moved on and the area seemed deserted. One week after a Vermont pair had cached 20 kg of suet, I searched approximately 5 ha but did not find raven tracks anywhere to indicate that any had been retrieved. Similarly, in Maine 12-14 days

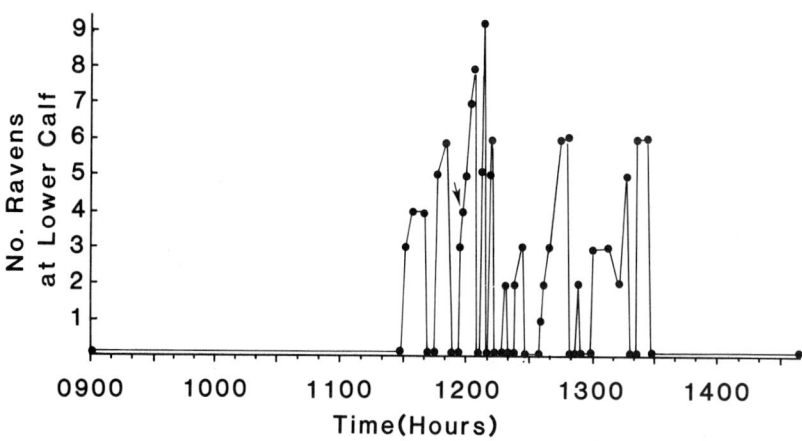

Fig. 5. *The number of ravens feeding at a calf carcass showing simultaneous flight from the carcass and simultaneous return. There were no apparent external stimuli that coincided with the birds' synchronous movements to and from the bait. Arrow indicates time when first feeding began.*

after a sheep was eaten by 40 birds (and 3-5 days after a fresh snow) I examined approximately 5 ha surrounding all sides of the bait site. There were no tracks to indicate that a raven might have attempted to retrieve any cache. Instead, tracks clearly indicated that coyotes had dug out nine presumptive raven caches. Only one known cache was untouched by coyotes.

G.   Olfaction?

Snowstorms of over 30 cm are common in western Maine and would remove most carcasses from sight. A bird which comes back to a feeding site the day after a storm faces the choice: to dig or not to dig? Since scavengers could have removed food in the meantime, the bird could waste considerable time and effort in fruitless digging if it could not determine whether or not the food was still present.

The ravens were able to retrieve meat under snow, provided it was located at the precise spot where they had been feeding previously. Do they detect the outline of the bait through the snow, smell it, or do they rely exclusively on memory of the site?

I tested whether or not ravens would find or retrieve meat buried in snow at other than the expected location. Two fresh pork chop bones, one freshly killed red squirrel, 2 kg unfrozen sheep entrails, and one run-over cat were buried in 5-10 cm of freshly fallen snow at different locations within 1-2 m where approximately 20 ravens had been feeding at any one time on the remains of a sheep. The sheep itself was then removed. It was snowing lightly when the birds returned, and several of them pecked into the snow at the site where the sheep had been, making sideways sweeps of the bill. After much apparent random motion one bird contacted a foot of the cat, and the whole cat was then pulled up and eaten. None of the other baits were located in the next two days, and the ravens left.

I also buried three lamb haunches under light snow within 1 to 2.5 m of where the birds were picking at bare deer ribs sticking out of the snow. None of these buried baits was retrieved in the two weeks that they were available. These results do not exclude the possibility that the birds can smell decaying meat (my captive ravens refused to eat decaying meat), but they indicate that in the winter food that is out of sight is usually unavailable, unless it is dug up by other animals, such as coyotes in Maine.

IV.   DISCUSSION

At the time of European settlement, the raven was found over most of New England (Bull 1974). Along with the wolf, it was virtually extirpated by the settlers who waged relentless warfare upon it (Furbush 1927). Samuels (1872) said "This bird is an extremely rare resident in New England. I have never heard of it breeding here." However, in the past 20 years *Canis latrans* var., a coyote-wolf hybrid, has appeared over

most of New England (Hilton 1978), apparently taking up the niche formerly occupied by timber wolves, *Capis lupus,* and ravens have returned
as well. The raven is now found over most of Maine, New Hampshire and
Vermont (Laughlin and Kibbe 1985).

The raven is now found over most of Maine, New Hampshire and
Ravens are well known to be associated with carnivores. Allen
(1979, p. 288) reports that "In winter a traveling wolf pack is normally
accompanied by a half-dozen or more ravens." Similarly, Mech (1970, p.
279) writes, "Flocks of ravens routinely follow wolf packs from kill to
kill and dine on the leavings of the wolf pack." In addition, Mech (1970,
p. 287) wrote, "When following wolves, ravens fly ahead of them, land in
trees, await the passing of the wolves, and then repeat the process . . .
besides following wolves directly, ravens often track them . . . flying
directly over the trail."   Ravens are also attracted to wolf howling
(Harrington 1978) and are reputed to have gathererd at the report of a
gun in the heyday of bison hunting (Prince Maximilian, as reported by
Allen [1979, p. 288]).   Tinbergen (1958, p. 38) writes of Eskimos in
Greenland who say that ravens follow polar bears on their wanderings
over the outer pack ice and live off the remains of seals killed by bears.
Similarly, Holzworth (1930, p. 370) observed ravens following Grizzly
bears. "The ravens pick at what the bears leave . . . they will fly about
and sit in the tree tops where the bears are . . ." Ravens have a similar
relationship with coyotes, as well as with other carnivores such as
humans where they are not persecuted. Bent (1964, p. 184) describes
ravens "as tame as hens" at Eskimo villages in the Aleutian Islands.

The birds' social behavior, as well as the source and nature of their
food supply, is relevant to a consideration of the evolution of sharing
behavior.   There are frequent reports in the literature of "flocks" or
"swarms" of ravens, principally at feeding sites (Mylne 1961; Dorn 1972;
Coombes 1948; Hauri 1958; Scheven 1955; Meier 1956; Schmidt 1957).
But the nature and mechanisms of the aggregations have previously not
been investigated.

American crows are well-known to have roosts that may number in
the millions (Madson 1976), and common ravens are also known to roost
in large numbers (Cushing 1941; Stiehl 1981). As indicated in the present
study, however, the Maine and Vermont birds were not reliably present
either in any one roost or in the nearby area, probably because of the
vagaries of the food supply, the carcasses. Generally roosts contain several dozen (Mylne 1961) to over a hundred birds (Hutson 1945; Cushing
1941; Lucid and Conner 1974). But nearly a thousand birds have been
reported in a single roost (Stiehl 1981; Young et al. 1985). The birds
from a roost disperse widely during the day (Dorn 1972; Hauri 1958), and
some individually identified birds have been seen to forage as much as 45
km away (Stiehl 1978, 1981). Work with individually marked birds at
Jackson Hole, Wyoming, USA shows that the birds of one feeding flock
are distinct from those of another (Dorn 1972).

The "swarms," which are often nomadic, may persist throughout the
years; numbers fluctuate as juveniles are recruited after the breeding
season in the summer and mated pairs leave in late winter in their third
year of life to establish permanent year-round residences. A single pair

may forage over a territory as large as 250 km^2 (Hauri 1956), although it is unlikely to do so if a reliable food source is locally available. For example, at a permanent feeding site in Nova Scotia extensive tagging studies indicate that most birds do little traveling (Coldwell 1972).

The above conclusions of raven natural history from Europe and North America are consistent with my observations. At both study sites a pair of ravens was present throughout the fall, winter and spring of two years, but groups of birds were present only sporadically. Given that several dozen to several hundred birds roost together, the possibility for information exchange exits (Ward and Zahavi 1973; Krebs 1974; Knight and Knight 1983; Rabenold 1983). However, it does not imply that successful foragers voluntarily share information. Successful foragers could still attempt to keep their finds a secret and fight others off from any food they find. Unsuccessful birds might simply follow other birds and parasitize their information of food localities.

Before sharing by **active** recruitment could evolve, its benefits must exceed its costs. The costs of sharing include the energy expenditure of recruitment and increased competition at the bait. How much does an individual give up by sharing? The carcass of a large game animal could probably provide a raven enough food for an entire winter. However, in nature ravens rely on social carnivores to provide carcasses; although I scented carcasses to keep carnivores away, their presence at the carcass is normal, and the meat should be quickly disposed of by them. Therefore, by sharing with other ravens a bird may be giving up little, i.e., it does not gain by defending the resource from other ravens because it cannot defend it from bears, wolves, and so on.

The primary question I wished to answer in this study was whether or not ravens actively recruit. Since the discoverers of baits were single birds or pairs in similar proportion to free-flying birds, and since groups of birds come in primarily **after** food discovery (Fig. 1), it is clear that the build-up of birds at baits involves social interactions.

In a general way, the pattern of how birds come to the bait (Fig. 2) over time reflects some aspects of the underlying mechanisms (Fig. 6) whereby the gathering is accomplished. A linear increase is expected if there is no recruitment from an infinite pool of birds, whereas the numbers arriving would level off for a finite population. Accumulation by nonindependent discovery could assume a variety of forms. If the "target" is increased (as when birds already at the bait serve as attractant), then this local enhancement should result in an upward inflection of numbers with time. A sudden increase in numbers could mean recruitment and whether the increase occurs before or after feeding begins should distinguish whether or not the birds share the food or the risk of feeding on a new and potentially dangerous food. The observed build-up of birds (Figs. 2 and 3) in relation to feeding initiation conforms to that expected if the birds share. Furthermore, since the vocalizations at the bait are not only loud but also food specific, and since they are a powerful attractant to other ravens, I conclude that ravens do, indeed, actively recruit.

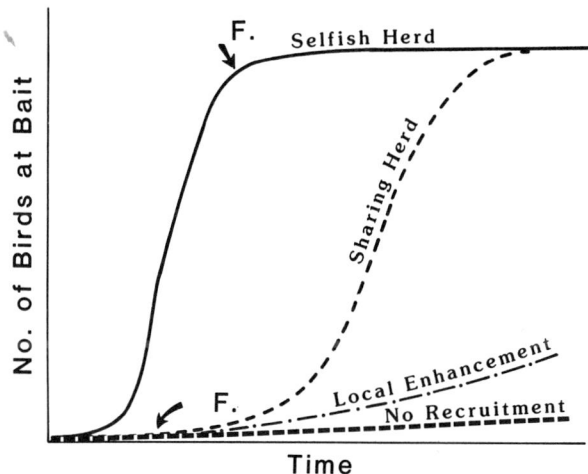

Fig. 6. *Expected build-up of birds at a food bonanza as a function of time of availability of the food and the beginning of feeding (F) depending on different contingencies.*

In view of the above, one may consider several hypotheses for the evolution of active recruitment to food bonanzas by ravens.

A.   Kin Selection Hypothesis

Cooperation could evolve by kin selection within a stable population that shows little dispersal and strong territoriality (Sherman 1977). Similarly, it might be expected in a mobile group whose members are composed along kinship lines. However, ravens appear to be highly vagrant in the winter, and roosts consist of birds originating from isolated nests that can be hundreds of kilometers apart in a population that is apparently panmictic (Stiehl 1978). The population shifts fluidly between roosts, with marked individuals showing no strict roost fidelity (Young et al. 1985). The roost-mates, which sometimes include other species of corvids (Hauri 1958) are unlikely to be exclusive kin-mates; and if the roost serves as a focus for recruitment, then it is difficult to envision how kin could be recruited selectively out of it without non-kin who are present parasitizing the message. Similarly, since the loud vocalizations near the bait act to recruit locally, it is also difficult to envision how non-kin could be excluded there, especially since vagrants could come by at any time. Even if non-kin are somehow excluded because they might not reciprocate in information exchange, then one must still presume that cheating is beneficial. But as indicated previously, the payoff to cheating is low under natural conditions if the food bonanza (a carcass) is controlled and ultimately eaten by a pack of carnivores. The benefits of sharing and the low costs involved should occur regardless of kinship structure and kin selection. Although kin selection could be a large contributing factor to the initial evolution of cooperation, it is unlikely at the present time to be of primary significance.

B. Improved Vigilance

Bird flocks may function in increasing the vigilance for hawks or carnivores while feeding (Caraco 1981; Kenward 1978; Elgar 1986). If the hypothesis applies to corvids foraging at a bait then one would expect recruitment to food in weaker animals such as jays or crows and lack of recruitment in strong ones such as ravens who have little to fear from hawks. In reality, however, the reverse applies. In my observations at baits, the ravens showed little apparent alarm to red-tailed hawks or goshawks, the strongest aerial predators available. Previous observations also indicate that ravens are relatively immune to harm from large predatory birds (Williamson and Rausch 1956; Hutson 1945; Hauri 1958; Cade 1960). To my knowledge there are no published reports of them being harmed by mammalian carnivores at a carcass, although it is not clear whether or not this simply reflects a paucity of detailed observations. Nevertheless, the idea that the birds do not settle on a bait because they fear carnivores in the vicinity is weak because when two baits were placed almost side-by-side, the birds fed for many hours or days on one or the other but not on both at the same time. Predator vigilance should be equally effective while feeding from either bait, and it may also be a contributing but not a prime factor in raven sharing behavior.

C. Selfish Herd

Perhaps there is danger associated with the bait itself. Recruitment could conceivably function to share the perceived risk of and at the bait by diluting the danger (Hamilton 1981). But what is that risk?

The possibility that the site itself is the main source of fear can be ruled out, because if a new bait is placed within several meters of the old that has been consumed then the new one is treated with nearly the same apparent caution as any other bait. After a bait has been fed from it can be moved, but the birds do not hesitate for long to resume feeding. I conclude that the birds fear primarily the bait itself and not potential dangers associated with the bait location.

It is likely that shyness is an innate fear of novel stimuli. Gwinner's (1964) ravens who were raised from nestlings in captivity were interested in but feared every "conspicuous, unknown object," and regarded such from the safety of their perch (in the cage) for one hour or more. They then approached it cautiously. When they were outside the cage the birds flew closer and closer to the object, finally landing briefly to hit it with their bill, to then fly up again and repeat the process until the shyness gradually left them. Gwinner (1964) speculated that their caution was concerned with testing an object for good. My observations in the wild are similar to Gwinner's except that the caution I observed was more exaggerated.

Gwinner (1964) observed the caution particularly expressed in juvenile birds. An adult dominant bird of the group eventually took on the task of checking the strange object, after which the others freely visited it. If the alpha-bird was absent and the object was not contacted, however, then the other birds were for days nervous and unwilling to land on the ground in their cage. My own observations with two groups of captive ravens are entirely consistent with Gwinner's observations: contrary to the behavior of most other birds, in ravens it is the juvenile birds who are the most skittish. Only the dominant ones among them were the leaders who tested the bait.

The ravens' caution and curiosity may reflect the possibility that the object might be either a dead animal to eat or a sleeping and/or dying but still dangerous predator. One way for a raven to be sure that the object is both harmless and palatable is for a predator to be feeding on it. Since ravens in my area probably normally feed on the carcasses of deer attended by coyotes, it may be unusual for them to find unattended cattle and sheep carcasses like the ones I provided. The difference in prefeeding time between native and non-native carcasses strongly supports the hypothesis that they were afraid of the food itself, not predators near it.

Regardless of the reason for their fear, the selfish herd hypothesis is unlikely to be the main reason for recruitment. This hypothesis suggests the birds should recruit as soon as they find food, when they still fear it. However, recruitment usually occurred only when feeding was well under way (Figs. 2, 3) after the birds appeared to have lost their fear.

D. Scavenger Alert

Although the ravens have strong bills they were unwilling or unable to penetrate the hides of frozen as well as unfrozen moose and deer. (It is possible that they can penetrate decayed carcasses, but they do not eat decaying meat.)

Has their vocal behavior at a bait evolved to bring in hide-cutting scavengers, with other birds then parasitizing the message? If so, then vocalizations and recruitment should be restricted to unopened carcasses where the birds cannot feed, rather than vice versa. However, my ravens showed little interest in unopened carcasses and did not recruit to them; no ravens accumulated for months at carcasses that had been discovered by ravens but which were not cut open. I therefore suggest that although it is possible that carnivores could parasitize the ravens' message, it is unlikely that ravens attempt to call in carnivores or scavengers.

E. Gangs

In the jackdaw, *Corvus munedula*, flocking may function to secure the bait from the much larger crows who are dominant to the jackdaws individually (Röell 1978). Ravens could potentially recruit in order to repel the carnivores holding a bait. However, there are no reports of

ravens excluding large carnivores from the kill, and my data indicate
that the agonistic interactions at a bait are generally not between specific potentially competing sub-groups. Nevertheless, the possibility
remains that a group of ravens may recruit to take away and to hold kills
from resident adult pairs.

F.   Reciprocal Altruism

The recruitment of others might be based on a return of a favor to
others who are individually recognized (Trivers 1971), and possibly
ravens recognize each other as individuals in their semi-permanent
"swarms" since many corvids show varying degrees of individual recognition (J.L. Brown 1963, 1964; Gwinner 1964; Thompson 1969; Berger and
Ligon 1977; Röell 1978; E.D. Brown 1985). However, for reciprocal
altruism to occur a successful raven would have to pick out to show food
to those specific individuals of the roost who have aided it in the past.
On the other hand, perhaps a bird could broadcast its message to as
many roost-mates as possible in order to have a large number of future
reciprocators. But opportunities to reciprocate are likely to be very
rare, so that a reciprocator will not likely be differentiated from cheaters. As in the kinship hypothesis, it is doubtful that the recruitment can
be selective either at the roost or in the vicinity of the bait itself, because the ravens are highly vocal and conspicuous within several kilometers of a profitable bait, and reciprocal altruism also appears to be an
unlikely explanation for the recruitment.

G.   Food Preparation and Maintenance

I saw no evidence that the birds used coordinated effort or even
multiplication of effort to help them in tearing open a carcass. However, they may require multiplication of effort to keep it from being
snowed under. In the winter the food supply can disappear in a few hours
during a snowstorm. Since the birds apparently cannot smell buried
meat in the winter, how could they solve the frequently recurring problem of their food supply being buried? They might rely on mammalian
scavengers to locate and dig out a carcass. More interesting is the hypothesis that snow-removal is accomplished by ravens acting in a group.
In one instance I observed 20 ravens feeding on a sheep carcass in the
early morning, just as a severe snowstorm was starting. By nightfall,
when the snowstorm had deposited over 30 cm of snow, the sheep carcass still lay fully exposed; the birds working constantly on the carcass
throughout the day had kept it cleared of accumulated snow.
The adaptive significance of recruitment under this hypothesis may
not be **directly** linked to snowfall, because snowstorms are random
events that cannot be predicted when a food bonanza is discovered.
Since there was nearly one snowstorm per week at my Maine study site
during the 1985-1986 winter, it might be advantageous to recruit regardless of whether or not it is snowing, on the possibility that it might snow
soon.

I conclude that snow removal could be a contributing factor in raven recruitment in that snow exaggerates the ephemeral nature of the food supply and reduces the advantage of cheating. In effect the mutual snow shoveling at a carcass is a reciprocation or team work, because the effort of one bird is immediately repaid by the efforts to another of the same group. However, this hypothesis is weakened by the observation that there were apparently no more presumptive recruitment calls per unit time during a snowstorm than before or after one.

H.   Status Enhancement

In many birds part of the male's mating display involves either direct food offering to the female being courted, or display (such as flight performance) that demonstrates vigor and capability to secure food. In ravens the primary limitations in getting food are in finding it and possibly in testing whether it is safe. One possibility (for which I have no evidence either for or against) is that ravens who recruit others (after they themselves have been recruited, or who have originally discovered and tested the food) thereby gain status and an advantage in securing mates from among the others of the swarm. Additionally, the bait could be an arena where the agonistic interactions result in honest advertisement of male quality. A linkage between recruitment and mate evaluation seems to me to be a promising possibility for further investigation. If correct, it would be analogous to the mate-acquisition techniques used by some dung beetles at elephant dung piles in Africa; these beetles use a large ephemeral food resource as a focus for finding and recruiting mates (Heinrich and Bartholomew 1978). Rather than incurring substantial costs that are commonly associated with most other sexual selection, recruitment to an ephemeral food bonanza would not only demonstrate individual fitness at little risk to the displayers, but it would also provide direct benefits to those displayed to.

I.   "Hopeful" Reciprocity

One of the greatest benefits of sharing, given the ephemeral nature of the food bonanzas is de facto reciprocation. I loosely define "hopeful" reciprocation as a long-term reciprocity in which direct individual paybacks for favors given are not assured. Rather, indiscriminate sharing among members of a long-term flock or among individuals that occupy the same geographical area would result in an informal sharing without any intent or direct selective pressure for it. Non-assured reciprocity is probably feasible only if there is huge asymmetry between costs and benefits and a low advantage to cheating. Reciprocity of information exchange between many independently searching birds, who together cover a vast territory, would increase enormously the availability of food on a more regular basis, and it might lower the threshold for the evolution of sharing by some other mechanism, such as in G and H above. By itself, however, the behavior would not likely evolve due to invasion of cheaters.

V. A POTENTIAL SCENARIO

Given the above nine hypotheses in the context of our present knowledge of raven natural history and behavior, I construct the following
tentative scenario for the evolution of the sharing behavior. The key is
in the great advantage of reciprocity given the birds' specialization on
ephemeral and scarce food bonanzas in fall and winter. American crows,
who have much larger roosts than ravens (Madison 1976) never recruited
in my study. But although they feed on carcasses, this food is not their
specialty.

The association of the young with the parents until fall and into the
winter could result in information exchange to spread the risk of not
finding sparsely distributed food bonanzas. However, raven clutches
from any one year are often only 2-3 individuals, and information exchange restricted to those family members that the birds grow up with
would be limited. The young could enormously increase their information pool, as well as their ability to exploit patchy resources, by joining a
larger group. This group also serves as their source of future mates.

The carnivores that provide the food resources to the group are
likely to be the major consumers of that food as well (but this remains to
be investigated), and a bird that shares a find probably gives up relatively little, and it can gain little by cheating and not sharing. Thus one
might expect that there is a relatively low threshold for the evolution of
sharing behavior. But how can the sharing behavior evolve?

I view both group selection and kin selection as highly unlikely explanations in view of the birds' now well-established dispersed population
structure, vagrancy and apparent lack of cohesive or exclusive group
maintenance. Reciprocal altruism is also an unlikely explanation. Individuals who share with many other individuals may indeed buy future
favors, but the behavior for reciprocity should not spread if that for
cheating has equal fitness. It is unlikely that cheaters can be discriminated against in the present system because they will not very likely be
identified; only a few of the 40 or more birds that could share a carcass
will ever be fortunate enough to find another carcass to share so that no
matter how hard genotypic sharers try to reciprocate they may be phenotypically identical to cheaters. Thus, although a bird may share with
many hopeful reciprocators, it would also share with the potential
cheaters. In both kin selection and reciprocal altruism the strategy is
open to invasion of cheaters. However if the sharing acts in status enhancement, and if the bait can serve as an arena where status can unam-
biguously be evaluated, then the benefits could accrue directly to the
giver of information. I therefore tend to favor this hypothesis as one of
the more promising for future focused experimental investigation. The
roles of males vs. females and high-status vs. low-status males in food
discovery could provide some interesting insights to this hypothesis.

Vultures face a similar situation as ravens with regard to food distribution. Have they evolved a similar foraging system? Most vultures
roost communally, and circumstantial evidence strongly suggests that
roosts of black and turkey vultures serve as information centers in food

finding (Rabenold 1983). Similarly, some species of vultures find food by watching others descend to food (Krunk 1967; König 1983). However, there is no evidence so far that any vulture **voluntarily** recruits others to share food. To the contrary, the evidence suggests that information exchange is by mutual parasitism.

In terms of ecological adaptation and evolution, however, the difference between mutual parasitism and "voluntary" reciprocity may be trivial. Because of the open habitat and prevalance of thermals, soaring vultures are perhaps singularly open to information parasitism. Like ravens, they fight at carcasses, but sharing results nevertheless. Whether intended or not, in effect the information parasitism in vultures results in sharing, and as long as the birds remain in the same geographical area it would result in reciprocity with the effect that the temporal patchiness of food is reduced.

The system evolved by the ravens may be closely analogous to that of the !Kung Bushmen of the Kalahari Desert. Each band of Bushmen, living off wild plant food (Veldkos, in Africaans) and large game, occupies a large territory. There are no social obligations to share the veldkos. However, the meat of large game animals is distributed throughout the band (Marshall 1965). The band's bivouac is like the ravens' roost from which hunting parties of one to five individuals go out in different directions. Like ravens searching for large food bonanzas, it is more advantageous for !Kung trying to find sparsely distributed game to cover a wide area with many small parties, rather than to search in one direction with one large party. The game animal, after being tracked down and killed is a temporary food bonanza. It must be processed quickly before it is consumed by leopards, hyenas, jackals, maggots and bacteria. The hunter may satisfy his hunger, but he is obligated to share the spoils with the rest of the band. He has little to lose by sharing in the present, and he has much to gain for the future. For one thing, a young hunter gains status; he must demonstrate his skill as a hunter to be eligible to marry. After a hunter makes a kill, a system of successive waves of sharing ensues until the carcass has been subdivided into ever-smaller portions that are distributed throughout the whole band. The pattern satisfies kin, as well as enmeshing non-kin in future obligations for insurance for the future. In arctic areas high unpredictability of food supply has also favored mutual food sharing in humans (Durham 1976). Ravens, who face highly unpredictable, ephemeral and large food bonanzas are possibly also sustained by a system of mutual sharing. The results of the present ongoing study are consistent with this hypothesis, but they raise more questions than they answer.

VI.   SUMMARY

Ravens breed and forage solitarily, and they are relatively rare and sparsely distributed in New England. However, up to 40 birds sometimes showed up within two days after a large food bonanza had been discovered by one of them. Playback experiments show that the vocal behavior of birds at the bait attract others, and circumstantial evidence suggests that there is long-range recruitment, possibly from a communal roost. The apparent recruitment behavior reduces the great patchiness of ephemerally super-abundant food that results from the birds' commensalism with large mammalian carnivores. Nine hypotheses to explain the evolution of the foodsharing behavior are considered, and an interaction of several factors is considered likely.

ACKNOWLEDGMENTS

Steven Smith and Gus Verderber helped in the field. Vernon Adams, Bernie Gaudette, and Mike Pratt helped in providing animal carcasses. Dana Eames gave logistical support. Joan Herbers made many helpful comments on the manuscript. I thank them all.

REFERENCES

Allen, D.L. 1979. *Wolves of Minong*. Boston:Houghton Mifflin.

Bent, A.C. 1964. *Life histories of North American jays, crows, and titmice*. Part I. New York:Dover Publishing, Inc..

Berger, L.R. and J.D. Ligon. 1977. Vocal communication and individual recognition in the pinon jay, *Gymnorhinus cyanocephalis*. *Animal Behaviour* 35:567-584.

Brown, J.L. 1963. Social organization and behavior of the Mexican jay. *Condor* 65:126-153.

Brown, J.L. 1964. The integration of agonistic behavior in the stellar's jay *Cyanocitta stelleri* (Gmelin). *University of California Publications in Zoology* 60:223-323.

Brown, R.N. 1974. Aspects of vocal behavior of the raven *Corvus corax* in interior Alaska. M.S. thesis. Fairbanks:University of Alaska. 139 pp.

Brown, E.D. 1985. The role of song and vocal imitation among common crows (*Corvus brachyrhynchos*). *Zeitschrift für Tierpsychologie* 68:115-136.

Bull, J. 1974. *Birds of New York State*. Garden City, N.Y.:Doubleday/ Natural History Press.

Cade, T.J. 1960. Ecology of the peregrine and gyrafalcon populations in Alaska. *University of California Publications in Zoology* 63:151-290.

Caraco, T. 1981. Energy budgets risk and foraging preferences in dark-eyed juncos (*Junco hyemalis*). *Behavioral Ecology and Sociobiology* 8:213-217.

Carpenter, R.L. and R.E. MacMillan. 1976. Threshold model of feeding territoriality and test with a Hawaiian honeycreeper. *Science* 194:639-642.

Coldwell, C. 1972. Raven banding in Nova Scotia. *Bird-Bander* 43:288.

Coombes, R.A.H. 1948. The flocking of the raven. *British Birds* 41:290-294.

Craighead, F.C. Jr. 1979. *Track of the grizzly*. San Francisco:Sierra Club Books.

Cushing, J.E. Jr. 1941. Winter behavior of ravens at Tamales Bay, California. *Condor* 43:103-107.

Dorn, J.L. 1972. The common raven in Jackson Hole, Wyoming. M.S. thesis. Laramie:University of Wyoming.

Durham, W.H. 1976. Resource competition and human aggression, Part I: a review of primitive war. *Quarterly Review of Biology* 51:385-415.

Elgar, M.A. 1986. House sparrows establish foraging flocks by giving chirrup calls if the resource is divisible. *Animal Behaviour* 34:169-174.

Emlen, S.T. and S.L. Vehrencamp. 1985. Cooperative breeding strategies among birds. pp. 359-374. In: B. Hölldobler and M. Lindauer, eds. *Experimental behavioral ecology and sociobiology*. Suttgart and New York:G. Fisher Verlag.

Furbush, E.H. 1927. *Birds of Massachusetts and other New England states*. Vol. 2. Boston: Massachusetts Department of Agriculture.

Gill, F.B. and L.L. Wolf. 1975. Economics of feeding territoriality in the Golden-winged sunbird. *Ecology* 56:333-345.

Gwinner E. 1964. Untersuchungen über das Ausdrucks- und Sozialverhalten des Kolkraben (*Corvus corax corax L.*). *Zeitschrift für Tierpsychologie* 21:657-748.

Goodwin, D. 1976. *Crows of the world*. Cornell University Press.

Hamilton, W.D. 1971. Geometry of the selfish herd. *Journal of Theoretical Biology* 31:295-311.

Harlow, R.C., R.G. Hooper, R.D. Chamberlain, H.S Crawford. 1975. Some winter and nesting season food of the common raven in Virginia. *Auk* 92:298-306.

Harrington, F.H. 1978. Ravens attracted to wolf howling. *Condor* 80:236-237.

Hauri, R. 1956. Beiträge zur Biologie des Kolkraben (*Corvus corax*). *Ornithologisch Beobachten* 55:156-168.

Heinrich, B. and G.A. Bartholomw. 1978. Roles of endothermy and size in inter- and intraspecific competition for elephant dung in an African dung beetle, *Scarabeus kevistriatus*. *Physiological Zoology* 52:484-494.

Hilton, H. 1978. Systematics and ecology of the eastern coyote. pp. 209-228. In: M. Bekoff, ed. *Coyotes*. New York:Academic Press.

Holzworth, J.M. 1930. *The wild grizzlies of Alaska*. New York:G.P. Putnam and Sons.

Hutson, H.P.W. 1945. Roosting procedure of *Corvus corax laurenci* Hume. *Ibis* 87:455-459.

Johnson, L.K. and S.P. Hubbel. 1974. Aggression and competition among stingless bees: field studies. *Ecology* 55:120-127.

Kenward, R.E. 1978. Hawks and doves: Factors affecting success and selection in goshawk attacks on woodpigeons. *Journal of Animal Ecology* 47:499-460.

Knight, S.K. and R.L. Knight. 1983. Aspects of food finding by wintering bald eagles. *Auk* 100:477-484.

König, C. 1983. Interspecific and intraspecific competition for food among old world vultures. pp. 153-171. In: S.R. Wilbur and J.A. Jackson, eds. *Vulture Biology and Management*. Berkeley:University of California Press.

Krebs, J.R. 1974. Colonial nesting and social feeding as strategies for exploiting food resources in the Great Blue Heron *(ardea nerodias)*. *Behaviour* 51:99-134.

Krebs, J.R. and Davies, N.B. 1981. *An introduction to behavioural ecology*. Oxford:Blackwell.

Laughlin, S.B. and Kibbe, D.P. eds. 1985. *The atlas of breeding birds of Vermont*. Honover:University Press of New England.

Lucid, V.J. and Conner R.N. 1974. A communal common raven roost in Virginia. *Wilson Bulletin* 86:82-83.

Madson, J. 1976. The dance on Monkey Mountain. *Audubon* 78:52-60.

Marshall, L. 1965. The !Kung Bushmen of the Kalahari Desert. pp. 243-278. In: J.L. Gibbs (ed.). *Peoples of Africa*. New York:Holt, Rinehart & Winston.

Mech, L.D. 1970. *The wolf: The ecology and behavior of an endangered species*. Garden City, New York:The Natural History Press.

Meier, H. 1956. Kolkrabensammlungen im Mai. *Ornithologisch Beobachten* 53:16.

Michener, C.D. 1985. From solitary to eusocial: need there be a series of intervening species? pp. 293-305. In: B. Hölldobler and M. Lindauer, eds. *Experimental behavioral ecology and sociobiology*. Stuttgart and New York:G. Fisher Verlag.

Mylne, C.K. 1961. Large flocks of ravens at food. *British Birds* 54:206-207.

Nelson, A.L. 1934. Some early summer food preferences of the American raven in southeastern Oregon. *Condor* 36:10-15.

Rabenold, P.P. 1983. The communal roost in black and turkey vultures - an information center? pp. 303-321. In: S.R. Wilbur and J.A. Jackson, eds. *Vulture Biology and Management*. Berkeley:University of California Press.

Röell, A. 1978. Social behaviour of the jackdaw, *Corvus monedula*, in relation to its niche. *Behaviour* 64:1-124.

Samuels, E.A. 1872. Birds of New England and adjacent states. 6th ed. Boston:Noyes, Holmes, and Co.

Scheven, J. 1955. Ein Kolkrabenschwarm. *Vogelwelt* 76:212-216.

Schmidt, G. 1957. Geselligkeit beim Kolkraben, insbesondere in schleswig-Holstein. *Ornithologischy Mitteilung* 9:121-126.

Sherman, P.W. 1977. Nepotism and the evolution of alarm calls. *Science* 197:1246-1253.

Stiehl, R.B. 1978. Aspects of the ecology of the common raven in Harney Basin, Oregon. Ph.D. dissertation. Portland:Portland State University.

Stiles, F.G. 1975. Ecology, flowering phenology, and hummingbird pollination of some Costa Rican *Haliconia* species. *Ecology* 56:285-301.

Stiles, F.G. and L.L. Wolf. 1970. Hummingbird territoriality at a tropical flowering tree. *Auk* 87:465-492.

Temple, S.A. 1974. Winter food habits of ravens on the Arctic slope of Alaska. *Arctic* 27:41-46.

Thompson, N.S. 1969. Individual identification and temporal patterning in the cawing of common crows. *Communications in Behavioural Biology* 4:29-33.

Thorne, B.L. 1985. Termite polygyny: the ecological dynamics of queen mutualism. pp. 325-31. In: B. Holldobler and M. Lindauer, eds. *Experimental behavioral ecology and sociobiology*. Stuttgart and New York:G. Fisher Verlag.

Tinbergen, N. 1958. *Curious naturalists*. New York:Doubleday, New York.

Trivers, R.L. 1971. The evolution of reciprocal altruism. *Quarterly Review of Biology* 46:35-37.

Ward, P. and A. Zahavi. 1973. The importance of certain assemblages of birds as information centers for food finding. *Ibis* 115:517-534.

Williamson, F.S.L. and R. Rausch. 1956. Interspecific relations between goshawks and ravens. *Condor* 58:165.

Wolf, L.L., F.G. Stile, and F.R. Hainsworth. 1976. Ecological organization of a tropical highland hummingbird community. *Journal of Animal Ecology* 45:349-379.

Woolfenden, G.E. and J.W. Fitzpatrick. 1984. *The Florida scrub jay: Demography of a cooperative-breeding bird*. Princeton:Princeton University Press.

Young, L.S., K.A. Engel, A. Brody and R. Bowman. 1985. Implications of communal roosting by common ravens to operation and maintenance of the Malin to Midpoint 500 KV transmission line. pp. 33-72. *Snake River Birds of Prey Research Project annual report 1985*. U.S. Department of the Interior. Boise, Idaho:Bureau of Land Management.

V. ECOLOGY OF SOCIAL ARTHROPODS

Chapter 13

SOCIAL COMPETITION UNDER MANDATORY GROUP LIFE

Gregory B. Pollock[1]

School of Social Science
University of California, Irvine 92717

Steven W. Rissing[2]
Department of Zoology
Arizona State University, Tempe 85287

[1]Supported by NSF grant IST-8305819 and a MacArthur Foundation
Fellowship in International Peace and Security granted by the Social
Science Research Council.

I. INTRODUCTION

Under the logic of kin selection, beneficiaries of altruism are essentially passive participants in social interaction at its emergence, for a beneficiary cannot solicit aid until the capacity to give aid (and violate the logic of classical fitness [*sensu* West Eberhard 1975]) already exists. Although potential beneficiaries may try to induce reproductive inferiority in relatives, such induction assumes that the latter are already phenotypically capable of altruism. Here, at the emergence of altruism, a beneficiary is unable to induce altruism; if he acts to limit the reproductive capacity of relatives he must do so for reasons unrelated to their potential altruistic behavior. Once altruism is evolutionarily stable, selection operates on beneficiaries to manipulate relatives to maximize the former's inclusive fitness. In this sense altruism must be evolutionarily prior to the *manipulation* of altruism, i.e., dominance.

Recently an alternative interpretation of the relationship between dominance and altruism has been made. Dominance contests for the control of resources--i.e., 'social competition'--are seen as inducing altruism on the part of subordinate losers. Having little hope of reproduction without access to lost resources, subordinates offer aid to dominants in order to remain near these resources. In this view the degree of relatedness between dominant and subordinate is of secondary importance; in principle, subordinate altruism may be regularly expressed among non-relatives (Vehrencamp 1983a; West Eberhard 1982).

For social competition to occur, resources must be mediated by social interaction (West Eberhard 1979, 1981, 1983). Selection for group life thus establishes the background for social competition (Emlen 1982a, 1984). When group life is mandatory, social interaction is inescapable, and one might expect socially induced altruism to be at its extreme (Vehrencamp 1983a; West Eberhard 1982). Here we examine the logic of social competition under mandatory group life. Rather than induce subordinate altruism, mandatory group life often selects against the expression of dominance; this will be shown as a consequence of the essentially group selective dynamics of TIT FOR TAT (Axelrod 1981, 1984; Axelrod and Hamilton 1981). We then present a further condition for the expression of subordinate altruism under group life: that the existence of the potential subordinate is not necessary for the continued reproduction of the 'dominant.' Violation of this simple condition helps explain patterns of cooperation among cofounding ant queens and argues against the evolution of a reproductive division of labor among non-relatives in the Hymenoptera. Further, satisfaction of the condition accounts for a primitive form of subordinance among social spiders and is consistent with patterns of subordinance in territorial birds and a territorially monogamous New World monkey.

II. MANDATORY GROUP LIFE AND SUBORDINANCE

A. Dominance, Group Life, and Kin Selection

Primitively eusocial wasps often establish colonies pleometrotically, i.e., with multiple foundresses. Such queens usually derive from the same parental nest (West Eberhard 1969; Klahn 1979; Jeanne 1972; Noonan 1981; Strassmann 1983), are often related (Gamboa et al. 1986), and exhibit a reproductive division of labor, with some cofoundresses largely foregoing oviposition to specialize as foragers (Pardi 1948; West Eberhard 1967, 1969; Jeanne 1972; Litte 1977, 1979; Noonan 1981; Strassmann 1981, 1983). Although this concurrence of relatedness and altruism is compatible with kin selection, this need not imply that all cofoundress interaction is a function of relatedness. Related queens often contest for positions of dominance on starting nests. Unless dominance contests are the proximate mechanism for gauging phenotypic quality among relatives, a process supplementary to kin selection seems necessary to explain why individuals appear to force relatives into a subordinate, fitness-limited role of forager-specialist. West Eberhard (1967, 1969) has suggested that adult nests might produce gynes of varying phenotypic quality to limit the options of the relatively inferior, predisposing them toward subordinance once aware of their relative inferiority via dominance contests. In this view the fitness of parental colonies is enhanced when sexual offspring co-found, selecting for parental manipulation of offspring nutrition to induce subordinance among siblings. When siblings of similar quality encounter each other they separate, searching for other nests or establishing their own; an inferior, however, remains to aid her superior relative. Here dominance proximally creates group life by permitting a female to judge her relative phenotypic quality. Dominants do not force altruism from subordinates but rather inform them of their phenotypic quality, thus allowing the subordinate to enhance its inclusive fitness via altruism.

Recent evidence questions this view. Under the above logic, phenotypic variance (as measured by body size) among cofoundresses on a given nest should be maximal; in *Polistes annularis*, however, females from the same parental nest assort randomly with respect to body size when founding new nests (Strassmann 1983; Sullivan and Strassmann 1984); further, larger foundress associations are not composed of smaller females. Hence dominance contests do not seem a consequence of parental manipulation during development in this species. If dominance contests do not mediate the creation of groups, then selection for group life among queens must exist as such (Strassmann 1983; Sullivan and Strassmann 1984). Dominance is then contingent on the prior existence of group life. This view, however, has lead to the unexpected conclusion that subordinate cofoundresses behave altruistically for reasons unrelated to kin selection (West Eberhard 1978b, 1979, 1981, 1982, 1983).

Suppose selection operates on queens to cofound, perhaps in defense against behavioral parasites (Michener 1958, 1974; Lin 1964; Lin and Michener 1972; West Eberhard 1978a) or to recover rapidly from nest predation (Gibo 1978; Strassmann 1981). Queens will exhibit an environmentally induced variance in body size (West Eberhard 1978a), providing natural competitive advantages to some in aggressive encounters. Larger females may then try to steal the provisions or eat the eggs of their weaker nestmates. If group life is highly advantageous, the latter may be unable to flee to another location, creating de facto forager specialists among losers (Lin and Michener 1972; West Eberhard 1978a). Although relatedness might make subordinate losers more readily altruistic, assuming the phenotypic capacity for altruism has already evolved via kin selection, in principle such 'altruism' could exist among non-relatives (Vehrencamp 1983a,b). If so, kin selection need not be involved in the evolution of hymenopteran sociality at all (West Eberhard 1982). The apparent failure of kin selection to account for dominance as exhibited in primitively eusocial wasps thus jeopardizes its import for eusociality as a whole.

B.   TIT FOR TAT and Dominance Contests

The preceding analysis of the possible irrelevance of kin selection rests on a simple yet incompletely explored assumption: that a subordinate-loser has no option in response to an aggressor save escape and concomitant loss of advantageous group life. By modeling aggressive encounters between cofounding wasps as an iterated Prisoner's Dilemma this assumption will be shown wanting.

Consider a female returning with prey. A nestmate may either attempt to steal this prey or not. If not, the females live a cooperative, communal life. Otherwise one female stands to gain at the expense of the other. If the returning female relinquishes her prey she behaves as a subordinate altruist. If she resists rather than submit, she potentially limits her future fitness, for she is unable to continue provisioning; this also, however, induces a cost for the aggressor, as the latter cannot provision so long as she meets resistance. A form of the Prisoner's Dilemma results where cooperation is the absence of aggression or resistance while defection indicates their presence (Fig. 1).

Resistance may end in three ways. First, the forager may retain possession of her prey. If so, the aggressor experiences a reduction in fitness, for she has expended energy without payoff and must still provision her eggs. Second, the forager may lose the contest but remain alive. The loser, nonetheless, retains a retaliatory option: refuse to forage further. If the winner forages, the loser may simply parasitize the provisioning mass of the former while she is away; this enhances the fitness of the latter at a cost to the former. If the winner refuses to forage, the fitness of both suffer.

Finally, resistance may end in death for the resister. But then the aggressor suffers a reduction in group size. Since group life is advantageous, this incurs a cost. When aggression is common, females should distribute themselves into groups to enhance their relative value, increasing the cost of their loss to an aggressor; whenever existence of the

INDIVIDUAL B

	COOPERATES	DEFECTS
COOPERATES	COMMUNAL LIFE	**A** SUBORDINATE TO **B**
DEFECTS	**B** SUBORDINATE TO **A**	COLONY FITNESS DECLINES

Figure 1. Social competition among cofoundresses as an iterated Prisoner's Dilemma. When subordinate resistance sufficiently depresses group fitness, inter-group competition makes dominance a sub-optimal strategy within the population.

'subordinate' resister is necessary for the aggressor, the potential dominance of the latter may be constrained.

When the continued existence of colonymates is itself an advantage, refusing to forage after aggression, or simply resisting that aggression, establishes a TIT FOR TAT strategy (Azelrod 1981, 1984; Axelrod and Hamilton 1981): forage in proportion to one's oviposition rate but resist aggression and/or refuse to forage when the subject of aggression. Since repeated foraging bouts are necessary, this strategy is played under an iterated Prisoner's Dilemma; given sufficiently low mortality while foraging, this establishes resistance as an evolutionarily stable strategy (Pollock, in review).

Under TIT FOR TAT, groups with cooperating females outproduce those in internal conflict. A potential dominant seeks to skew intra-colony fitness in her favor (Vehrencamp 1983a,b). This strategy is effective, however, only if the colony's relative population fitness is not unduly affected by dominance contests. By resisting aggression a female controls colony foraging effort, depressing colony fitness and thus leaving a 'dominant' at a disadvantage in the population (Pollock, in review). Inter-colony selection prevents or depresses the expression of intra-colony aggression.

C.   Cooperation among Claustral, Cofounding Ant Queens

The above argues for the impossibility of a phenomenon--a reproductive division of labor among unrelated, cofounding wasps. Since wasp cofoundresses exhibit both a reproductive division of labor and relatedness, such species are supportive but not demonstrative of the present thesis; one could hold that both kin selection and social competition produce a reproductive division of labor among cofounding wasps. Some ant species also establish nests pleometrotically (Rissing and Pollock in press[a]). Unlike social wasps, ant cofoundresses readily associate with non-relatives (Bartz and Hölldobler 1982; Tschinkel and Howard 1983; Mintzer and Vinson 1985; Ross and Fletcher 1985; Rissing and Pollock 1986; Rissing et al. in press). Here kin selection may be excluded in an analysis of cofoundress behavior, and a reproductive division of labor among ant queens would provide clear evidence for the logic of social competition. Below we review data on the behavior of cofounding queens of the desert seed harvester ant *Veromessor pergandei* supporting the thesis that social competition yields evolutionarily stable aggression only when subordinates are necessary for the reproduction of dominates.

1.   Selection for pleometrosis in ants
Species of ants regularly founding colonies pleometrotically share a tendency to aggregate starting nests. Starting nests of *Solenopsis invicta* are grouped on microtopographically elevated areas (Tschinkel and Howard 1983), a possible adaptation to frequent flooding in their native flood-plain habitat. Starting nests of *V. pergandei* are grouped in ravine bottoms where they may have closer access to groundwater (Pollock and Rissing 1985; Rissing unpub. data). Finally, starting nests of the desert leaf-cutter ant *Acromyrmex versicolor* are clumped under the canopy of large trees where soil temperatures are low enough to apparently permit successful establishment (Rissing et al. in press).
Adult colonies of each of these pleometrotic species display distinct territoriality (*S. invicta*: Wilson et al. 1971; *V. pergandei*: Went et al. 1972, Wheeler and Rissing 1975, Byron et al. 1980, Ryti and Case 1984; *A. versicolor*: Gamboa 1974); thus, from each clump of starting nests, only one adult nest prevails. Under these conditions, starting colonies rapidly display territorial behavior in the form of brood raiding among nests (Bartz and Hölldobler 1982; Tschinkel and Howard 1983; Rissing and Pollock in press[b]). Pleometrotic ant colonies generally produce more brood in less time than their haplometrotic (i.e. founded by a single queen) counterparts (Waloff 1957; Stumper 1962; Markin et al. 1972; Taki 1976; Mintzer 1979; Bartz and Hölldobler 1982; Tschinkel and Howard 1983; Rissing and Pollock in press[b]; Rissing, unpub. data for *V. pergandei*). In *V. pergandei* there is direct evidence that young colonies with larger worker forces enjoy a competitive advantage over their smaller rivals. Postclaustral colonies engage in reciprocal brood raiding, leading to eventual abandonment of brood-depleted nests. When pleometrotic colonies were pitted against haplometrotic ones, the former overcame the latter in 16 of 19 trials (Rissing and Pollock in press[b]). Starting colonies of *V. pergandei* are clumped in the field, often less than a meter apart (Pollock and Rissing 1985; Rissing unpub. data). Such a population structure among immature nests, when coupled with brood raiding, selects for pleometrosis via the production of a superior raiding force (Rissing and Pollock in press[b]).

2.   Queen aggression and the value of group life
Like most higher ants, *V. pergandei* queens establish nests claustrally, rearing their first brood from stored energy reserves in isolation.

Claustral queens tend and contribute to brood equally, with no evidence of aggression or dominance (Rissing and Pollock 1986). Once workers eclose and forage, however, queens become hostile to one another, engaging in prolonged fights which lead to an increase in queen mortality and a trend toward monogyny (Rissing and Pollock in press[b]).

While claustral, colonies are closed energy systems. Since queens rear brood from their own energy reserves, investing in brood potentially decreases a queen's longevity. When queens cofound, the potential for exploitation exists; by investing less in brood than a colonymate, a queen increases her chances of surviving claustrality at the expense of those relatively overinvesting. Let cooperation be defined as investing in brood as needed. Define defection as investing less than this. If one queen defects while another cooperates, the former benefits at the expense of the latter. Mutual defection reduces brood quality and thus the benefits of group life. This defines a persistent, or interated, Prisoner's Dilemma (Fig. 1). If queens are able to monitor their stress level, increasing stress may indicate defection by a colonymate. To retaliate, a queen need merely cease brood care, conserving her remaining energy. This retaliatory strategy is similar to TIT FOR TAT: care for brood equally, but resist attempts to induce asymmetries in this care between queens. Resistance disrupts the selective value of pleometrosis, for colonies in internal conflict invest less in brood and must thus produce an inferior raiding/defense force relative to those without conflict. By refusing to invest further the victim of defection controls colony fitness, making the defection of her colonymate an unviable option within the population. Brood raiding under a dense population structure of starting nests also occurs in *Myrmecocystus minicus* and *Solenopsis invicta*; both species also exhibit pleometrosis and cooperative group life while claustral (Bartz and Hölldobler 1982; Tschinkel and Howard 1983).

Once workers forage, colony fitness is no longer completely contingent on queen cooperation. Brood raiding has already begun, and an independent resource now exists for which queens may compete, for the value of the foraging force is invariant to queen aggression (Rissing and Pollock 1986). Furthermore, at the end of claustrality queens are under considerable energetic stress. Eliminating fellow queens channels worker investment toward the aggressor, maximally reducing her stress. 'Subordinate' control over colony fitness is limited, and queens now have divergent preferred allocations of worker effort; it is precisely at this point that queen aggression leading to mortality occurs (Rissing and Pollock in press[b]).

An analogue of the above analysis of *V. pergandei* exists in the tropical wasp *Metapolybia aztecoides* (M.J. West Eberhard, pers. comm.). This species founds nests in swarms with workers and usually several queens. Initially several queens produce worker-destined eggs, permitting a rapid buildup of the worker force. Queen conflict increases as the colony matures until it is monogynous. Subordinated queens either leave with a worker swarm, presumably to establish a new colony, disappear, or become foraging workers (West Eberhard 1978b, 1981). That dominance conflict increases as a colony produces its own worker force suggests that intra-queen aggression is inversely related to the value of cofoundresses as producers of workers. Since an independent resource, the swarming worker force, exists at colony foundation and queens are related, some degree of subordinance is possible at colony initiation.

D. Competition at the end of mandatory group life

1. Selection for pleometrosis and oligogyny

The early colony ontogenies of only a few claustral, pleometrotic ants have been documented to date. Rather than engage in direct queen aggression for control of the worker force, some species exhibit 'oligogyny' or multiple, dispersed queens in a single nest. If selection for pleometrosis is largely through an enhanced worker force, oligogyny has the potential to cancel this advantage. Pleometrotic colonies often produce fewer workers/queen than haplometrotic colonies (references above). If queens divide the pre-worker brood and establish independent units, a queen should, on average, have fewer workers than a haplometrotic counterpart. Queens of the subterranean ant *Lasius flavus* do exactly this (Waloff 1957). Queens become oligogynous prior to first worker eclosion and continue to drift apart as their colonies mature, for Waloff and Blackith (1962) note young field colonies sometimes bud into still smaller ones. If workers of this species exhibit site fidelity to their area of eclosion (and hence to the nearby queen), as is known for other polygynous/oligogynous ant species (*Pseudomyrmex venefica*, Janzen 1973; *Iridomyrmex purpureus*, Hölldobler and Carlin 1985), oligogyny effectively reduces the worker force of a given queen. Nonetheless, the other major benefit of pleometrosis, faster production of workers, remains. The natural history of *L. flavus* suggests that this may be the primary advantage of pleometrosis in this species. Colonies farm aphids and seem able to live off them completely (Pontin 1978; Brian 1983). When workers farm under ground rather than forage above ground, fewer seem necessary for the survival of a young colony. Rapid worker production, however, would still reduce demand on a queen's energy reserves. In contrast to *V. pergandei*, ergonomic stress at the end of claustrality is likely brief and moderate in *L. flavus*, permitting a smaller worker force per queen and oligogyny as a solution to queen competition.

Brood raiding among immature colonies, however, may preclude oligogyny completely. To the extent that oligogyny establishes their independence, queens are more vulnerable to brood raiding after oligogyny. In such cases queens are forced to hold their worker force in common. Indeed, rather than exhibit oligogyny, *V. pergandei* queens spend more time together tending a common brood as the eclosion of workers approaches (Rissing and Pollock in press[b]). Queens are unable to divide their communally created resource; when coupled with energetic stress at the end of claustrality, a climate ripe for queen aggression results.

An analogue of the above exists in the Brazilian honeybee which occasionally produces afterswarms with several queens. These queens fight, or the swarm divides into smaller monogynous ones. Since worker number is correlated positively with swarm/colony survival, there must be a lower bound where fighting is preferred over splitting (Cosenza 1972, in Michener 1974). Similarly, an oligogynous ant queen must produce some minimum number of workers to survive.

2.   Queen conflict through parasitism and oligogyny

Competition may take forms other than queen aggression. When queens coexist in a nest they must share workers to some extent. Maintenance and growth of the worker force potentially limit a queen's production of sexuals. Any behavior which reduces a queen's share of worker production while not affecting colony size permits her to increase her production of sexuals. Given the opportunity, queens should parasitize the worker force of others (Janzen 1973). Oligogyny, when coupled with worker site fidelity, is, however, an effective means of preventing such parasitism. If workers care for brood primarily where they themselves eclosed, brood care as a resource is not communal; queens must produce their own workers. Worker site fidelity exists in monogynous and secondarily polygynous species of *Pseudomyrmex* (Beulig and Janzen 1969; Janzen 1973), the meat ant *I. purpureus* (Hölldobler and Carlin 1985), and foragers of several ant species (Carroll and Janzen 1973; Davidson 1977; Herbers 1977). If workers specializing in brood care also exhibit such fidelity, possibly to regulate colony efficiency (see Wilson 1985), site fidelity in brood care can evolve prior to oligogyny. Coexisting queens can then gain exclusive access to brood care simply by spacing themselves apart and producing brood at a single location.

III.   DOMINANCE WHEN SUBORDINATE CONTROL OF GROUP FITNESS IS LIMITED

A.   Subordinance as a Consequence of Asymmetry in the Value of Group Life

The greatest harm a subordinate may do to group fitness may be measured by her absence. When a dominant is subject to subordinate resistance the possibility of expelling the subordinate exits. When dominants have this ability, expulsion should occur when subordinate resistance provides the greater decrement to a dominant's population-wide fitness. Through resistance, subordinates depress their value from the perspective of those dominant. If dominants may reproduce on their own, there is a lower bound on the degree of fitness suppression a subordinate may induce via resistance.

The degree of resistance a subordinate will show should also be a function of her fitness outside of the group. When her relative extra-group viability is high, resistance should be pronounced (Vehrencamp 1983a,b; Emlen 1982b, 1984). Effective subordination should occur when the extra-group viabilities (here 'group' is defined by inclusion of the potential subordinate) of 'dominant' and 'subordinate' are highly asymmetric. That is, subordinate 'altruism' among non-relatives should occur when the subordinate values group life more than the dominant. In such cases a subordinate's altruism buys her tenure in the group through the tolerance of the dominant. When subordinates must live in groups, but are of limited value to dominants, the former may beg for the opportunity to be 'altruistic.'

The Florida scrub jay is obligatorily territorial; no individual has been observed to survive without tenure on some territory. Territories saturate the habitat. An individual losing its territory must thus gain access to another, already occupied, if it is to survive. Monogamous pairs, however, often breed without the presence of others (Woolfenden and Fitzpatrick 1984). The value of group life is thus asymmetric between a resident monogamous pair and an extra-territorial individual. As a consequence, such an individual will beg a pair for the privilege of becoming subordinate in return for territorial residence.

That subordinate scrub jays are often offspring of the breeding pair does not invalidate this analysis. Territoriality does not require that only one pair breed within a territory; other territorial birds regularly exhibit multiple breeders (references below). Although kin selection may act on offspring to increase their tolerance to subordination, dominance itself may be interpreted as a consequence of the asymmetric value of the presence of third parties for breeders relative to the necessity of territorial residence for non-breeders. That individuals remain subordinate on their natal territory even after one or both parents are supplanted supports this view (Woolfenden and Fitzpatrick 1984).

More direct evidence of a secondary role for relatedness between subordinate and breeder occurs in the tamarin *Saguinus oedipus*. This New World primate's breeding system is similar to that of the Florida scrub jay: individuals are territorial, with a single breeding pair and several non-breeders on a territory. Considerable immigration into established territories exists, however, especially among immatures (Dawson 1978). Juveniles disperse to other territories by accepting subordinate positions; in return for residence, subordinates act as sentinels while others feed. Such dispersal reduces relatedness between subordinates and those dominant. Nonetheless, the former remain non-breeding. To reproduce, a tamarin must colonize a territorial opening. Immigration permits a hopeful breeder to monitor a larger number of territories. Admittance to an adjacent territory likely has a higher value for immigrants than for residents admitting them. In the Florida scrub jay, flight permits monitoring without immigration. Individuals may then remain on their parental territory while subordinate, enhancing their indirect fitness (*sensu* Brown and Brown 1981; Emlen 1982a, 1984).

B.   Communal Breeding and Inter-group Competition

When a territorial breeding pair is unable to hold on to a territory adequate for reproduction, group life becomes advantageous, and possibly mandatory, for residents as well as joiners. Dominance should then be incomplete (Craig 1980a). Breeding pairs of the pukeko *Porphyrio porphyrio* have a greater fledgling success than their communal counterparts; communal breeding is, nonetheless, more common (Craig 1980b).

This species exhibits prolonged territorial battles. Territories with extra males are able to expand at the expense of those composed of a single breeding pair, often to the point of dissolution of the latter (Craig 1984). Although territories are necessary for breeding, they are not necessary for survival (Craig 1979). This places a potential joiner in a favorable bargaining position, for here a reproductive resident needs group life perhaps more than a completely subordinated, functionally sterile, joiner. Joining should then be contingent on some degree of mating success on the part of the joiner when group size is small (Craig 1980a). Mating is promiscuous in *P. porphyrio* (Craig 1980a) although not equitable (Craig pers. comm. in Vehrencamp et al. 1986). The degree of skew (*sensu* Vehrencamp 1983a,b) should be a function of the relative value of group life for resident and joiner (Emlen 1982b, 1984).

Group competition need not be direct (as in *P. porphyrio* and *V. pergandei*) for subordinates to possess leverage insuring them a share of group fitness. Subordinates may themselves generate inter-group competition by suppressing group fitness (see Section IIB). The communal breeding groove-billed ani, *Crotophaga sulcirostris*, is mildly territorial, with only the most dominant male displaying to other groups (Vehrencamp 1978). Subordination among females is incomplete (Vehrencamp 1977; Vehrencamp et al. 1986). Indeed, there is evidence that females will not tolerate complete subordinance.   Entire clutches may be destroyed, presumably by nesting females, forcing re-nesting and associated egg-tossing dominance to begin anew (Vehrencamp et al. 1986) Such intra-group induced nest loss is greater than loss through predation and clearly consistent with the thesis that individuals will depress group fitness as a tactic in social competition. Some groups in internal conflict fail to reproduce at all and, apart from destruction of eggs, females may affect group fitness by emigrating (Vehrencamp et al. 1986). Similar nesting conflict occurs in newly formed groups of *P. porphyrio* when no territorial resident exists prior to group formation (Craig 1980a,b). Established groups, however, exhibit little overt conflict.   Since group life is more ubiquitous in *P. porphyrio* (Craig 1979, 1980b, 1984) relative to the groove-billed ani (Vehrencamp 1978) this might be interpreted as evidence for rapid intra-group negotiation when individual (or breeding pair) reproduction is largely excluded.

Final evidence for subordinate leverage in *Crotophaga sulcirostris* should be found in the advantage of group life for dominants. All nocturnal incubation is done by the single, dominant male in a group. Such males are subject to higher mortality through predation (Vehrencamp 1978; Vehrencamp et al. 1986). Subordinate males, generally pair bonded to subordinate females, gain some reproductive success without the cost of nocturnal incubation. Without a compensating benefit, a dominant male is in fact altruistic. That subordinated females may emigrate, often with their mate (Vehrencamp et al. 1986) opens the possibility that subordinates are being 'paid' to remain in groups. Within pasture habitat, nest predation decreases with group size (Vehrencamp 1978), providing an advantage for nocturnal incubators with subordinates; this benefit does not, however, replicate in marsh habitat. Under the above logic, some such benefit must exist whenever dominants tolerate reproducing subordinates.

C. Application to Social Spiders

For the depression of group fitness to be an effective retaliatory strategy to dominance a multi-group population structure must exist. Groups must compete with one another in that colonies with cooperators outproduce those in conflict. Effective subordinate resistance is possible in a subdivided population, where intra-group fitness need not be equivalent to population-wide fitness (c.f. Wilson and Colwell 1980). The existence of a group does not, however, imply a subdivided population. Rather, the population may be a single group. In such a case intragroup fitness is equivalent to population fitness, and the logic behind TIT FOR TAT cannot apply. Here the logic of social competition follows without qualification: when a population forms a single, essentially homogenous (*sensu* Wilson 1980) group, competitively superior individuals can force their inferiors into subordinate roles (West Eberhard 1979). We suggest some communal spiders come close to satisfying this population structure.

1. Kin selection and dispersal in communal spiders

Females of the neo-tropical spider *Anelosimus eximius* live in large, long-lived colonies without internal territories. Colony foundation occurs through budding, communal emigration, and the dispersal of lone foundresses (Overal and Ferreira da Silva 1982; Vollrath 1982). The latter are often joined by conspecifics, most likely from the same parental colony (Vollrath 1982). Females mate with colonymates prior to dispersal, implying that inbreeding is common (Vollrath 1982, 1986; Smith 1986). *Achaeranea wau*, which similarly lacks dispersal among juveniles, is homozygous at multiple loci within a locality but not between localities (Lubin and Crozier 1985). Similar dispersal trends are evident in other highly communal spiders, leading several authors to suggest that here high degrees of relatedness may provide fertile ground for kin selection (Lubin 1974, 1980; Fowler and Levi 1979; Lubin and Robinson 1982; Smith 1983; Vollrath and Rohde-Arndt 1983).

Kin selection, however, cannot generate altruism when interactions are between same generation conspecifics and ultimate dispersal from parental colonymates is precluded (Pollock 1983). Since an altruism coding allele necessarily declines in relative frequency *within* a group after altruistic interactions, such an allele may increase in frequency within a population only through the differential production of groups, i.e., intergroup competition (Wade 1980, 1982, 1985). When cofoundresses are of the same generation and offspring of the same parental nest, such competition is precluded. The relative intra-group frequency of an altruism-coding-allele ultimately translates into its populational frequency, driving the allele to extinction (Pollock 1983).

Non-territorial communal spiders generally do not establish an exclusive parent/offspring bond, but rather care for spiderlings indiscriminately (Burgess 1978; Buskirk 1981; Kraft 1982; Christenson 1984; for an exception see Jacson and Joseph 1973). Without offspring altruism directed towards their mother, the extreme population viscosity (*sensu* Hamilton 1964) exhibited by communal spiders provides limited scope for kin selection as a factor in the evolution of a reproductive division of labor. Social spiders, especially those with limited parent/offspring overlap (as in *Anelosimus eximius* [Aviles 1986]), represent a natural system for the analysis of social competition without concurrent kin selection.

2. Social competition in *Anelosimus eximius*

Solitary *Anelosimus eximius* females exhibit low survivorship relative to communal conspecifics, the former being more prone to predation (Vollrath 1982; Christenson 1984) and the latter more effective at capturing larger prey (Nentwig 1985). When subject to food stress in laboratory colonies, females returning with prey to a retreat are pushed from their feeding position by those more vigorous (Vollrath and Rohde-Arndt 1983). Heavier females win in such contests. Monopolizing food permits a female to maintain her weight advantage as well as lay eggs. Since the process of capturing prey exposes an individual to predation (Brach 1975; Vollrath 1982; Christenson 1984), competitive losers are functionally altruistic.

In contrast to pleometrotic wasps and ants, the dispersal strategy of *A. eximius* may prevent effective subordinate resistance through the manipulation of group fitness. Many generations pass before females emigrate or colony fissioning occurs. Group competition is thus delayed, potentially raising the value of intra-group fitness in the calculation of populational values. Over several generations, the intra-group advantages of dominance may outweigh any subordinate resistance when the group component to selection is sufficiently moderate. Vollrath (1986) suggests that colonies undergo periodic food stress such that available prey cannot always support a colony's entire population. If so, a periodic negative correlation between group fitness and the feeding of subordinate losers would result.

A subordinate refusing to capture prey will eventually die. Her heavier nestmates should outlive her, foraging for themselves. Although group capture efficiency may decline with the loss of a subordinate, this need not translate into a loss in group fitness when food is sufficiently limited. Under such conditions a subordinate continuing to retrieve prey may potentially remain alive by occasional feeding, hoping for more abundant prey and a chance at reproduction in the future (c.f. West Eberhard 1978b, 1979, 1981)

IV. CONCLUSION: THE PLASTICITY OF SUBORDINANCE

Under the logic of social competition, a functionally sterile subordinate must retain the possibility of reproduction; that is, its phenotype must remain plastic. That primitively eusocial bees and wasps often exhibit such plasticity is consistent with the possibility of social competition (West Eberhard 1979, 1981, 1982). Kin selection, however, may permit plasticity as well. When benefits and costs associated with altruism vary significantly in an individual's lifetime, selection may operate to retain the option of abandoning altruism.

Unlike social competition, kin selection also permits an irreversible decision point. There is evidence that such decision points exist in primitively eusocial hymenoptera. Upon removal of their queen, freshly eclosed females of the neo-tropical wasp *Mischocyttarus drewseni* behave aggressively toward colonymates, exhibiting dominance and enlarged ovaries (Jeanne 1972). Females subordinate prior to queen removal, however, remain so, suggesting an irreversible decision point shortly after eclosion. Females of the eusocial wasps *Ropalidia cyathiformis* and *R. marginata* divide into three behavioral types: sitters, fighters, and foragers (Gadagkar and Joshi 1982, 1983, 1984, 1985) with only one behavioral type (dependent upon species) exhibiting oviposition. Gadagkar and Joshi (1984:27) suggest that an individual's behavioral repertoire is adopted early and "retained for most of its lifetime." Such decision points may differentiate the operation of kin selection from social competition, with the latter operating only among a subset of individuals on a nest.

Eusociality is traditionally defined as a reproductive division of labor among overlapping generations with communal brood care (Wilson 1971; Michener 1974). This definition clearly subsumes the Florida scrub jay as well as the social Hymenoptera and Isoptera, and might also include species such as the groove-billed ani if the reproductive division of labor may be synchronously incomplete. Further documentation of the possible existence of irreversible decision points on the part of 'helpers' in the Hymenoptera would argue for a refinement in this definition, excluding reversible helper phenotypes. Such a distinction also permits a discussion of the simultaneous operation of social competition and kin selection while clearly demarcating their effects. Although kin selection may be unnecessary for the evolution of helping, its import for such a refined eusociality may remain essential.

ACKNOWLEDGEMENTS

D.R.R. Smith provided needed cautions in the interpretation of spider sociality. Work reported here was supported by National Science Foundation grant DEB-8207052 to SWR. GRP was also supported by NSF grant IST-8305819 (through the kindness of L. Narens) and a MacArthur Foundation Fellowship in International Peace and Security granted by the Social Science Research Council.

REFERENCES

Aviles, L. 1986. Sex-ratio bias and possible group selection in the social spider *Anelosimus eximus. American Naturalist* 128:1-12.

Axelrod, R. 1981. The emergence of cooperation among egoists. *American Political Science Review* 75:306-318.

Axelrod, R. 1984. *The Evolution of cooperation*. New York:Basic Books, Inc.

Axelrod, R. and W.D. Hamilton. 1981. The evolution of cooperation. *Science* 211:1390-1396.

Bartz, S.H. and B. Hölldobler. 1982. Colony founding in *Myrmecocystus mimicus* Wheeler (Hymenoptera:Formicidae) and the evolution of foundress associations. *Behavioral Ecology and Sociobiology* 10:137-147.

Beulig, M.L. and D.H. Janzen. 1969. Variation in behavior among obligate acacia-ants from the same colony (*Pseudomyrmex nigrocincta*). *Journal of the Kansas Entomological Society* 42:58-67.

Brach, V. 1975. The biology of the social spider *Anelosimus eximius* (Areneae:Theridiidae). *Bulletin of the Southern California Academy of Science* 74:37-41.

Brian, M.V. 1983. *Social Insects*. London:Chapman and Hall.

Brown, J.L. and E.R. Brown. 1981. "Kin selection and individual selection in babblers." pp. 244-256. In: R.D. Alexander, and D.W. Tinkle, eds. *Natural selection and social behavior*. New York:Chiron Press.

Burgess, J.W. 1978. Social behavior in group-living spider species. *Symposium of the Zoological Society of London* 42:69-78.

Buskirk, R.E. 1981. "Sociality in the Arachnida." pp. 281-367. In: H.R. Hermann, ed. *Social insects*. Vol. II. New York: Academic Press.

Byron, P.A., E.R. Byron and R.A. Bernstein. 1980. Evidence of competition between two species of desert ants. *Insectes Sociaux* 27:351-360.

Carroll, C.R. and D.H. Janzen. 1973. Ecology of foraging by ants. *Annual Review of Ecology and Systematics* 4:231-257.

Christenson, T.E. 1984. Behaviour of colonial and solitary spiders of the theridiid species *Anelosimus eximius*. *Animal Behaviour* 32:725-734.

Craig, J.L. 1979. Habitat variation in the social organization of a communal gallinule, the Pukeko, *Porphyrio porphyrio melanotus*. *Behavioral Ecology and Sociobiology* 5:331-358.

Craig, J.L. 1980a. Pair and group breeding behaviour of a communal gallinule, the Pukeko, *Porphyrio p. melanotus*. *Animal Behaviour* 28:593-603.

Craig, J.L. 1980b. Breeding success of a communal gallinule. *Behavioral Ecology and Sociobiology* 6:289-295.

Craig, J.L. 1984. Are communal pukeko caught in the prisoner's dilemma? *Behavioral Ecology and Sociobiology* 14:147-150.

Davidson, D.W. 1977. Foraging ecology and community organization in desert seed-eating ants. *Ecology* 58:725-737.

Dawson, G.A. 1978. "Composition and stability of social groups of the tamarin, *Saguinus oedipus geoffroyi*, in Panama: Ecological and behavioral implications." pp. 13-22. In: D.G. Kleiman, ed. *The biology and conservation of the Callitrichidae*. Washington, D.C.:Smithsonian Institution Press.

Emlen, S.T. 1982a. The evolution of helping. I. An ecological constraints model. *American Naturalist* 119:29-39.

Emlen, S.T. 1982b. The evolution of helping II. The role of behavioral conflict. *American Naturalist* 119:40-53.

Emlen, S.T. 1984. "Cooperative breeding in birds and mammals." pp. 305-339. In: J.R. Krebs and N.B. Davies, eds. *Behavioural ecology: An evolutionary approach*. Sunderland, Mass.:Sinauer Associates, Inc.

Fowler, H.G. and H.W. Levi. 1979. A new quasisocial *Anelosimus* spider (Araneae, Theridiidae) from Paraguay. *Psyche* 86:11-18.

Gadagkar, R. and N.V. Joshi. 1982. Behaviour of the Indian social wasp *Ropalidia cyathiformis* on a nest of separate combs (Hymenoptera: Vespidae). *Journal of Zoology, London* 198:27-37.

Gadagkar, R. and N.V. Joshi. 1983. Quantitative ethology of social wasps: Time-activity budgets and caste differentiation in *Ropalidia marginata* (Lep.) (Hymenoptera:Vespidae). *Animal Behaviour* 31:26-31.

Gadagkar, R. and N.V. Joshi. 1984. Social organization in the Indian wasp *Ropalidia cyathiformis* (Fab.) (Hymenopotera:Vespidae). Zeitschrift für Tierpsychologie 64:15-32.

Gadagkar, R. and N.V. Joshi. 1985. Colony fission in a social wasp. Current Science 54:57-62.

Gamboa, G.J. 1974. Surface behavior of the leaf-cutter ant *Acromyrmex versicolor versicolor* Pergande (Hymenoptera: Formicidae). Masters thesis, Arizona State University, Tempe.

Gamboa, G.J., H.K. Reeve and D.W. Pfennig. 1986. The evolution and ontogeny of nestmate recognition in social wasps. *Annual Review of Entomology* 31:431-454.

Gibo, D.L. 1978. The selective advantage of foundress associations in *Polistes fuscatus*: A field study of the effects of predation on productivity. *Canadian Entomologist* 110:519-540.

Hamilton, W.D. 1964. The genetical theory of social behaviour, II. *Journal of Theoretical Biology* 7:17-52.

Herbers, J.M. 1977. Behavioral constancy in *Formica obscuripes* (Hymenoptera:Formicidae). *Annals of the Entomological Society of America* 70:485-486.

Hölldobler, B. and N.F. Carlin. 1985. Colony founding, queen dominance and oligogyny in the Australian meat ant *Iridomyrmex purpureus*. *Behavioral Ecology and Sociobiology* 18:45-58.

Jacson, C.C. and K.J. Joseph. 1973. Life history, bionomics and behaviour of the social spider *Stegodyphus sarasinorum* Karsch. *Insectes Sociaux* 20:189-204.

Janzen, D.H. 1973. Evolution of the polygynous obligate acacia-ants in Western Mexico. *Journal of Animal Ecology* 42:727-750.

Jeanne, R.L. 1972. Social biology of the neotropical wasp *Mischocyttarus drewseni*. *Bulletin of the Museum of Comparative Zoology, Harvard University, Cambridge* 144:63-150.

Klahn, J.E. 1979. Philopatric and non-philopatric foundress associations in the social wasp *Polistes fuscatus*. *Behavioral Ecology and Sociobiology* 5:417-424.

Kraft, B. 1982. "Eco-ethology and evolution of social spiders." pp. 73-84. In: P. Jaisson, ed. *Social insects in the tropics*. Paris:Universite Paris-Nord.

Lin, N. 1964. Increased parasitic pressure as a major factor in the evolution of social behavior in halictine bees. *Insectes Sociaux* 11:187-192.

Lin, N. and C.D. Michener. 1972. Evolution of sociality in insects. *Quarterly Review of Biology* 47:131-159.

Litte, M. 1977. Behavioral ecology of the social wasps, *Mischocyttarus mexicanus*. *Behavioral Ecology and Sociobiology* 2:229-246.

Litte, M. 1979. *Mischocyttarus flavitarsis* in Arizona: Social and nesting biology of a polistine wasp. Zeitschrift für *Tierpsychologie 50:282-312*.

Lubin, Y.D. 1974. *Adaptive advantages and the evolution of colony formation in Cyrtophora* (Araneae:Araneidae). *Zoological Journal of the Linnean Society* 54:321-339.

Lubin, Y.D. 1980. Population studies of two colonial orb-weaving spiders. *Zoological Journal of the Linnean Society* 70:265-287.

Lubin, Y.D. and R.H. Crozier. 1985. Electrophoretic evidence for population differentiation in a social spider *Achaearanea wau* (Theridiidae). *Insectes Sociaux* 32:297-304.

Lubin, Y.D. and M.H. Robinson. 1982. Dispersal by swarming in a social spider. *Science* 216:319-321.

Markin, G.P., H.L. Collins and J.H. Dillier. 1972. Colony founding by queens of the red imported fire ant, *Solenopsis invicta*. *Annals of the Entomological Society of America* 65:123-124.

Michener, C.D. 1958. The evolution of social behavior in bees. *Proceedings of the Xth International Congress of Entomology, Montreal, 1956*, 2:441-447.

Michener, C.D. 1974. *The social behavior of the bees*. Cambridge, Mass.:Belknap Press of Harvard University Press.

Mintzer, A. 1979. Colony founding and pleometrosis in *Camponotus* (Hymenoptera:Formicidae). *Pan-Pacific Entomologist* 55:81-89.

Mintzer, A. and S.B. Vinson. 1985. Cooperative colony foundation by females of the leafcutting ant *Atta texana* in the laboratory. *Journal of the New York Entomological Society* 93:1047-1051.

Nentwig, W. 1985. Social spiders catch larger prey: A study of *Anelosimus eximius* (Araneae:Theridiidae). *Behavioral Ecology and Sociobiology* 17:79-85.

Noonan, K.M. 1981. "Individual strategies of inclusive-fitness-maximizing in *Polistes fuscatus* foundresses." pp. 18-44. In: D.R. Alexander and D.W. Tinkle, eds. *Natural selection and social behavior*. New York:Chiron Press.

Overal, W.L. and R. Ferreira da Silva. 1982. "Population dynamics of the quasisocial spider *Anelosimus eximius* (Araneae: Theridiidae)." pp. 181-182. In: M.E. Breed, C.D. Michener and H.E. Evans, eds. *The biology of social insects*. Boulder, Colorado:Westview Press.

Pardi, L. 1948. Dominance order in *Polistes* wasps. *Physiological Zoology* 21:1-13.

Pollock, G.B. 1983. Population viscosity and kin selection. *American Naturalist* 122:817-829.

Pollock, G.B. and S.W. Rissing. 1985. Mating season and colony foundation of the seed-harvester ant, *Veromessor pergandei*. *Psyche* 92:125-134.

Pontin, A.J. 1978. The numbers and distribution of subterranean aphids and their exploitation by the ant *Lasius flavus* (Fabr.). *Ecological Entomology* 3:203-207.

Rissing, S.W., R.A. Johnson and G.B. Pollock. In press. Natal nest distribution and pleometrosis in the desert leaf-cutter ant *Acromyrmex versicolor* (Pergande) (Hymenoptera:Formicidae). *Psyche.*

Rissing, S.W. and G.B. Pollock. 1986. Social interaction among pleometrotic queens of *Veromessor pergandei* (Hymenoptera: Formicidae) during colony foundation. *Animal Behaviour* 34:226-233.

Rissing, S.W. and G.B. Pollock. In press (a). "Pleometrosis and polygyny in ants." In: R.L. Jeanne, ed. *Interindividual behavioral variability in social insects*. Boulder, Colorado:Westview Press.

Rissing, S.W. and G.B. Pollock. In press (b). Queen aggression, pleometrotic advantage and brood raiding in the ant *Veromessor pergandei* (Hymenoptera:Formicidae). *Animal Behaviour.*

Ross, K.G. and D.J.C. Fletcher. 1985. Comparative study of genetic and social structure in two forms of the fire ant *Solenopsis invicta* (Hymenoptera:Formicidae). *Behavioral Ecology and Sociobiology* 17:349-356.

Ryti, R.T. and T.J. Case. 1984. Spatial patterns and diet overlap between colonies of three desert ant species. *Oecologia* 62:401-404.

Smith, D.R.R. 1983. Ecological costs and benefits of communal behavior in a presocial spider. *Behavioral Ecology and Sociobiology* 13:107-114.

Smith, D.R.R. 1986. Population genetics of *Anelosimus eximius*. *Journal of Arachnology* 14:201-217.

Strassmann, J.E. 1981. Evolutionary implications of early male and satellite nest production in *Polistes exclamans* colony cycles. *Behavioral Ecology and Sociobiology* 8:55-64.

Strassmann, J.E. 1983. Nest fidelity and group size among foundresses of *Polistes annularis* (Hymenoptera:Vespidae). *Journal of the Kansas Entomological Society* 56:621-634.

Stumper, R. 1962. Sur un effet de groupe chez les femelles de *Camponotus vagus* (Scopoli). *Insectes Sociaux* 9:329-333.

Sullivan, J.D. and J.E. Strassmann. 1984. Physical variability among nest foundresses in the polygynous social wasp, *Polistes annularis*. *Behavioral Ecology and Sociobiology* 15:249-256.

Taki, A. 1976. Colony founding of *Messor aciculatum* (Fr. Smith) (Hymenoptera:Formicidae) by single and grouped queens. *Physiological Ecology, Japan* 17:503-512.

Tschinkel, W.R. and D.F. Howard. 1983. Colony founding by pleometrosis in the fire ant, *Solenopsis invicta*. *Behavioral Ecology and Sociobiology* 12:103-113.

Vehrencamp, S.L. 1977. Relative fecundity and parental effort in communally nesting anis, *Crotophaga sulcirostris*. *Science* 197:403-405.

Vehrencamp, S.L. 1978. The adaptive significance of communal nesting in groove-billed anis (*Crotophaga sulcirostris*). *Behavioral Ecology and Sociobiology* 4:1-33.

Vehrencamp, S.L. 1983a. A model for the evolution of despotic versus egalitarian societies. *Animal Behaviour* 31:667-682.

Vehrencamp, S.L. 1983b. Optimal degree of skew in cooperative societies. *American Zoologist* 23:327-335.

Vehrencamp, S.L., B.S. Bowen and R.R. Koford. 1986. Breeding roles and pairing patterns within communal groups of groove-billed anis. *Animal Behaviour* 34:347-366.

Vollrath, F. 1982. Colony foundation in a social spider. *Zeitschrift für Tierpsychologie* 60:313-325.

Vollrath, F. 1986. Eusociality and extraordinary sex ratios in the spider *Anelosimus eximius* (Araneae:Theridiidae). *Behavioral Ecology and Sociobiology* 18:283-287.

Vollrath, F. and D. Rohde-Arndt. 1983. Prey capture and feeding in the social spider *Anelosimus eximius*. *Zeitschrift für* Tierpsychologie 61:334-340.

Wade, M.J. 1980. Kin selection: its components. *Science* 210:665-667.

Wade, M.J. 1982. The effect of multiple inseminations on the evolution of social behaviors in diploid and haplodiploid organisms. *Journal of Theoretical Biology* 95:351-368.

Wade, M.J. 1985. Soft selection, hard selection, kin selection, and group selection. *American Naturalist* 125:61-73.

Waloff, N. 1957. The effect of the number of queens of the ant *Lasius flavus* (Fab.) (Hym.:Formicidae) on their survival and on the rate of development of the first brood. *Insectes Sociaux* 4:391-408.

Waloff, N. and R.E. Blackith. 1962. The growth and distribution of the mounds of *Lasius flavus* (F.) (Hym.:Formicidae) in Silwood Park, Berkshire. *Journal of Animal Ecology* 31:421-437.

Went, F.W., G.C. Wheeler and J. Wheeler. 1972. Feeding and digestion in some ants (*Veromessor* and *Manica*). *Bioscience* 22:82-88.

West, M.J. 1967. Foundress associations in polistine wasps: Dominance hierarchies and the evolution of social behavior. *Science* 157:1584-1585.

West Eberhard, M.J. 1969. The social biology of polistine wasps. *Miscellaneous Publications of the Museum of Zoology, University of Michigan* 140:1-101.

West Eberhard, M.J. 1975. The evolution of social behavior by kin selection. *Quarterly Review of Biology* 50:1-33.

West Eberhard, M.J. 1978a. Polygyny and the evolution of social behavior in wasps. *Journal of the Kansas Entomological Society* 51:832-856.

West Eberhard, M.J. 1978b. Temporary queens in *Metapolybia* wasps: Non-reproductive helpers without altruism? *Science* 200:441-443.

West Eberhard, M.J. 1979. Sexual selection, social competition, and evolution. *Proceedings of the American Philosophical Society* I, II, III:222-234.

West Eberhard, M.J. 1981. "Intragroup selection and the evolution of insect societies." pp. 3-17. In: R.D. Alexander and D.W. Tinkle, eds. *Natural selection and social behavior.* New York:Chiron Press.

West Eberhard, M.J. 1982. "Communication in social wasps: Predicted and observed patterns, with a note on the significance of behavioral and ontogenetic flexibility for theories of worker 'altruism'." pp. 13-36. In: A. de Haro, ed. *Communication chez les insectes sociaux.* Barcelona:Les Societes D'Insectes Colloque International, UIEIS Section Francaise.

West Eberhard, M.J. 1983. Sexual selection, social competition, and speciation. *Quarterly Review of Biology* 58:155-183.

Wheeler, J. and S.W. Rissing. 1975. Natural history of *Veromessor pergandei.* II. *Behavior. Pan-Pacific Entomologist* 51:303-314.

Wilson, D.S. 1980. *The natural selection of population and communities.* Menlo Park, CA:The Benjamin/Cummings Publishing Company, Inc.

Wilson, D.S. and R.K. Colwell. 1981. Evolution of sex ratio in structured demes. *Evolution* 35:882-897.

Wilson, E.O. 1971. *The insect societies.* Cambridge, Mass.:Belknap Press of Harvard University Press.

Wilson, E.O. 1985. The sociogenesis of insect colonies. *Science* 228:1489-1495.

Wilson, N.L., J.H. Dillier and G.P. Markin. 1971. Foraging territories of imported fire ants. *Annals of the Entomological Society of America* 64:660-665.

Woolfenden, G.E. and J.W. Fitzpatrick. 1984. *The Florida scrub jay. Demography of a cooperative-breeding bird.* Princeton, New Jersey:Princeton University Press.

Chapter 14

RELATIONSHIPS BETWEEN ECOLOGY AND SOCIAL BEHAVIOR IN COCKROACHES

J. Y. Gautier
P. Deleporte
C. Rivault

Laboratoire Ethologie, CNRS UA 373,
Université de Rennes, France

I.   INTRODUCTION

When the field of socioecology was in its early stages, ethologists tended to view social structures either as environmentally determined or as the direct result of genetic programming. In the years that followed, socioecologists oscillated between a critical analysis of the adaptive value of each social characteristic and trust in optimization theory. They tended to regard social structure as an optimal context in which individuals perform their vital functions which are then analyzed in terms of costs and benefits (Crook et al. 1976). Economic concepts invaded biology.

Later on these interpretations were toned down. Ecological conditions were no longer considered to directly determine social structure but they were regarded as facilitating the development of one particular type of social organization. Genetic inheritance was no longer thought to induce stereotyped social behavior, but rather to set the range of behavioral responses to environmental factors. According to this point of view, social systems are controlled by both specific features and environmental factors. Jarman's (1974) work perfectly demonstrates this approach to socioecology.

For several years socioecology emphasized the search for correlations between habitat parameters and social characteristics. Clutton-Brock and Harvey (1977a,b), working with primates, successfully took this approach. Studies such as theirs indicate that species living in the same type of habitat and with the same general diet have similar social structures (e.g., group size, number of males in a group).

Such interspecific comparisons are limited for two primary reasons: they do not always allow us to establish clear correlations between ecology and some social characteristics such as group composition. Furthermore, the correlations which have been established do not provide any proof of causal relationships. Despite these limitations, this approach is a useful one that can yield important results.

Using this correlational method, we analyzed relationships between ecology and social behavior in cockroaches, and investigated whether a given social system corresponded to one type of habitat. In this paper we define the social system by two main factors: spacing and social interactions. Cockroaches provide a good model for this purpose because they have invaded many types of environments in which they exhibit different social organizations. Furthermore, separate phyletic lines can differ greatly in some characteristics such as their type of reproduction (oviparity or ovoviviparity). In our study, we analyzed social variation within species.

II.   COCKROACH SOCIOECOLOGY

We set out in our study to answer two major questions: (1) Do species with close phyletic relationships (i.e., those belonging to the same genus or family) have more behavioral traits in common than more distantly related species? We referred to MacKittrick's (1964) phyletic tree (Fig. 1). (2) Do species which colonize the same habitat tend to share the same social characteristics and have a particular social system different from that of other species living in another type of habitat?

A.   Phyletic Relatedness and Behavioral Similarities

Comparisons between species within the same genus or the same subfamily reveal behavioral similarities associated with phyletic relatedness. In several species of *Blaberus* (*B. atropos, B. colosseus, B. craniifer*) (Gautier 1974a), social relationships are all characterized by complex agonistic interactions between males (Fig. 2) who establish stable hierarchies. Subordinate males present submissive postures and dominant males hold territories. Similar characteristics are observed in two

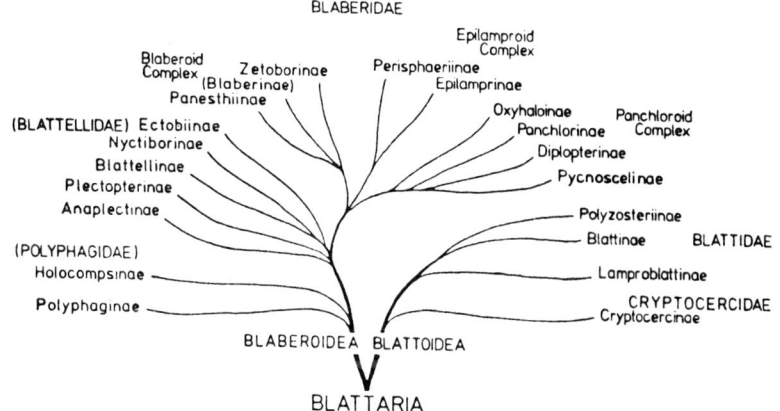

Fig. 1. *The phyletic tree of cockroaches (according to MacKittrick 1964).*

species of a closely related genus, *Eublaberus* (*E. posticus*) (Gorton et al. 1979), and *E. distanti* (Gautier and Foraste 1982).

Intense aggressive behavior (reciprocal lunge and kick) is also exhibited by territorial adult males in two *Gromphadorhina* species studied under laboratory conditions, *G. brunneri* (Ziegler 1972) and *G. portentosa* (Rivault 1985). Species of three genera, *Blatella* (Breed et al. 1975),

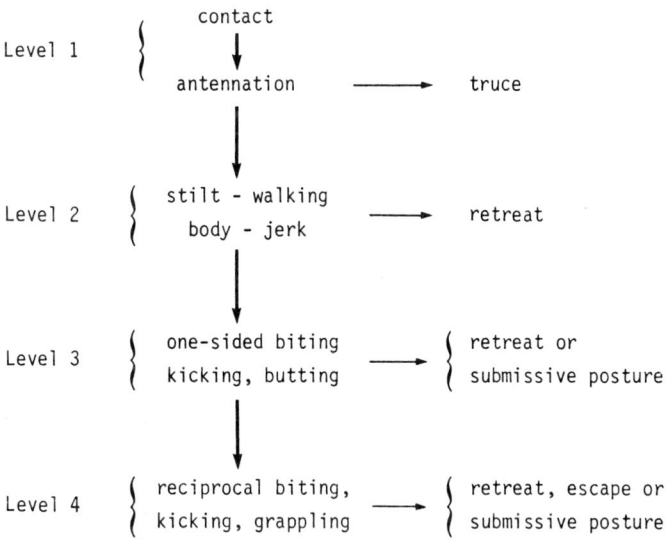

Fig. 2. *Sequences of agonistic interactions with increasing level of aggressiveness (from Breed et al. 1975).*

Parcoblatta (Gorton 1979a,b; Schal pers. comm.), and *Shawella* (Gorton et al. 1983), exhibit very short and low-level agonistic encounters. Furthermore, *P. pennsylvanica* and *S. couloniana* do not produce aggregation pheromones. Aggregation pheromones have been identified in many *Blaberus* species (Roth and Cohen 1973; Brossut 1980). In both *Xestoblatta* species, *X. cantralli* and *X. hamata*, receptive females adopt calling postures at the same time as they emit a volatile sex attractant (Schal and Bell 1985). In Oxyhaloinae (Sreng 1984), sexual attractants and calling postures are male attributes.

These examples would make us think that phylogenetically related species share similar social behaviors and communication systems. However, it is important to note that other related species have developed completely different social characteristics. For example, the larvae and adults of *Arenivaga investigata* are dispersed in the desert sand, while *A. apacha* individuals are grouped around food stored in rodents' burrows (Edney et al. 1974; Cohen and Cohen 1981).

Related species belonging to the Panesthiinae subfamily have developed different social structures. In *Panesthia cribrata* (Rugg and Rose 1984), clusters of animals can be observed under natural conditions. Approximately 50 percent of these groups are composed of several adult females and larvae, while the other half are solely made up of larvae. In *Salganea taiwanensis* (Matsumoto 1986) most groups are composed of one couple and their larvae.

Although some closely related species of cockroaches tend to have very similar social organizations, some species which are not closely related also have developed very similar behaviors. For example, similar female calling postures are observed in one species of Blattidae, *Periplaneta americana*, and in one species of Blaberidae, *Capucina patula* (Schal and Bell 1985). Monogamous territorial families are observed in species belonging to two different superfamilies, Blaberoidea (*Salganea taiwanensis*) and Blattoidea (*Cryptocercus punctulatus*) (Seelinger and Seelinger 1983).

These data provide a partial answer to the first question posed above. Phyletic relatedness is not enough in itself to explain behavioral similarities. The nature and origin of behavioral similarities can better be understood through synchronous studies of habitat and social biology.

B. Types of Habitat and Social Systems

In order to examine the influence of habitat on social systems, we studied several cockroach species. Several parameters were analyzed during the course of the study. The following habitat characteristics and their utilization were measured in the laboratory and in the field: (1) resting sites and nocturnal foraging areas; (2) food resources and food intake; and (3) predators and anti-predatory behavior. The biological traits analyzed were: (1) reproductive processes (ovoviviparity vs. oviparity); (2) social structure; and (3) social behavior. We compared a number of species within the same habitat and between habitats in an attempt to analyze the influence of habitat on social systems.

1. Tropical forest

Our first comparison involves *Lamproblatta albipalpus* (Lamproblattinae) (Gautier and Deleporte 1986) and *Cariblatta imitans* (Plectopterinae) (Schal 1982, 1983; Schal et al. 1984). These species will be designated as *L.a.* and *C.i.* in the following paragraphs. Both species inhabit humid, tropical forests in French Guyana and Costa Rica, respectively, where they find abundant but scattered food resources (fungi for *L.a.* and algae, leaf trichomes, moss, liverwort and fungi for *C.i.*). Army ants, spiders and ant birds actively prey on these two species. The cockroaches try to avoid their predators by concealing themselves under dead leaves, in folded leaves, or under rotten logs and branches.

The main differences between these two species are that *L.a.* rests in dead trunks and branches lying on the ground during the day, while *C.i.* uses refuges such as rolled dead leaves, loose bark or roots and leaves of epiphytes. In addition, *L.a.* individuals maintain a constant home range of a few square meters which includes resting sites and foraging places on the ground and in rotten logs. Occasionally some *L.a.* individuals climb up stumps to heights of 0.6 m. *C.i.*, on the contrary, is a truly arboreal species. All life stages remain in the vegetation both day and night (Fisk 1983) at a mean perch height of 1 m, although adult males tend to remain higher up in the foliage than females. *L.a.* females deposit their oothecae in dead trunks and branches while *C.i.* females lay them in the vegetation. Several egg cases have been found in flower inflorescences of *Dieffenbachia* sp.

The most likely explanation for the difference in vertical stratification between these two species relates to the fact that *L.a.* adults are completely apterous, whereas *C.i.* individuals possess fully developed wings which allow them to fly in and out of the foliage and thus escape predation.

In spite of dissimilarities in their vertical stratification within the habitat, the movements and foraging behavior of the two species are solitary. Both species can be defined as solitary, although adult *L.a.* males and females sometimes tend to be found together in diurnal resting sites. Observations under laboratory conditions demonstrate that *L.a.* individuals share the same diurnal shelters, but without any real aggregation. Usually the distance between individuals is greater than twice the length of the legs in resting posture, and therefore the animals' bodies are not in direct contact.

Our second comparison contrasts *Lamproblatta albipalpus* (*L.a.*) to another oviparous species, *Xestoblatta hamata* (*X.h.*) (Blattelinae) (Schal 1982, 1983; Schal et al. 1984). This latter species inhabits the humid tropical forests of Costa Rica, and finds resting sites under leaf litter or in underground refuges. Young larvae live in the leaf litter; older larvae go higher in the foliage to a level almost as high as the mean perching heights of adults (0.4 m). Adult males tend to perch higher in the vegetation than females.

Gut contents have revealed that *X.h.* individuals eat wood chips, leaf bits, arthropods, fungi and algae. Adults have also been observed feeding on fruits, flowers and epiphylls. Large aggregations of both sexes occur near food plants, particularly on shed bark under an inga tree (*Inga coruscans*). The sex ratio is skewed toward females. Individual females appear under an inga tree where they call, following an 8-10 day cycle that corresponds to the length of the gonadotrophic cycle. Females tend to be more site specific than males during a given night, but move at least as much as males between consecutive nights. Aggressive interactions are not very common but females tend to defend a perch which they have occupied for some length of time.

A trait common to the *L.a.* and *X.h.* species is that adult males and females gather on spatially predictable sites. In the case of *X.h.*, the site has only a trophic value, whereas it also is used for shelter and oothecae deposition by *L.a.* In both cases, the aggregations are loose, with no social structure of their own. These clusters seem to have the unique function of facilitating meeting between sexes.

It is important to note that the three species discussed above live in the same habitat with scattered resources. All three species are either solitary or they aggregate only in very loose clusters. None of them is gregarious.

2.   Temperate forest

Our comparison can now be extended to species living in temperate climates, such as *Ectobius* (Ectobiinae) and *Parcoblatta* (Blattellinae). Four *Ectobius* species, *E. lapponicus*, *E. lividus*, *E. sylvestris* and *E. panzeri* are found in European forests and moorlands (Morvan 1972a,b; Brown 1973). Five *Parcoblatta* species, *P. uhleriana*, *P. lata*, *P. pennsylvanica*, *P. virginica*, and *P. bolliana* inhabit North American forests and grasslands (Gorton 1980).

Although the forest, moorland and grassland habitats are diverse, their resources are all dispersed. Given this similarity, an important question to ask is whether or not the cockroach species inhabiting these areas share similar behavioral characteristics.

None of the nine temperate species seems strictly attached to any particular habitat. Resting sites are situated under dead leaves in the litter, under loose bark, or at the foot of Gramineae plants. Clusters have never been observed at these sites.

Litter, bushes, ferns and grasses are used as foraging areas. Vertical stratification is restricted for *Ectobius sylvestris* brachypterous females and for four out of the five *Parcoblatta* species mentioned. The diets of the *Ectobius* species are completely unknown but *Parcoblatta* species have been observed feeding on mushrooms, mammalian and bird feces, sap on tree wounds, flowers, and mosses.

Most of our knowledge of social behavior in these species thus comes from laboratory observations which indicate that *Parcoblatta* species exhibit low levels of agonistic interactions, including biting and kicking upon occasion. Male-female and female-female agonistic encounters tend to be greater in intensity than male-male contacts, with females being the more aggressive. *Parcoblatta uhleriana* individuals have been observed fighting each other, and *Parcoblatta pennsylvanica* males are known to defend a food source against other *Parcoblatta* species (Schal pers. comm.). In *Ectobius*, male-male interactions are very short and usually end with the departure of one of the protagonists after antennal fencing and occasional kicking.

All of these nine temperate species are solitary, regardless of whether they are arboreal or terrestrial. Individuals meet and withdraw after short agonistic encounters of low intensity. Some species even lack aggregation pheromones. These species are similar to the tropical species discussed above. Although the tropical species *L.a.* and *X.h.* tend to aggregate loosely, such groups have no social cohesion or organization, and therefore, these animals are functionally solitary.

Most of these solitary species belong to the Blattidae and Blatellidae, which are oviparous. A few ovoviviparous Blaberidae species such as *Hyporhicnoda reflexa*, some *Epilampra* and some *Panchlora*, which all live in tropical forests, can be included in this group.

3. Caves

Caves constitute a very different type of habitat from tropical or temperate forests. Conditions within caves tend to be very stable. The first species that we will consider is *Blaberus colosseus* (Blaberinae), an ovoviviparous species studied in Trinidad, West Indies (Gautier 1974a,b). Contrary to the species discussed previously, members of this one are large, with adults measuring up to 8 cm. Individuals feed mainly on guano, fruits and seeds brought back by bats, and on dipteran and lepidopteran larvae living in guano.

Diurnal resting sites of adults and last-instar larvae are located in holes, galleries, crevices, niches, or on ledges. Until the seventh instar, larvae remain concealed in the guano and in organically rich soil on the cave floor. They are absent from the zones of dry soil, stones and pebbles located within the cave. Young larvae forage at night on the guano surface, while older larvae and adults forage on the cave walls, ledges and niches. Predators are numerous, perhaps because the cockroach species is relatively concentrated. Predators include *Morion* beetles, *Tarantula* spiders, terrestrial crabs, *Anolis* geckos, *Bufo* toads, *Phyllostomus* bats and opossums (Gautier 1974a; Brossut 1980). Young larvae avoid predation by burrowing into the guano while older larvae and adults escape and hide in crevices.

Unlike the solitary forest, grassland and moorland species, the cave species of cockroaches tend to be gregarious. Social groups consisting of 20 to 50 adult males, females and older larvae gather at particular sites such as tunnels and niches. In these groups, adults come into close contact and even climb onto one another. A few isolated animals or small groups of three or four adults are usually found nearby. Laboratory observations reveal that adult males establish rather rigid hierarchies, while the highly gregarious females develop only a few aggressive acts which are very rarely intense (Fig. 2). Females come down onto the ground at the end of the gestation period and give birth on the guano. Then the females, which are again receptive, become very active.

Other species of Blaberinae, including *Eublaberus distanti* (Gautier 1974a; Brossut 1980) and *Eublaberus posticus* (Darlington 1968), have been studied in caves in Trinidad. They were observed to form social groups in sometimes heavily populated places within the caves. In the field, *Eublaberus distanti* respond to an olfactory signal enabling them to recognize the group to which they belong (Brossut 1980). Laboratory experiments confirm that the males establish dominance hierarchies, although the precise modalities used to obtain and hold a high rank vary from one species to another (Gautier and Foraste 1982).

The second example of a gregarious, cave-dwelling species in Trinidad, *Periplaneta americana* (Blattinae), often lives in very heavily populated areas consisting of several thousand individuals (Deleporte 1976). Resting sites of adults and older larvae are located mainly in crevices and holes in the rocky wall up to 2.5 m high. They take refuge in these crevices when threatened by a predator. The predators that prey on them are the same ones feeding on *Blaberus colosseus*. Most of the young larvae occupy galleries that they dig in the layers of clay at the foot of the wall. On the ground, individuals of this species eat guano and fruit brought back by bats. Some of these animals are sedentary and exhibit homing behavior between their resting sites and feeding areas. Others are wanderers. Adult females deposit their oothecae in the guano or clay.

Small groups of variable size are found at the resting sites (e.g., one male, two females, and three old larvae; nine males, seven females, and eight third or fourth-instar larvae). Aggressive behavior on the ground near food or resting sites has been observed sites between adults, between adults and older larvae, and between older larvae. Laboratory observations show that aggressive behavior is frequent during all the larval stages, but that it does not induce stable hierarchies.

Other *Periplaneta* spcies, *P. australasiae* and *P. brunnea,* have the same type of crowded spatial distribution in caves. According to Roth (1980), *Symploce cavernicola* (Blatellinae) also swarms in great numbers on boulders near guano beds.

The cave-dwelling species of cockroaches that were studied all seem to have developed a similar social system. They cluster together in groups that are characterized by a hierarchical organization. It seems likely that the cave habitat, with its stable climatic conditions and abundant but patchy food resources, facilitates the development of this type of social system.

4. Discussion

As demonstrated in the previous sections, those cockroach species discussed that inhabit grasslands, moorlands and forests where resources are dispersed, tend to be solitary. Individuals of the species *Lamproblatta albipalpus* tend to be found together at diurnal resting sites such as hollow logs, however. Although they are relatively close to one another, they do not cluster together.

The species living in these habitats belong to two large phyletic groups, and consequently have different biological characters. Nevertheless, they share the same social organization. This similarity could be explained by convergence in that they acquired the same adaptation to one type of environment independently.

Quite a different situation exists among the cave-inhabiting species of cockroaches such as the Blaberinae that have been studied. Within the caves, climatic conditions tend to be stable and resources clumped. Bat guano, for instance, on which many of them feed, is highly concentrated. In this situation, members of the cockroach species cluster together. Males have developed hierarchical organizations, whereas females tend to be merely gregarious. Some Blattidae (*Periplaneta* spp.) and possibly some Blattellidae (*Symploce cavernicola*) exhibit in caves a social organization resembling that of Blaberidae. It differs, however, in that the dominance hierarchies are less stable, and include adult males and females as well as larvae. The similarities in social organization observed in Blaberidae, Blattidae and Blattellidae could be the result of a convergence phenomenon in response to the same ecological pressures to which these different species are subject.

The scattered resources and unstable conditions that prevail in tropical and temperate forests, as well as grasslands, seem to make it impossible for individuals to group tightly together and develop a social organization such as that which occurs among cave-dwelling Blaberidae males especially.

In socioecological studies, it is very important to be able to have enough detailed data available about different species in order to be able to compare them. For example, animals described as gregarious can represent different kinds of gregariousness. Non-obligatory gregariousness, without any close contact between animals, for instance, is very different from that observed in Blaberidae where crowding is induced by a pheromone. Precise descriptions of interactions between males in several species have led to the definition of several agonistic modalities (Bell et al. 1979; Gautier and Foraste 1982). Consequently, it is possible to understand how some hierarchical rankings are stable in some species, but not in others. This type of analysis needs to be continued under laboratory conditions with stable groups of animals in a complex environment.

It is not sufficient to describe the ethogram of just one sex or age class in characterizing a particular species. In some species of cockroaches, males and females exhibit the same social behaviors, while in others, males and females have very different ones; i.e., clustered females and dispersed or territorial males, as in *Gyna maculipennis* (Gautier 1980), and in *Gromphadorhina portentosa* (Rivault 1985). Differences between sexes and age classes are often as important as differences between species.

Although we are able to localize precisely the different cockroach species which live in a given environment, we are rarely able to understand why, for example, they are found settled in a particular vertical stratum of the vegetation. This may be due to microclimatic conditions which correspond to their requirements, or to competition between two or more species. Further studies of microclimatic conditions and physiological resistance potentialities should help us to solve this problem.

III. MONOGAMOUS TERRITORIAL FAMILIES

In cockroaches this type of social organization is observed only in xylophagous species. The species that have been studied most extensively are *Cryptocercus punctulatus* (Cryptocercinae) (Seelinger and Seelinger 1983; Nalepa 1984), *Salganea taiwanensis* and *S. esakii* (Panesthiinae) (Matsumoto 1986). These animals live permanently in galleries dug in dead logs where climatic factors are buffered and where moisture is always high. These species eat wood, which can be considered a permanent and abundant resource of low nutritive value. No predators or antipredator acts are known.

The most frequently encountered groups found in the field are composed either of one pair of adults or one pair and its larvae. The link between adults and larvae is long lasting, and in *Cryptocercus* can last the whole larval life. *Cryptocercus* mated pairs defend the gallery against conspecific intruders and exhibit alarm behavior, thus leading to social isolation of the family units.

These cockroaches harbor a gut cellulose-digesting symbiotic fauna composed of flagellates and bacteria (*Cryptocercus*), or only of bacteria (Panesthiinae). Symbionts are transferred from adults to larvae through proctodeal trophallaxis in *Cryptocercus* and probably by means of trophallaxis or feces ingestion in Panesthiinae.

Wood-eating cockroaches (Panesthiinae and Cryptocercinae) are not closely related according to MacKittrick's phyletic tree (1964) (Fig. 1). *Cryptocercus* is frequently referred to as a primitive cockroach. According to a review of anatomical characters, *Cryptocercus* probably descended from an early emerging stem of the phyletic tree (Fig. 3) (Deleporte unpub. data). The association with flagellates could then be explained by the Grasse and Noirot hypothesis (1959). Flagellates inherited from a presumed xylophagous common ancestor of Blatteria and Isoptera, perhaps corresponding to the hypothetical *Archisoptera* (Martynov 1938), have been retained in *Cryptocercus* and lost in other Blatteria species which are no longer xylophagous, nor do they still have a family structure. It is interesting to note that flagellates have also been lost in Termitidae which are considered to be evolved species with a eusocial organization. The appearance of a family structure in Panesthiinae (Blaberidae) is therefore a true convergence.

The similarities in the social organization of non-closely related wood-feeding cockroaches argue in favor of a correlation between habitat and social structure. This argument would be even more convincing if the same kind of correlation occurred in other wood-eating species not related to Blattaria. Seelinger and Seelinger (1983) and Nalepa (1984) point out the similarities between *Cryptocercus* and termites. The *Cryptocercus* territorial family resembles the foundation phase of a termite colony where the larvae stay with their parents. Symbiont transfer is made through proctodeal trophallaxis in both cases. Termites have developed eusociality with differentiated sterile castes, polyethism, cooperation in nest building and egg care (Wilson 1975), which *Cryptocercus* have not.

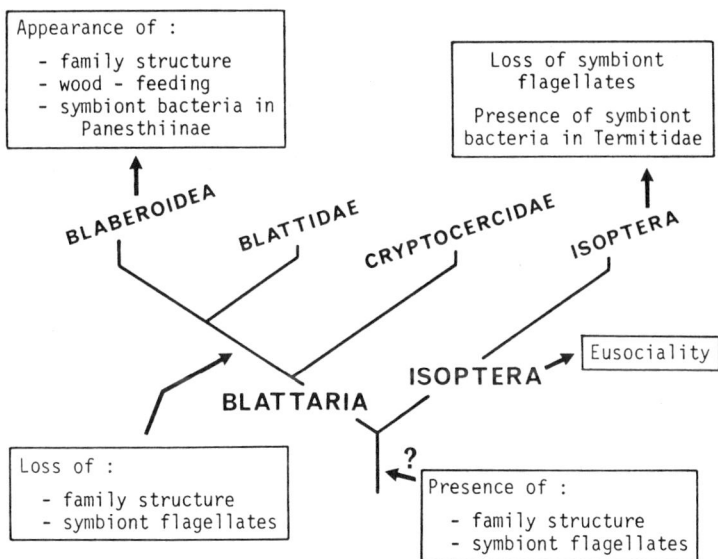

Fig. 3. Evolutionary development of socialization in Dictyoptera from a probable flagellate-harboring common ancestor.

Thus, different insect groups have developed a social organization related to a particular trophic niche. Family structure can be interpreted as an adaptation which allows symbiont transfer and trophic exchanges from adults to larvae. In *Cryptocercus*, even though gut fauna is complete at the age of about one month (Cleveland et al. 1934), larval feeding from proctodeal liquid lasts for several years (Nalepa 1984).

IV. SOCIAL PLASTICITY

In the previous section, we discussed the broad relationship between habitat types and social organization. However, it is important to note that even within a particular habitat type such as caves, the social organization of a species may be fairly plastic. This social plasticity in some cockroaches came to light when several populations of the cave-dwelling species, *Blaberus colosseus*, were studied in depth. Similar plasticity has been reported for other *Blaberus* species including *B. craniifer* and *B. giganteus*, and *Eublaberus* species such as *E. distanti* (Gautier 1974a) and *E. posticus* (Gorton et al. 1979). In all the species investigated, the males proved to be gregarious; however, the type and extent to which they were social varied greatly. Three degrees of gregarious social organization were observed, including the hierarchical, territorial and intermediate types.

In a hierarchical situation, adult males and females as well as older larvae form groups of a few dozen animals. The dominant male within the group exhibits aggressive behavior, while subordinate males adopt submissive postures. Such a situation creates a dominance hierarchy. Only one population displaying this type of organization has been observed in the deeper part of a cave covered by thick guano and heavily populated with young larvae. Many older larvae and adults gather on a few sites where density is especially high. Only one predator, a bufo toad, was seen.

Under territorial conditions, males hold sites such as rock ledges of 0.5 to 2 square meters in area. Density of cockroaches in such a situation is relatively low. Within any one population, there may be ten males, an equal number of females, and a few larvae. Territorial males can be site specific for up to 45 days, while others search for an unoccupied site. Some males actually remain on their larval site. Populations exhibiting this type of social organization have been found in caves where food is abundant but patchy. Predators of all sizes prey on them, particularly the opossum.

In the intermediate type of social organization, adult populations consist of several dozen individuals. Females live in small groups. An average of five males inhabit a site of 1 to 3 square meters, a surface area similar to that held by one territorial male in the previous case. Some males can be observed on the same site for 10 to 27 days, others for only 3 to 5 days. The remaining males move from site to site. In the absence of females, male interactions show a low level of aggression, and encounters are of short duration. Most encounters end with either the withdrawal of one individual, or the separation of both. Males have overlapping home ranges and aggression seems to be reduced by frequent encounters between neighbors. Populations with this type of sociality have been observed under conditions where food is abundant and large predators are absent.

For a better understanding of the relationship between population density and social organization, we supplemented the observations taken under natural conditions with laboratory experiments. When density is low, interactions between adults or between adults and older larvae are rare, whereas such interactions are frequent at high densities. Our experiments measured the development of male agonistic behavior in relation to the number and quality of their interactions with their partners.

Isolation during the end of larval life and the beginning of adult life promotes the development of male aggression. Encounters between these young males are characterized by a high level of aggressive behavior resulting in the acquisition of dominant or subordinate status. Dominant males chase away subordinates. After several days of isolation, subordinates recover their aggressive behavior. Young males that have experienced larval grouping cluster with subordinate males after their imaginal molt. They avoid dominant males whose agonistic behavior inhibits their own. Adult male sexual displays encourage early manifestation of aggressive behavior among young males and promote its suppression in mature subordinate males.

From these laboratory data and the preceding field observations, we can see that the isolation of young males induces the development of aggressive behavior, enabling them to become territorial. Isolation of subordinate males allows them to recover a dominant status and to occupy a vacant site. This situation is often encountered in low-density populations. Grouping leads to the inhibition of aggression in most males, especially young and subordinate ones. It also results in the development of a hierarchical system with a few top-ranking males. This situation is frequently found in high-density populations.

Receptive females can modify these patterns of social organization in males by inducing sexual displays. When the population density is high, this can lead to reversals of male ranking even among subordinate males. At intermediate densities, some males become temporarily territorial. When population densities are low, territorial males follow receptive females and fight with other males as they move from one territory to another. In caves, this situation produces extensive modifications in male spatial distribution.

Two main factors thus seem to influence the size of a population: food, which is rarely limiting, and the impact of predators, which can be high in all age classes. These two factors, together with the abundance and distribution of favorable sites, control the spatial distribution of *Blaberus colosseus*. As the population density increases, the gregarious social organization varies from territorial to hierarchical, through an intermediate type, thereby demonstrating that the specific social organization of *Blaberus colosseus* males is density dependent.

It is now possible to describe a series of events ranging from the control of population size and local density by ecological factors to the differentiation of social behavior through individual ontogeny which leads to plasticity in the social organization of a given species. Figure 4 summarizes the relationships which link three levels of organization--the population-environment system, the social system, and the individual.

V. CONCLUSIONS

According to Baldwin and Baldwin (1979), an adequate analysis of the relationship of behavior to ecology requires data on a species' phylogenetic history as well as individual ontogenetic development. We are in complete agreement with these authors, and based on the data presented within this paper, we can envision the following three different social histories in cockroaches: (1) some species such as *Cryptocercus* developed a monogamous family life, dwelling inside wood and feeding on it; (2) other species settled in forests, grasslands and meadows, and became solitary; and (3) yet other species living in habitats such as caves with abundant, clumped resources became social.

Based on the data we have presented earlier in this paper, it seems clear to us that social organization in cockroaches is better correlated with environmental parameters than with phyletic classification. Different social ontogenies occur in species having different lifestyles. For

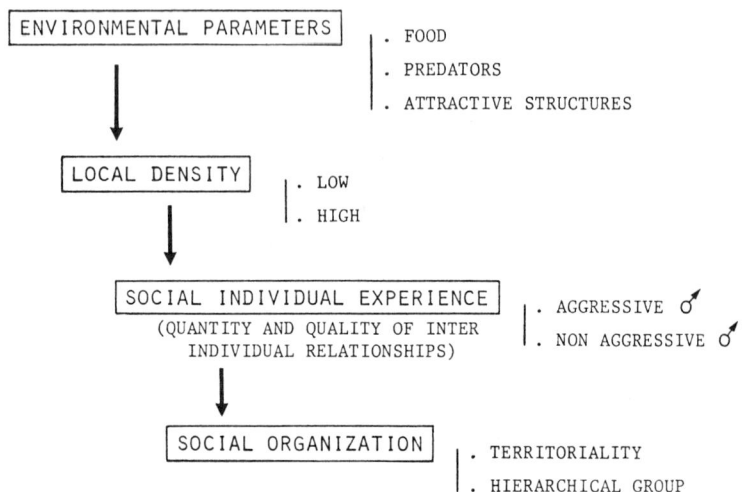

Fig. 4. Series of events leading to social plasticity in a natural population of Blaberus.

example, aggressive behavior develops early in *Periplaneta americana* first-instar larvae living in galleries within caves, while it never develops among young larvae of Blaberidae living in guano which is a very abundant, clumped resource. Aggression develops at a later developmental stage when the cockroaches live on cave walls. Several types of gregariousness ontogeny are described in different Blaberidae sp. in Brossut (1979).

Even in insects, often considered genetic automatons, individual experience can be an important source of social variation. The part played by experience has to be integrated into a more general study of the different conditions and mechanisms favoring social plasticity. Responses to supra-normal stimuli used in ethological studies for a long time have led to the conclusion that animals rarely express behaviors which are at the limits of their capabilities. This unused capacity allows the organism some plasticity, opportunism and adaptation to environmental changes. When discussing behavioral plasticity, however, we must not forget that some specific characters resist changes. In cockroaches, social plasticity and conservation are observed respectively in Blaberidae and *Cryptocercus.* There is the need for new socioecological theoretical models integrating the antagonism between innovation and conservation at the social level.

REFERENCES

Baldwin, J.D. and J.I. Baldwin. 1979. "The phylogenetic and ontogenetic variables that shape behavior and social organization." pp. 89-116. In: I.S. Bernstein and E.O. Smith, eds. *Primate ecology and human origins: Ecological influences on social organization.* New York and London:Garland STPM Press.

Bell, W.J., R.E. Gorton Jr., M.K. Tourtelot and M.D. Breed. 1979. Comparison of male agonistic behavior in five species of cockroaches. *Insectes Sociaux* 26:252-263.

Breed, M.D., C.M. Hinkle and W.J. Bell. 1975. Agonistic behavior in the German cockroach *Blatella germanica. Zeitschrift für Tierpsychologie* 39:24-32.

Brossut, R. 1979. "Gregarism in cockroaches and in *Eublaberus* in particular." pp. 237-246. In: F.J. Ritter, ed. *Chemical ecology: Odour communication in animals.* Amsterdam:Elsevier and North-Holland Biochemical Press.

Brossut, R. 1980. La communication chimique chez les blattes. Thesis, Université de Dijon.

Brown, W.K. 1973. Aspects of the reproductive biology of three species of *Ectobius* (Dictyoptera:Blattidae). *Entomologia Experimentalis et Applicata* 16:213-222.

Cleveland, L.R., S.R. Hall, E.P. Sanders and J. Collier. 1934. The woodfeeding roach, *Cryptocerus*, its protozoa, and the symbiosis between protozoa and roach. *Memoirs of the American Academy of Science* 17:185-342.

Clutton-Brock, T.H. and P.H. Harvey. 1977a. Primate ecology and social organization. *Journal of Zoology* (London) 183:1-39.

Clutton-Brock, T.H. and P.H. Harvey. 1977b. "Species differences in feeding and ranging behaviour in primates." pp. 557-584. In: T.H. Clutton-Brock, ed. *Primate Ecology.* London:Academic Press.

Cohen, A.C. and J.L. Cohen. 1981. Microclimate, temperature and water relations of two species of desert cockroaches. *Comparative Biochemistry and Physiology* 69:156-167.

Crook, J.H., J.E. Ellis and J.D. Goss-Custard. 1976. Mammalian social systems: Structure and function. *Animal Behaviour* 24:261-274.

Darlington, J. 1968. Bioenergetics of *Eublaberus posticus.* Thesis, University of Trinidad, W.I.

Deleporte, P. 1976. L'organisation sociale chez *Periplaneta americana* (Dictyopteres): Aspects éco-éthologiques, ontogenese des relations interindividuelles. Theses 3é Cycle, Université de Rennes.

Edney, E.B., S. Haynes and D. Gibo. 1974. Distribution and activity of the desert cockroach *Arenivaga investigata* (Polyplagidae) in relation to microclimate. *Ecology* 55:420-427.

Fisk, F.W. 1983. Abundance and diversity of arboreal Blattaria in moist tropical forests of the Panama canal area and Costa Rica. *Transactions of the American Entomological Society* 108:479-489.

Gautier, J.Y. 1974a. Processus de différenciation de l'organisation sociale chez quelques especes de blattes du genre *Blaberus*: Aspects
écologiques et éthologiques. These Université de Rennes.

Gautier, J.Y. 1974b. Etudes comparées de la distribution spatiale et
temporelle des adultes de *Blaberus atropos* et *Blaberus colosseus*
(Dictyopteres) dans quatre grottes de Trinidad. *Revue du Comportement Animal* 9:237-258.

Gautier, J.Y. 1980. Distribution spatiale et organisation sociale chez
Gyna maculipennis (insecte dictyoptere) dans les cavernes et gale-
ries de la region de Belinga au Gabon. *Acta oecologica Oecologia
generalis* 1:347-358.

Gautier, J.Y. and P. Deleporte. 1986. "Behavioural ecology of a forest
living cockroach, *Lamproblatta albipalpus* in French Guayna." pp.
17-22. In: L.C. Drickamaer, ed. *Behavioral Ecology and Population
Biology*. Private, I.E.C., Toulouse. (Readings from the 19th International Ethological Conference, No. 4).

Gautier, J.Y. and M. Forasté. 1982. Etude comparée des relations interindividuelles chex les males de deux expeces de blatttes, *Blaberus
craniifer* (Burm.) et *Eublaberus distanti* (Kirby): Phénomene de dominance et plasticité de l'organisation sociale. *Biology of Behaviour*
7:69-87.

Gorton, R.E. Jr. 1979a. Behavior and ecology of the cockroaches of
northeastern Kansas. Ph.D. thesis, University of Kansas.

Gorton, R.E. Jr. 1979b. Agonism as a function of relationship in a cockroach *Shawella couloniana* (Dictyoptera:Blatellidae). *Journal of the
Kansas Entomological Society* 52:438-442.

Gorton, R.E. Jr. 1980. A comparative ecological study of the wood cockroaches in northeastern Kansas. *University of Kansas Science Bulletin* 52:221-230.

Gorton, R.E. Jr., K.G. Colliander and W.J. Bell. 1983. Social behaviour
as a function of context in cockroach. *Animal Behaviour* 31:152-159.

Gorton, R.E. Jr., J. Fulmer and W.J. Bell. 1979. Spacing patterns and
dominance in the cockroach, *Eublaberus posticus* (Dictyoptera:Bla-
beridae). *Journal of the Kansas Entomological Society* 52:334-343.

Grassé, P.P. and C. Noirot. 1959. L'évolution de la symbiose chez les
Isopteres. *Experientia* 15:375-372.

Jarman, P.J. 1974. The social organisation of antelope in relation to
their ecology. *Behaviour* 48:215-266.

McKittrick, F.A. 1964. Evolutionary studies of cockroaches. *Cornell
University Agricultural Experimental Station Memoirs* 389:1-197.

Martynov, A.B. 1938. Etudes sur l'histoire géologique et de phylogénie des
ordres des Insectes Ptérygotes. I. Palaeoptera, Neoptera, Poly-
neoptera. *Travaux de l'Institut Paléontologique de l'Académie des
Sciences de l'URSS* 7, fasc. 4.

Matsumoto, T. 1986. Colony compositions of the monogamous wood-
feeding cockroach, genus *Salganea* Roth (Blattariae, Blaberidae,
Panesthiinae). *Xth Congress of the International Union for the
Study of Social Insects Abstracts*, 129.

Morvan, R. 1972a. Etude du cycle biologique de quatre especes d'Ectobius (Blattaria: Blatellidae) de la region de Paimpont. *Bulletin de la Société Scientifique de Bretagne* 47:257-274.

Morvan, R. 1972b. Etude comparative des relations intra et interspécifiques au cours du comportement sexuel chex cinq especes d'*Ectobius* Steph. (Blattaria: Blatellidae). *Revue du Comportement Animal* 6:245-253.

Nalepa, C.A. 1984. Colony composition, protozoan transfer and some life history characteristics of the woodroach *Cryptocercus punctulatus* Scudder (Dictyoptera:Cryptocercidae). *Behavioural Ecology and Sociobiology* 14:273-279.

Rivault, C. 1985. Rythme circadien de comportement: synchronisation par l'environnement physique et social chez deux especes de blattes. These, Université de Rennes.

Roth, L.M. 1980. Cave dwelling cockroaches from Sarawak, with one new species. *Systematic Entomology* 5:97-104.

Roth, L.M. and S. Cohen. 1973. Aggregation in Blattaria. *Annals of the Entomological Society of America* 66:1315-132.

Rugg, D. and H.A. Rose. 1984. Intraspecific association in *Panesthia cribrata* (Sauss.) (Blattodea:Blaberidae). *General and Applied Entomology* 16:33-35.

Schal, C. 1982. Intraspecific vertical stratification as a mate finding mechanism in tropical cockroaches. *Science* 215: 1405-1407.

Schal, C. 1983. Behavioral and physiological ecology and community structure of tropical cockroaches (Dictyoptera:Blattaria). Ph.D. thesis, University of Kansas.

Schal, C. and W.J. Bell. 1985. Calling behavior in female cockroaches (Dictyoptera:Blattaria). *Journal of the Kansas Entomological Society* 58:261-268.

Schal, C., J.Y. Gautier and W.J. Bell. 1984. Behavioural ecology of cockroaches. *Biological Review* 59:209-254.

Seelinger, G. and U. Seelinger. 1983. On the social organisation, alarm and fighting in the primitive cockroach *Cryptocercus punctulatus* (Scudder). *Zeitschrift für Tierpsychologie* 61:315-333.

Sreng, L. 1984. Morphology of the sternal and tergal glands producing the sexual pheromones and the aphrodisiacs among the cockroaches of the subfamily Oxyhaloinae. *Journal of Morphology* 182:279-294.

Wilson, E.O. 1975. *Sociobiology: The new synthesis*. Cambridge, Mass.: The Belknap Press of Harvard University Press.

Ziegler, R. 1972. Sexual- und Territorialverhalten der Schabe *Gromphadorhina brunneri* Butler. *Zeitschrift für Tierpsychologie* 31:531-541

Chapter 15

RISK SENSITIVITY AND FORAGING IN COLONIAL SPIDERS

George W. Uetz[1]

Department of Biological Sciences
University of Cincinnati, Ohio 45221-0006

[1]*The research reported here was supported by the National Geographic Society, the American Philosophical Society, and the University of Cincinnati Research Council.*

THE ECOLOGY
OF SOCIAL BEHAVIOR

I. INTRODUCTION

Much attention recently has been focused on the ecological costs and benefits of sociality for a variety of animals. A number of advantages may result from group living, including defense against predators and parasites, increased foraging efficiency, and exploitation of new resources or habitats (Wilson 1971, 1975; Alexander 1974; Brown 1975; Oster and Wilson 1978; Wittenberger 1981; Hermann 1982; Pulliam and Caraco 1984). Although many studies have considered the benefits of social foraging in providing increased food intake, only recently have investigators noticed its importance in reducing starvation risk by influencing variance in food intake (Thompson et al. 1974). Models generated from risk-sensitive foraging theory may have considerable relevance to our understanding of the ecology of social behavior (Pulliam and Millikan 1982; Pulliam and Caraco 1984).

In this chapter, I will focus on risk-sensitivity in social foraging in a group of animals for which sociality is quite rare--the spiders. Spiders are regarded by many as ideal animals with which to investigate questions in behavior and ecology (Wise 1984), and research on social behavior in certain spider species has recently received much attention (see Buskirk 1981; Uetz 1986a). Following a review of spider social behavior, I will examine how coloniality may be considered as a response to variability in prey intake. I will then speculate on the role of risk-sensitive foraging in the evolution of sociality in spiders.

II. SOCIAL BEHAVIOR AND FORAGING

A. Social Foraging and Increased Foraging Efficiency

There are a number of ways in which members of social groups may benefit from group foraging. The rate at which food is discovered may be influenced by group foraging, and this could be especially important in environments where food is patchily distributed. Krebs et al. (1972) demonstrated that captive birds were more successful in finding food than solitary ones, because when one individual located a patch containing food, others were attracted to the same location. The process by which individuals learn of foraging locations from the behavior of others has been called "local enhancement" (Crook 1965).

The importance of social interaction in foraging may be seen in a number of similar processes. The "information centre" hypothesis (Ward and Zahavi 1973; deGroot 1980) proposes that in colonial roosting birds, individuals learn of foraging locations by observing and following successful individuals. Another mechanism, referred to as "social facilitation" may occur when individuals watch others feeding and copy their feeding behavior, thereby increasing their own efficiency at finding or handling food items (Murton 1971).

Foraging in groups may also influence the availability of food to individuals as the result of combined effort or cooperation. Rand (1954)

and Morse (1970) have shown that insectivorous birds in groups more effectively flush prey from hiding, and thereby shorten search time. Wittenberger (1981) calls this the "beater effect," and suggests that it is a
passive form of cooperation. Similar processes have been seen in schooling fish (Major 1978), crocodiles (Pooley and Gans 1976) and a variety of
birds (Emlen and Ambrose 1970). Studies have also shown that active
cooperation in foraging serves to increase individual food intake, at least
in small groups like lion prides (Caraco and Wolf 1975) or wolf packs
(Nudds 1978).

B.   Social Foraging and Risk-Sensitivity

An important consideration in analyses of social foraging involves
not only the effect of groups on the rate of food intake but the variance
as well. Many animals may be confronted with the risk of starvation of
reduced reproductive success if the variance in their food intake over
time is high. Recently, a sizeable amount of theory has been developed
based on the risk-sensitivity of foraging behavior (see review by Real
and Caraco 1986). Most of this work concerns the patterns of animal
foraging and the circumstances under which animals forage that serve
either to reduce variance in food intake (risk-averse foraging), or capitalize on that variance (risk-prone foraging).

Although most of the processes discussed above serve to increase
food intake in social foragers, some, like local enhancement or social
facilitation, may also affect the variance in individual food intake.
Thompson et al. (1974) developed a simulation model of foraging in
flocking birds based on prey detectability and bird movement patterns,
and found that one of the major benefits of foraging in flocks accrues
from minimizing risk (by minimizing variance in food intake) when
foraging in patchy environments. They further suggest that this benefit
may be more important in some cases than maximizing feeding rate.

Several recent models have been developed which suggest that the
decision to forage solitarily or in a group may be based on risk-sensitivity (Pulliam and Millikan 1982; Pulliam and Caraco 1984; Caraco and
Pulliam 1984). These models demonstrate that in environments where
food distribution is patchy, individuals in groups can reduce variance in
search time or variance in food acquisition over a given amount of time
by foraging in groups (see Figure 1A and B). In a habitat where the
average availability of food per patch or unit time exceeds that needed
to survive, an individual animal can reduce the probability of starvation
by foraging in a risk-averse manner and joining a group. On the other
hand, in a habitat where the average availability is equal to or less than
an individual's needs, survivorship will be proportional to the variance in
food availability. Under these circumstances, individuals are better off
foraging alone in a risk-prone manner on the chance that they will get a
higher rate of intake.

The role of risk-sensitivity in social foraging is currently under
investigation with a number of animal species (see Real and Caraco
1986). In this chapter, I report on research with social foraging behavior

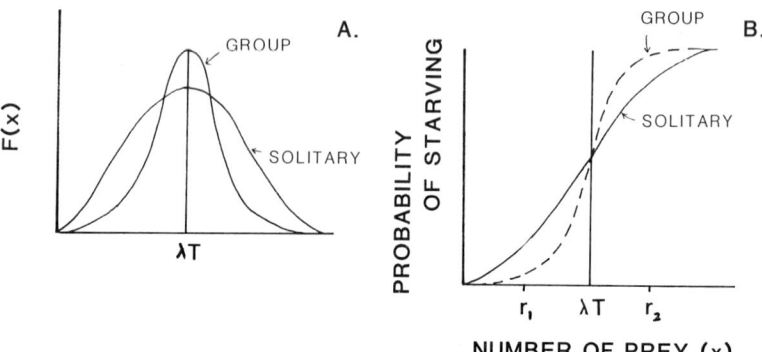

Fig. 1 A) The probability distribution of expected food intake (F_x) differs between solitary and group foragers. For the same average rate of food intake (λ) over a fixed time period (T), the variance is lower for grouped foragers. B) The probability of consuming less than (x) prey is the integral of the curves in (A). If the average expected rate of intake (λT) is greater than the amount of food required to avoid starvation (r_1), then individuals in a group will have a lower probability of starving. If the requirement is greater than the average expected rate (r_2), solitary individuals have a greater survival probability than individuals in a group. (From Caraco and Pulliam, ch. 10, In: P.W. Price, C.N. Slobodchikoff, and W.S. Gaud, Eds. A New Ecology: Novel Approaches to Interactive Systems, 1984 J. Wiley and Sons).

in colonial spiders. Since work on social behavior in spiders is in a relatively early stage (see Buskirk 1981; Uetz 1986a), and far less widely known than for other animal groups, the next section of the chapter will be devoted to providing the reader with a brief introduction to spider sociality and its evolution.

III.   SOCIAL BEHAVIOR IN SPIDERS

Most spiders are solitary predators and show aggression and even cannibalistic behavior toward other spiders (Riechert 1982). Among web-building spiders, a resource-based form of territoriality exists wherein the web is defended against intruders (Riechert 1978, 1981, 1982). These characteristics would appear to preclude the occurrence of social groups among the spiders, yet there are about 35 species (out of approximately 35,000 worldwide) which exhibit some form of social behavior (Shear 1970; Kullmann 1972; Burgess 1978; Buskirk 1981; Krafft 1982).

The major features of insect sociality (reproductive division of labor and morphological castes) are absent in the "social" spiders, as is the haplodiploid sex determination mechanism implicated in the evolution of most eusocial insects (Hamilton 1964, 1972; Wilson 1971, 1975; Oster and Wilson 1978). In entomological terms, these spiders are *presocial* in their level of organization, but have characteristics that make them difficult to place within a broad sociobiological classification scheme (Michener 1969; Cloudsley-Thompson 1982; Smith 1983).

Jackson (1978) proposed that spider sociality be classified into three forms: "solitary" spiders with non-overlapping webs and aggressive behavior toward conspecifics, "communal-territorial" spiders with interconnected but individually occupied (and defended) webs, and "communal-nonterritorial" spiders with a single web occupied by many spiders with little or no aggression. However, since this classification scheme is based predominantly on the web and living arrangements (Tietjen 1986b), it allows no distinction for other criteria of social behavior, e.g., degrees of cooperative behavior (Darchen and Delage-Darchen 1986), and has not been widely used. For example, some of the group-living spider species exhibit cooperation in all aspects of their social behavior (in prey capture, feeding and brood care), while others vary in the degrees of cooperation shown depending on a variety of factors (e.g., prey size), and yet others are strictly non-cooperative but share in some aspects of web framework construction (Tietjen 1986b; Uetz 1986b; Cangialosi and Uetz in press).

A. Evolution of Sociality in Spiders

Questions about the evolution of sociality in spiders are of particular interest because the diversity of their social behavior suggests that many selective regimes might account for its occurrence (Buskirk 1981; Smith-Trail 1982; Smith 1983). The origins of sociality in spiders can, to some extent, be seen in the behavior of many common species. Rudimentary social interactions take the form of maternal care (Rovern et al. 1973), territorial contests over webs (Riechert 1982), interactions of recently hatched spiderlings on communal webbing (Hill and Christenson 1981), and even establishment of social hierarchies (Aspey 1977; Rypstra 1979). It has often been suggested that preadaptation for sociality exists among the spiders, especially the web-builders (Buskirk 1981).

Two pathways of evolution from solitary to social existence have been suggested for spiders: the "subsocial" and "parasocial" routes (Shear 1970; Wilson 1971; Kullman 1972; Buskirk 1981). On the subsocial route, parental care might be extended, or the dispersal of spiderlings delayed, resulting in maternal/juvenile aggregations or prolonged aggregations of siblings, which in turn could lead to "subsocial" (also called "quasisocial") (Brach 1977; Nentwig and Christenson 1985) and ultimately "social" (Burgess 1976) spiders with cooperative prey capture, feeding, and brood care. An alternate route, the "parasocial" pathway, is where individual spiders of any age (possibly unrelated) could live together, aggregated in areas of high prey density (Rypstra 1983, 1986). If the aggregation provided ecological benefits such as increased prey capture efficiency or protection from predators, selection for communal habits might lead to a "colonial" form of sociality (Shear 1970; Buskirk 1981; Uetz et al. 1982; Uetz 1986b).

B. Examples of Spider Social Organization

An example of an evolutionary series on the subsocial pathway can be seen in species of the genus *Anelosimus* (Theridiidae), which build large communal sheetwebs on three branches. Subsocial (quasisocial) *A. studiosus* from Mexico and the southern U.S., and *A. jucundus* from Central America show a living arrangement where spiders live together as mother-offspring groups. Initially, the spiderlings are fed by their mother through regurgitation; then as they mature, individuals cooperate in prey capture and feed communally. Spiders also cooperate in web-building and nest-cleaning activities, and live together until the onset of sexual maturity, when colonies break up and females found new groups (Brach 1977; Nentwig and Christenson 1985). Social *A. eximius* live together in large colonies with overlap of generations, and show cooperation in web construction, nest cleaning, prey capture and brood care (Vollrath 1982; Vollrath and Rohde-Arndt 1983; Christenson 1984; Nentwig 1985).

The highest levels of sociality seen in spiders are seen among members of only a few species from the families Theridiidae (*Anelosimus eximius, Achearanea wau*), Agelenidae (*Agelena consociata*), Dictynidae (*Mallos gregalis*), and Eresidae (*Stegodyphus sarasinorum*). These spiders have in common the characteristics described above for *Anelosimus eximius*, and are all presumed to have arisen through similar subsocial evolutionary pathways. The probability of genetic relatedness among members of these kinds of groups is high, and thus kin selection is considered to be a potential driving force for cooperative behavior to evolve (Buskirk 1981; Smith 1986; Riechert et al., 1986). Interestingly, these spiders also show other characteristics--a highly female-biased sex ratio, and colony formation by fragmentation and/or swarming--which have led several researchers to suggest that group selection may also be involved (Lubin and Crozier 1985; Vollrath 1986a,b; Aviles 1986).

An example of an evolutionary series on the parasocial pathway is found in the orb-weaver genus *Cyrtophora* (Lubin 1974, 1980). These spiders build a non-sticky horizontal orb with a maze of barrier webbing above it, where the spider sits in a retreat. *Cyrtophora monufli* from Pacific Islands is a solitary species, while the widely distributed *C. citricola* may occur solitarily or (more often) in colonies of tens to hundreds of individuals (Lubin 1974, 1980; Rypstra 1979). *C. molluccensis* from New Guinea persist in large colonies for years at a single site (Lubin 1980). Prey capture for *Cyrtophora* is an individual affair, even when individuals all contribute to the construction and repair of the web framework supporting their webs. Similar living arrangements are seen in a number of orb weavers (Buskirk 1975, 1981; Tietjen 1986b; Uetz 1986b).

C.   Colonial Spiders as Foraging Flocks

As information on the various types of social organization seen in spiders accumulates, it may be possible to examine spiders in light of a number of current models, which consider social behavior as a consequence of individual foraging decisions (Pulliam and Caraco 1984; Caraco and Pulliam 1984). Because spider webs are not only habitations, but prey capture devices (Burgess and Witt 1976), groups of web-building spiders have been considered to be "foraging flocks" (Rypstra 1979) as well as "colonies." The behavior of colonial spider groups, however, is somewhat different from many other animal groups because the "foraging flock" is stationary, and distributed in three dimensions instead of two.

There are constraints on the social behavior of orb-weaving spiders (families Araneidae and Uloboridae) which make studies of their sociality quite intriguing. Unlike the sheet or tangle webbing of spiders in other families which exhibit communal web building (Theridiidae, Agelenidae, Dictynidae, Eresidae, Amaurobiidae, Oecobiidae, and Pholcidae), the orb web cannot be built by more than one spider. It is the result of a complex sequence of behaviors which are tightly controlled by the genetic "program" of individual spiders (Witt and Reed 1965; Witt et al. 1986), unlikely to be modified to include participation by others (Lubin 1974; Burgess and Uetz 1982). Thus, cooperative web construction, at least of the prey-catching portions of the web, is precluded. Most orb weavers are limited in their social organization to aggregations of individuals building and occupying their own webs within a shared web foundation (Lubin 1974; Buskirk 1975 a,b; Fowler and Diehl 1978; Uetz et al. 1982; Smith 1983). The result is a three-dimensional foraging flock in which the spatial arrangement of group members is somewhat fixed, yet may change every time webs are renewed (usually on a daily basis) depending on the outcome of aggressive interactions between individuals.

A question raised often with regard to colonial spider sociality is whether some actually are "social" groups, or in fact are "fortuitous" aggregations of individuals exhibiting an aggregative response to prey density (Readshaw 1973; Riechert 1974, 1978). Under such conditions, there would be a reduction in territory size in response to reduced levels of hunger (Riechert 1978; Buskirk 1981; Burgess and Uetz 1982). For example, recent studies by A.L. Rypstra with solitary spiders (1985, 1986) have shown that manipulation of food supply has profound effects on the dispersion of spiders. She was able to produce web "colonies" in areas of high prey density, and force those colonies to disperse by lowering prey availability. This could argue that sociality in spiders is an artifact of patchy resource distribution, and/or that tolerance of conspecifics is a result of reduction in aggression with increased food availability. On the other hand, others have argued that these fortuitous aggregations represent the earliest stages of spider sociality (Riechert 1974; Burgess 1978; Buskirk 1981; Rypstra 1986). In such situations, selection could favor individuals with increased tolerance of conspecifics, as this would reduce energy lost in frequent agonistic encounters, and/or allow individuals to benefit from increased foraging efficiency provided by grouping webs (Riechert 1978; Uetz et al. 1982; Rypstra 1979, 1986; Uetz 1986b). The social organization of most species of "colonial" or "communal/territorial" orb weaving spiders (Jackson 1978) is conceptually (and possibly evolutionarily) intermediate between the incipient sociality of fortuitous aggregations of otherwise solitary spiders, and that of more advanced social spiders (which exhibit some degree of cooperation) (Buskirk 1981; Burgess and Uetz 1982; Rypstra 1983, 1986; Uetz 1986b). These spiders must reconcile the demands of conflicting behavioral strategies--communal aggregation and defense of space--while at the same time contend with a mosaic of contiguous territories in three dimensions instead of two. The "territory" defended within the colony has indistinct boundaries, depending on the nature of silk connections between the retreat, the orb or catching spiral, and the space web of other individuals. In addition, this is a "multipurpose territory" (Davies 1978), in that it may include a foraging site (the orb), a habitation (the retreat), a mating site (orb and/or retreat), and a breeding/egg laying site (the retreat).

A number of studies have suggested that spiders in "foraging flocks" benefit from increased prey capture by grouping webs (Lubin 1974; Rypstra 1979; Uetz 1985, 1986b; Rypstra in press). Increased prey capture efficiency in colonial spiders may come about in two ways--through the process of "local enhancement" (Hinde 1959, 1961; Crook 1965) and/or as a consequence of the "richochet effect" (Uetz 1986b). Local enhancement occurs when the attention of one individual is drawn to a particular foraging location by the behavior of another. In spiders, such an effect may be the result of the concentration of silk lines (indicating the presence of other successful foragers), or from the increased perception of prey by vibrations through connection of silk lines in colonies (Gillespie in press). The ricochet effect (similar to the beater effect) is seen in web colonies when prey are captured after they bounce off several webs in succession (Uetz 1985, 1986b; Rypstra in press). These effects may in fact be inseparable, because the richochet of prey from one web to another alerts spiders to their presence, and causes individuals to move to a capture-ready position at the web hub (which undoubtedly improves capture efficiency). As a consequence, spiders in colonies capture more and larger prey than solitary spiders in the same habitat (Uetz 1986b).

Considering that most spiders are solitary predators, the behavior of the colonial spiders represents a major shift--from solitary to group foraging. Whether they are seen as the first stages in evolution of spider sociality through the parasocial pathway, or considered as permanent or temporary foraging flocks, colonial spiders provide an opportunity to examine how foraging decisions, in response to ecological factors, may shape the evolution of animal societies.

IV.   VARIATION IN GROUP FORAGING IN COLONIAL ORB WEAVERS

Colonial spiders of the genus *Metepeira* Pickard-Cambridge (Araneidae) found in central Mexico (Pickard-Cambridge 1903) show considerable variation in their social organization, ranging from solitary individuals to small colonies to huge groups of tens of thousands (Burgess and Witt 1976; Uetz and Burgess 1979; Burgess and Uetz 1982; Uetz et al. 1982). The wide variation in group size seen in these spiders affords an opportunity to address a number of questions concerning the ecological factors affecting group foraging in colonial spiders.

The web of individual *Metepeira* contains a three-dimensional space (or barrier) web with a retreat, and a sticky orb connected to the retreat by signal threads (Burgess and Witt 1976; Levi 1977). Although the sticky orb webs are taken down and renewed on a daily basis, the space web persists, and in colonies acts as a framework for the web-building activities of numerous individuals. Individuals maintain and defend orbs and retreats within the colony, and capture their own prey.

Previous work has demonstrated that benefits gained by *Metepeira* spiders from living in groups include increased foraging efficiency, increased predator detection, and occupation of web sites that could not be exploited by solitary spiders (Uetz 1983, 1985, 1986b). Costs of group-living include behavioral competition and energy lost in agonistic encounters (Benton and Uetz 1986), and increased vulnerability to predators and parasitoids (Uetz and Hieber unpubl.). Social spacing in *Metepeira* is influenced by prey availability, mediated by behavioral interactions, and partially determined by genetically based differences in social tolerance (Uetz et al. 1982; Uetz and Cangialosi 1986; Cangialosi and Uetz in press; Uetz et al. in press).

Group size in *Metepeira* varies geographically between and within species, largely in response to habitat and prey availability. Populations, once thought to be a single species, have been shown to be behaviorally distinct species (Uetz and Cangialosi 1986; Cangialosi and Uetz, in press). In the mesquite grasslands of San Miguel de Allende, Guanajuato, in north-central Mexico, where climatic conditions are typically harsh and prey availability is low (e.g., desert grasslands), *Metepeira atascadero* are predominantly solitary or in small groups, maintaining considerable distances between individuals. In the agricultural central valley of Mexico, near Mexico City, there are alternating rainy and dry seasons, and spiders there (*Metepeira spinipes*) live in groups of 5-150, with variable spacing. In mid-elevation rainforest/agricultural sites in Veracruz, where climate is benign all year and insect abundance is great, colony size of *Metepeira incrassata* is very large, ranging from hundreds to tens of thousands of closely spaced individuals.

A.   Prey Availability and Group Size: A Paradox

The observed relationship between spider group size and availability of prey is deceptively simple, and upon closer examination reveals a paradox with respect to group foraging. Not unexpectedly, spiders preferentially forage in areas of greater prey abundance (Turnbull 1964; Gillespie 1981; Morse and Fritz 1982; Rypstra 1985), and prey availability is a key factor permitting higher spider densities (Riechert 1978, 1981; Rypstra 1983). However, in colonial spiders, prey capture efficiency increases with colony size (Lubin 1974; Rypstra 1979; Uetz 1985). The paradox, then, is that given the advantage of group foraging with respect to prey capture efficiency, why don't spiders in prey-poor environments group webs together and take advantage of this? On the other hand, why would spiders in prey-rich environments need to forage together?

Given the importance of the "decision" to forage solitarily or in groups as a step in the evolution of sociality in spiders, this apparent paradox requires an explanation. A reasonable one might be that in some cases, patches providing space for webs and sufficient food for more than one spider coincide only rarely--as "oases in a desert" (Carr and MacDonald 1986). The number of spiders in a patch or colony will be determined by the amount of food and the compressibility of web "territories." This explanation assumes that there is some threshold value of prey availability in a patch which could permit the coexistence of several spiders, except that their territorial behavior would prohibit it. In this case, territory size would shrink over time, as selection would favor "tolerant" individuals who expend less energy fighting over unneeded space (Brown 1964; Riechert 1978). Once the threshold of prey availability permitting tolerance has been exceeded, spiders can take advantage of the enhanced prey capture afforded by grouping webs. In the case of M. incrassata, living in prey-rich tropical areas, it may be that other selection pressures (e.g., protection from predators) are more important in fostering social grouping.

One difficulty with this explanation is that tolerance must evolve first, before spiders can take advantage of the richochet effect. Spider territorial behavior appears to be tightly controlled genetically (Riechert in press) and opportunities allowing such a behavioral strategy to take hold in a population might be very rare. On the other hand, it would appear that once the step is taken, larger groups would be expected to occur far more frequently than they are seen in sites with intermediate levels of prey. Given this explanation, social behavior in spiders might well be expected to be more common than it is.

Another problem with this explanation is that it does not take into account temporal variability of prey resources. It is hard to envision how a tolerant behavioral strategy could take over in a population, as selection would not favor tolerant individuals when colonies experience periodic food shortages, or when group size grows to exceed the limits of local food supply. In this light, the variation seen might argue that aggregations of spiders are "fortuitous," and that all such spiders are merely facultative in their sociality. However, open field studies in the laboratory (Burgess and Witt 1976) and analyses of spatial distribution in the field (Uetz and Burgess 1978) argue that aggregation in Metepeira is based on interattraction and not environmental factors such as resource dispersion.

Recent studies provide somewhat contradictory information, rather than clarification of this paradox. In studies of *Metepeira spinipes* in intermediate habitats, changes in social spacing have been seen in response to short-term variation in prey (Uetz et al. 1982). Similarly, lab studies manipulating prey levels show a reduction in cannibalism and reduced nearest neighbor distance at higher levels of prey, suggesting flexibility in the degree of tolerance (Uetz et al. in press). Even so, minimum spacing limits appear to be genetically fixed in different populations (Uetz and Cangialosi 1986; Cangialosi and Uetz in press). Field observations of agonistic behavior suggest that the number of aggressive interactions per spider increases with colony size, and is highest in the large colonies of *M. incrassata* from prey-rich tropical sites (Uetz 1985). Laboratory studies of agonistic behavioral interactions in tropical *M. incrassata* and desert *M. atascadero* have shown that even under conditions of satiation levels of prey and equal colony size, tropical spiders show several times as many conflicts (Uetz and Cangialosi 1986). Although interactions among tropical spiders are less likely to escalate, the energy drain from increased encounter rates in large colonies may cancel gains from increased prey capture efficiency--resulting in lower egg production (Benton and Uetz 1986). Clearly, there are a number of problems with the initial explanation proposed for the apparent paradox of increased group size with increased prey availability.

B.   Risk-Sensitivity and Group Foraging
in *Metepeira*

An alternative argument may be one that considers the foraging decisions spiders make (solitary vs. group) and their relationship to the level and variability of prey in different habitats--are spiders making choices based on risk-sensitivity?   For sedentary predators like *Metepeira*, whose foraging sites occur as discrete, widely separated patches, the models of Pulliam and Millikan (1982) and Pulliam and Caraco (1984) have particular relevance. If grouping webs affects not only prey capture efficiency, but also reduces variability in the amount each individual captures, then spiders would be expected to conform to the predictions of these models.

Within a habitat, some patches have high average prey availability (because of prey attractants like flowers or animal feces--Riechert and Tracy 1975), while others have low prey. Variability of prey within a patch over time may also be high or low. In habitats where the overall average availability of prey is at or above the daily requirement of an individual, the best choice should be to adopt a risk-averse strategy that reduces the variance in prey, and forage in a group. On the other hand, in habitats where the average availability of prey is close to or below the individual's requirements, the best choice should be a risk-prone strategy of solitary foraging. Although the risk-prone strategy tends to increase variance in prey, a hungry spider would be better off taking that risk on the chance that variance will provide a higher prey reward, rather than be faced with almost certain starvation. Given these predictions, solitary spiders and smaller groups would be expected in prey-poor habitats like desert grasslands, and larger groups would be expected in prey-rich habitats like tropical forest/agricultural areas.

C. Does Group Foraging Affect Variance in Prey Capture?

Laboratory studies (Hollis and Uetz unpub.) were conducted with *Metepeira spinipes* hatched from field-collected egg sacs and raised under controlled environmental conditions (light/dark cycle 12 hrs/12 hrs; temperature 27° C; relative humidity 65-75%). Spiders were housed in clear plastic and screen cages (100 x 85 x 75 cm) in groups of 1, 4, 9, and 20 individuals (five replicates of each group size). Spiders were fed *Musca domestica* at a rate equal to that necessary to provide a daily maintenance requirement of approximately 16-20 mg dry wt/day (about three to five flies/spider/day). Prey capture rates were recorded daily for seven to ten days for each cage.

In general, spiders with grouped webs captured significantly more prey per individual than solitary spiders (Scheffe test, $p < 0.05$), and prey capture efficiency also increased with group size, supporting the assumptions of Pulliam and Millikan (1981) and Pulliam and Caraco (1984) (Fig. 2a). Variance in number of prey captured/spider decreases dramatically as a function of group size, as predicted by the above models (Fig. 2b). Given individual requirements, at the rate of prey capture seen in these laboratory cages, it is likely that about half of the solitary spiders would starve or experience limited growth and egg production, while most grouped spiders would survive and reproduce (Riechert and Tracy 1975).

Field observations also support these findings. Prey captured/spider and efficiency of capture increased across a range of colony sizes in *M. spinipes* (Uetz 1985, 1986b). In 25 hours of field observation of solitary and colonial spiders occurring side by side in the same habitat, the mean number of prey captured/individual was significantly greater for colonial individuals (1.80 prey/spider/hr; SD = 1.18) than for solitaries (1.30 prey/spider/hr; SD = 1.26). Coefficients of variation were also different--96.9% for solitaries and 65.6% for colonial spiders. By grouping webs together, *Metepeira* may increase their individual rate of prey capture and decrease variance in prey.

D. Solitary vs. Group: Does Prey Availability Affect Foraging Decisions?

Spiders were studied in three geographic regions in Mexico: near San Miguel de Allende, Guanajuato (*Metepeira atascadero*) in desert/mesquite grassland habitat; near Fortin de las Flores, Veracruz (*Metepeira incrassata*) in montane tropical rainforest/agricultural sites; and near Toluca and Tepotzotlan, Mexico (*Metepeira spinipes*) in agricultural areas. At each of these sites, spiders were surveyed in plots ranging in

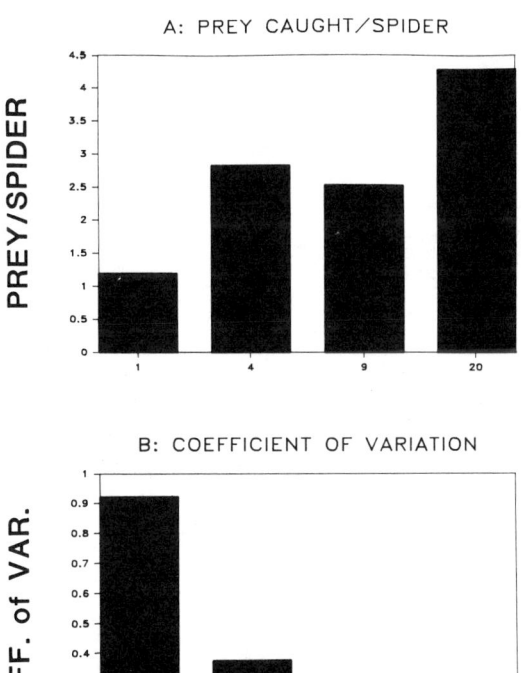

Fig. 2. Laboratory studies with caged spiders show: a) Prey capture rate increases with group size; and b) Variance in prey/spider (expressed here as a coefficient of variation) decreases with size of group.

size from 600 to 10,000 sq m (plot size used was determined by area required to sample >100 spider colonies).   Within each plot, all spiders were located and numbers of individuals and groups counted.

Prey availability in these sites was measured in two ways.   Prey abundance and variability were assessed in all sites by direct observation of insects flying into 1 m³ of colony web in hourly observation periods. In the desert/grassland and tropical forest/agriculture sites, prey insects were trapped on adhesive-coated artificial sticky web traps which approximated the size (250 cm²) of spiders' webs.   Daily prey biomass available was estimated from length-weight regressions for various taxa applied to insects captured by these traps.

Abundance and variability of prey differed among sites (Table I).   In the desert/mesquite grassland site, prey availability was lowest, while in the tropical forest/agriculture site, prey availability was highest.   Varia-

TABLE I. The frequency of spiders foraging in groups or solitarily varies with the average level of prey availability as well as with variation in prey availability.

Species	Site	X Prey/m^3 (1 hr obs.)	Coeff of Var	Freq. of Groups >1
Metepeira atascadero	San Miguel de Allende	10.3	.54	.21
M. incrassata	Fortin	59.7	.27	.99
M. spinipes	Toluca	12.0	.56	.67
	Tepotzotlan	18.7	.52	.76
	Tepotzotlan	29.2	.61	.93

bility in the desert site was about twice that of the tropical site. In central Mexico agricultural sites, prey abundance is intermediate between the above, but variability is equal to or higher than the desert site.

The frequency of spiders foraging solitarily or in groups varies--both between species habitats and within a species habitat between sites--with the average level of prey (Table I). In desert/mesquite grassland habitat, where average daily prey biomass (mean = 20.67 mg/250 cm^2; 16 mg SD; coeff. var. = 0.77) is less than or equal to individual needs and of relatively high variability, most *M. atascadero* are solitary foragers (risk-prone strategy) (Fig. 3). In montane tropical forest/agriculture habitats, where average daily biomass (mean = 58.8 mg/255 cm^2; 2.4 mg SD; coeff. var. = 0.43) is at least two to three times the daily individual requirement, and with relatively low variability, *M. incrassata* forage in large groups (risk-averse strategy) (Fig. 4).

In the Central Valley of Mexico, where prey variability is relatively high (similar to that of the desert/mesquite grassland habitat), yet varies in average abundance from site to site, *Metepeira spinipes* shows varying degrees of group foraging. Over a range of sites with levels of prey biomass from approximately equal (Toluca, Tepotzotlan agriculture) to greater than individual needs (Tepotzotlan feedlot, Guadalupe Lake), *M. sinipes* populations shift from a predominantly solitary (risk-prone) foraging strategy to increasing frequency of group (risk-averse) foraging (Fig. 5A-D). Results of field transplant experiments also show evidence of this shift: When relocated to prey-poor sites (e.g., in Tepotzotlan, from feedlot to agriculture), groups of *M. spinipes* decrease in size or break up completely; when prey are supplemented, spiders coalesce into groups (Uetz et al. 1982).

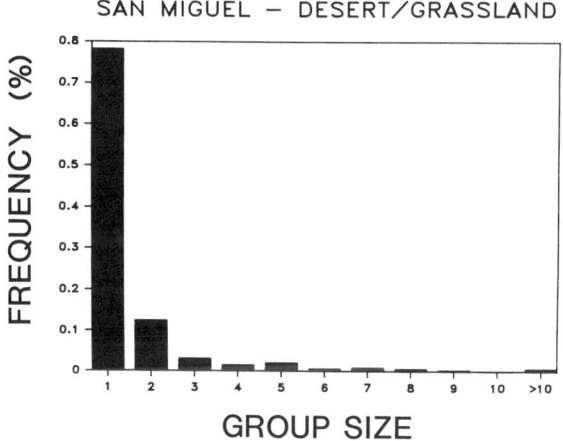

Fig. 3. *In desert/mesquite grassland habitat, where average daily prey biomass is less than or equal to individual needs, most Metepeira atascadero are solitary foragers (risk-prone strategy).*

Group foraging behavior observed in Metepeira fits the predictions of risk-sensitive foraging models proposed by Pulliam and Millikan (1982) and Pulliam and Caraco (1984). These findings help explain the paradox of earlier observations of a correlation between prey availability and colony size (Uetz et al. 1982). A similar explanation may also be offered

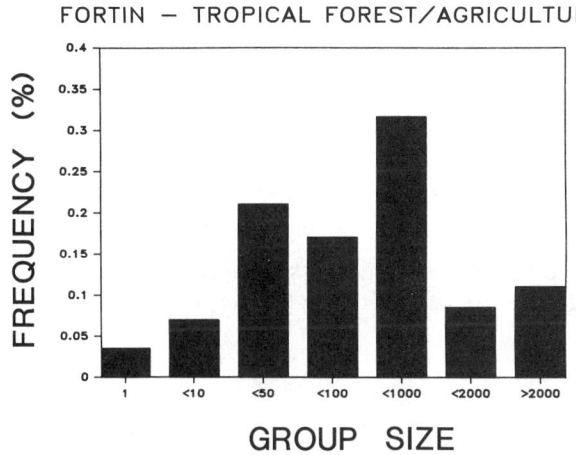

Fig. 4. *In tropical rainforest/agriculture sites, where average daily prey biomass is 2-3 x individual needs, Metepeira incrassata forage in large groups (risk-averse strategy).*

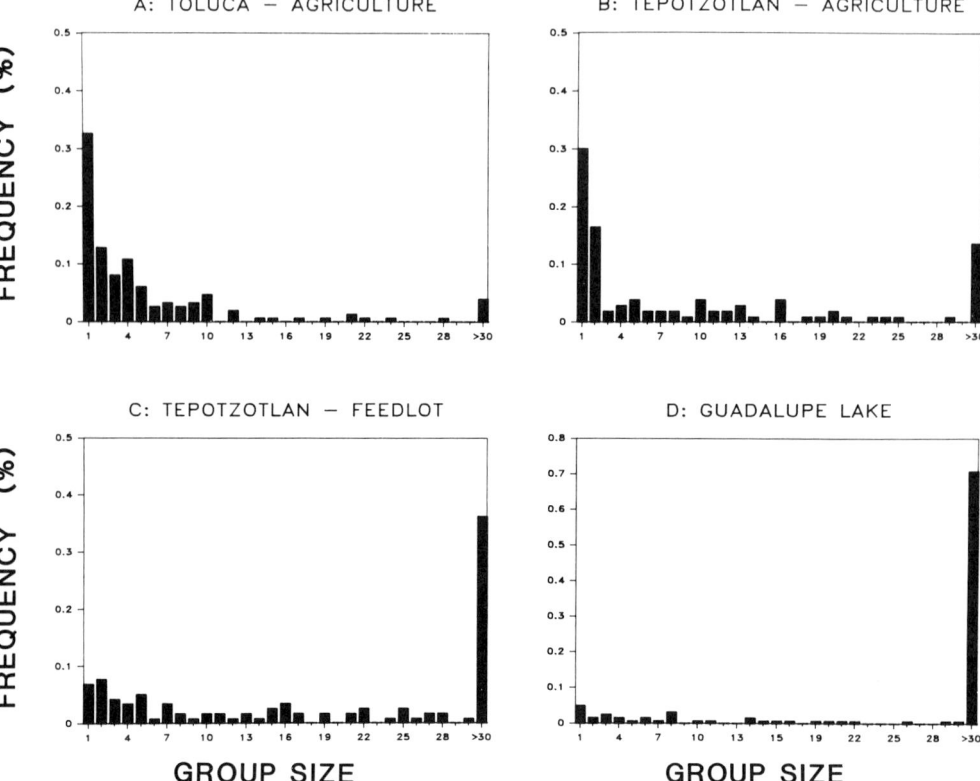

Fig. 5. *In central Mexico, where prey availability varies widely, Metepeira spinipes shows varying degrees of group foraging. Over a gradient of prey availability, from biomass equal to individual needs (A,B) to greater than individual needs (C, D), M. spinipes shows a shift from predominantly solitary (risk-prone) to predominantly group (risk-averse) foraging.*

for the tendency of a number of solitary spider species to form aggregations in areas of high prey density (Rypstra 1985; Gillespie in press).

V.   RISK-SENSITIVITY AND THE ORIGINS OF SOCIALITY IN SPIDERS

The data presented above raise new questions regarding the potential role of risk-sensitive foraging decisions as a critical element in the origin of sociality in spiders. Caraco and Gillespie (1986) have suggested that for certain orb-weaving spiders, the decision to change web sites or remain may be based on risk-sensitivity. Their model predicts that for spiders faced with spatial and temporal variation in prey, the risk-averse strategy is to adopt a mobile foraging mode, changing web sites frequently. It seems likely under these circumstances that spiders would move along gradients of increasing prey availability and end up aggregated in sites providing prey. This has been shown to be true for a number of spider species (Turnbull 1964; Honjo 1977; Morse and Fritz 1982; Gillespie 1981; Rypstra 1985; Riechert and Gillespie 1986).

The next foraging decision--to continue foraging solitarily or join with neighbors--would be based on the level of prey available. If the site provided prey at a level close to the individual requirement, then solitary foraging (the risk-prone strategy) would be the likely choice. Thus, for most spider species in most habitats, sociality may be precluded. Only where prey exceeds individual requirements is it likely that spiders will choose the risk-averse option of group foraging. It should come as no surprise that most species of colonial spiders (as well as all social species) occur in tropical environments where insect prey are abundant (Buskirk 1981; Riechert 1986).

Metepeira provides an illustration of the ecological circumstances under which sociality might arise on the parasocial pathway. These spiders only occur in habitat patches providing rigid support for their three-dimensional web (comprised of a barrier/space web and a catching orb). In most habitats where *Metepeira* are found, these "patches" are cacti, succulents and shrubs. The distance between patches may be great, and movement between them may not only be energetically costly, but life-threatening (desiccation is a real possibility, and a variety of predators abound). As a consequence, when spiders find a suitable web site, they tend not to move, and are faced with temporal variability in prey. When *Metepeira* are found aggregated in patches, they are not necessarily colonial--often a dozen or more may occupy the same shrub without connecting webs. When *Metepeira* do join webs into colonies, they do so by attaching barrier webs, which may contribute to increased prey capture and reduction in variance by local enhancement or the "ricochet effect." In habitats where the level of prey availability dictates a shift from risk-prone solitary foraging to risk-averse group foraging, selection may then act on the additional benefits provided (e.g., energy savings through a reduction in cost per individual of web construction), and sociality may evolve. The transition from solitary to colonial living seen among *Metepeira* species may reflect these initial steps toward sociality.

Recent work has shown that many species of web-building spiders have the capacity to shift from a solitary foraging mode to group foraging in sites providing a superabundance of prey (Burgess and Uetz 1982; Rypstra 1983, 1985, 1986; Gillespie in press). Rypstra (1986) has shown that *Achearanea tepidariorum* (Theridiidae) exhibits increased tolerance and a reduction in cannibalism when provided with a superabundance of prey, facilitating group foraging in limited space. Although *Achaearanea* in groups get approximately the same amount of prey as do solitary individuals, variance in prey intake is significantly reduced in groups (Rypstra in press). Gillespie (in press) has shown that in high-density aggregations in prey-rich sites, *Tetragnatha elongata* (Tetragnathidae) not only show an increase in tolerance, but also a reduction in individual web building, and indiscriminate use of pre-laid silk. These consequences of aggregation provide energetic cost savings from reduced silk production, as well as benefits from increased efficiency of prey detection over a larger area. Gillespie further suggests that this latter benefit is what serves to reduce variance in individual prey capture, since individual areas of prey detection will overlap and create a local enhancement effect. Thus, aggregative behavior in these otherwise solitary orb weavers may be a risk-averse foraging strategy.

VI. SUMMARY

Group foraging behavior observed in colonial *Metepeira* fits the predictions of risk-sensitive foraging theory (Pulliam and Millikan 1982; Pulliam and Caraco 1984). As group foraging reduces variance in prey capture/individual, spiders tend to forage solitarily where prey availability is less than or equal to their needs, and in groups only where prey availability is greater. These findings help explain the paradox of earlier observations of the tendency of *Metepeira* and a number of other orb-weaving spiders to build webs in groups only in areas of high prey availability (Uetz et al. 1982). If spiders in groups benefit from enhanced prey capture, why don't they do this in prey-poor environments? The answer is that the reduction in variance in prey intake which occurs as a consequence of group web building dictates that spiders in prey-poor habitats forage in a risk-prone manner and be solitary. It is better for them to take the risk of starvation on the chance that they will encounter a higher rate of prey, rather than forage in a group and increase the risk that they will starve.

In most patchy environments, solitary foraging is the appropriate strategy for spiders, because prey availability is likely to be close to individual requirements. This explains why sociality is so rare in spiders. Only in certain habitats or patches--those with superabundant prey--do spiders group webs together (Rypstra 1983, 1985; Gillespie in press), and reap the collective benefits of social foraging. In this light, risk-sensitive foraging may be seen as a critical element in the evolution of coloniality in spiders.

There are a number of as-yet unanswered questions regarding risk-sensitivity and decision making in spider foraging, and much research needs to be done. To fully understand the individual spider's decision to forage solitarily or in a group, it is necessary to clarify the factors impinging on that decision, as well as the consequences of that decision in terms of fitness. A number of studies have shown that birds, mammals and bees may adjust their foraging behavior in the short term in a risk-sensitive manner (Real and Caraco 1986). Within a number of spider species, there appears to be the capability of shifting from solitary to social foraging (and back) depending on prey availability. What are the proximate mechanisms that enable spiders to make such choices? The reduction in aggression and tolerance of conspecifics which occurs in many spiders as a result of prey abundance may be critical in allowing spiders to forage in groups. How this occurs is not well understood. Although it is clear that spiders are capable of reacting to hunger level, with what levels of resolution can they detect and respond to variance in food reward over time?

Perhaps more important is the question of the contribution of risk-sensitive foraging decisions to fitness. In most spiders there appears to be a direct relationship between food consumption, growth, and egg production (Kessler 1971; Wise 1975, 1979; Eberhard 1979; Palanichamy 1984; Fritz and Morse 1985). Spiders whose access to prey is limited by a variety of environmental factors suffer reduced fecundity (Riechert and Tracy 1975; Fritz and Morse 1985; Benton and Uetz 1986). Increased energy expenditure or reduced food intake result in slower growth and smaller size at adulthood, which may also affect egg production (Turnbull 1962; Anderson 1974; Palanichamy 1984; Suter 1985; Vollrath 1985; Benton and Uetz 1986). Other fitness-related parameters, especially juvenile survivorship, may be affected by foraging success, but are less known for spiders.

The research presented here provides evidence that colonial spiders forage in a risk-sensitive manner. It seems likely that risk-prone and risk-averse foraging strategies may differ between spider species, and even within species as a function of age and reproductive status (Caraco and Gillespie 1986). Future studies should reveal how individual foraging decisions influence fitness, and provide more insights regarding the role of risk-sensitive foraging in the evolution of sociality in spiders.

ACKNOWLEDGMENTS

The research reported here was conducted with support from the National Geographic Society, the American Philosophical Society, and the University of Cincinnati Research Council. I am grateful to a number of people who have contributed to this research in a variety of ways: for assistance in the field and laboratory, and for editorial comments, Debbie Fritz, Maggie Hodge, Criag Hieber, Kitty Uetz, Sonie Scheffer, Karen Cangialosi, Gail Stratton, Tom Kane, Bob Hollis, Mary Klus, Rosemarie Gillespie, and Ann Rypstra. Thanks also to Con Slobodchikoff for including me in this volume.

REFERENCES

Alexander, R.D. 1974. The evolution of social behavior. *Annual Review of Ecology and Systematics* 5:325-383.

Anderson, J.F. 1974. Responses to starvation in the spiders *Lycosa lenta* (Hentz) and *Filistata hibernalis* (Hentz). *Ecology* 55:576-585.

Aspey, W.P. 1977. Wolf Spider Sociobiology: I. Agonistic display and dominance-subordinance relationships in adult male *Schizocosa crassipes*. *Behaviour* 12:103-141.

Aviles, L. 1986. Sex-ratio bias and possible group selection in the social spider *Anelosimus eximius*. *American Naturalist* 128:1-12.

Benton, M.J. and G.W. Uetz. 1986. Variation in life-history characteristics over a clinal gradient in three populations of a communal orb-weaving spider. *Oecologia* 68:395-399.

Brach, V. 1977. *Anelosimus studiosus* (Araneae:Theridiidae) and the evolution of quasisociality in Theridiid spiders. *Evolution* 31:154-161.

Brown, J.L. 1964. The evolution of diversity in avian territorial systems. *Wilson Bulletin* 76:160-169.

Brown, J.L. 1975. *The evolution of behavior.* New York:Norton. 761 pp.

Burgess, J.W. 1976. Social spiders. *Scientific American* 234:100-106

Burgess, J.W. 1978. Social behavior in group-living spider species. *Symposia of the Zoological Society of London* 42:69-78.

Burgess, J.W. and P.N. Witt. 1976. Spider webs: design and engineering. *Interdisciplinary Science Review* 1:322-355.

Burgess, J.W., and G.W. Uetz. 1982. "Social spacing strategies in spiders." pp. 317-351. In: P.N. Witt and J.S. Rover, eds. *Spider communications, mechanisms and ecological significance.* Princeton, NJ:Princeton University Press. 440 pp.

Buskirk, R.E. 1975. Coloniality, activity patterns and feeding in tropical orb-weaving spider. *Ecology* 56:1314-1328.

Buskirk, R.E. 1982. "Sociality in the Arachnida." pp. 282-367. In: H.R. Hermann, ed. *Social insects*, vol. II. London:Academic Press.

Cangialosi, K.R. and G.W. Uetz. In press. Spacing in colonial spiders: effects of environment and experience. *Ethology.*

Caraco, T. and L.L. Wolf. 1975. Ecological determinants of group sizes of foraging lions. *American Naturalist* 190: 343-352.

Caraco, T. and H.R. Pulliam. 1984. "Sociality and survivorship in animals exposed to predation." pp. 279-309. In: P.W. Price, C.N. Slobodchikoff and W.S. Gaud, eds. *A new ecology: Novel approaches to interactive systems.* New York:John Wiley and Sons

Caraco, T. and R.S. Gillespie. 1986. Risk-sensitivity: foraging mode in an ambush predator. *Ecology* 67:1180-1185.

Carr, G.M. and D.W. MacDonald. 1986. The sociality of solitary foragers: a model based on resource dispersion. *Animal Behaviour* 34:1540-1549.

Christenson, T.E. 1984. Behavior of colonial and solitary spiders of the theridiid species *Anelosimus eximius*. *Animal Behaviour* 32:725-734.

Cloudsley-Thompson, J.L. 1981. On communal and social spiders. British Arachnological Society, *The Secretary's Newsletter*, No. 32, November, 1981.

Crook, J. 1970. Social organization and environment: Aspects of contemporary social ethology. *Animal Behaviour* 18:197-209.

Darchen, R. and B. Delage-Darchen. 1986. Societies of spiders compared to the societies of insects. *Journal of Arachnology* 14:227-238.

Davies, N.B. 1978. "Ecological questions about territorial behaviour." pp. 317-350. In: J.R. Krebs and N.B. Davies, eds. *Behavioural ecology: An evolutionary approach.* Sunderland, Mass.:Sinauer Assoc.

DeGroot, P. 1980. Information transfer in a socially roosting weaver bird (*Quelea quelea*:Ploceinae): An experimental study. *Animal Behaviour* 28:1249-1254.

Eberhard, W.G. 1979. Rates of egg production by tropical spiders in the field. *Biotropica* 11:292-300.

Emlen, S.T. and Ambrose, H.W. 1970. Feeding interactions of snowy egrets and red-breasted mergansers. *Auk* 87:164-165.

Fowler, H.G. and J. Diehl. 1978. Biology of a Paraguayan orb-weaver, *Eriophora bistriata* (Rengger) (Araneae:Araneidae). *Bulletin of the British Arachnological Society* 4:241-250.

Fritz, R.S. and D.H. Morse. 1985. Reproductive success and foraging of the crab spider *Misumena vatia*. *Oecologia* 65:194-200.

Gillespie, R.G. 1981. The quest for prey by the web building spider *Amaurobius similis* (Blackwell). *Animal Behaviour* 29:953-954.

Hamilton, W.D. 1964. The genetical theory of social behaviour. *Journal of Theoretical Biology* 7:1-52

Hermann, H.R. 1982. *Social insects.* I. New York:Academic Press.

Hill, E.M. and T.E. Christenson. 1981. Effects of prey characteristics and web structure on feeding and predators response of *Nephila clavipes* spiderlings. *Behavioral Ecology and Sociobiology* 8:1-5.

Honjo, S. 1977. Social behavior in *Dictyna follicola* (Araneae:Dictynidae). *Acta Arachnologie* 27:213-219.

Jackson, R.R. 1978. Comparative Studies of *Dictyna* and *Mallos* (Araneae:Dictynidae). I. Social organization and web characteristics. *Revue Arachnologique* 1:133-164.

Krafft, B. 1982. The significance and complexity of communication in spiders. In: P.N. Witt and J.S. Rovner, eds. *Spider communication, mechanisms and ecological significance.* Princeton, NJ:Princeton University Press. 440 pp.

Krebs, J.R., M.H. MacRoberts and J.M. Cullen. 1972. Flocking and feeding in the Great Tit, *Parus major* L.--an experimental study. *Ibis* 114:507-503

Kullmann, E. 1972. Evolution of social behavior in spiders (Araneae, Eresidae and Theriddidae). *American Zoologist* 12:419-426.

Levi, H.W. 1977. The orb weavers genera *Metepeira*, *Kaira*, and *Aculepeira* in America north of Mexico. *Bulletin of the Museum of Comparative Zoology, Harvard University* 148:185-238.

Lubin, Y.D. 1974. Adaptive advantages and the evolution of colony formation in *Crytophora* (Araneae:Araneidae). *Journal of the Linnean Society* 54:321-339.

Lubin, Y.D. 1980. Population studies of two colonial orb weaving spiders. *Journal of the Linnean Society* 70:265-287.

Lubin, Y.D. and R.H. Crozier. 1985. Electrophoretic evidence for population differentiation in a social spider *Achaeranea wau* (Theridiidae). *Insectes Sociaux* 32:297-304.

Major, P. 1978. Predator-prey interactions in two schooling fishes, *Caranx ignobilis* and *Stolephorus purpureus. Animal Behaviour* 26:760-777.

Michener, C.D. 1969. Comparative social behavior of bees. *Annual Review of Entomology* 14:299-342.

Michener, C.D. 1974. *The social behavior of bees.* Cambridge, Mass:Harvard University Press.

Morse, D.H. 1970. Ecological aspects of some mixed-species foraging flocks of birds. *Ecology Monographs* 40:119-168.

Morse, D.H. and R.S. Fritz. 1982. Experimental and observational studies of patch choice at different scales by the crab spider *Misumena vatia. Ecology* 63:172-182.

Murton, R.K. 1971. Why do some bird species feed in flocks? *Ibis* 113:534-536.

Nentwig, W. 1985. Social spiders catch larger prey: a study of *Anelosimus eximius* (Araneae:Theridiidae). *Behavioral Ecology and Sociobiology* 17:79-85.

Nentwig, W. and T. Christenson. 1985. Natural history of the nonsolitary sheet-weaving spider *Anelosimus jucundus* (Araneae:Theridiidae).

Nossek, M.E. and J.S. Rovner. 1984. Agonistic behavior in female wolf spiders (Araneae:Lycasidae). *Journal of Arachnology* 11:407-422.

Nudds, T.D. 1978. Convergence of group size strategies by mammalian social carnivores. *American Naturalist* 112:957-960.

Oster, G.F. and E.O. Wilson. 1978. Caste and ecology in the social insects. *Monographs in Population Biology*, No. 12. Princeton:Princeton University Press.

Palanichamy, S. 1984. Relation between food consumption, growth and egg production in the tropical spider *Cyrtophora cicatrosa* (Araneidae, Araneae) under experimental conditions of food abundance and ration. *Proceedings of the Indian Academy of Science* 3:599-615.

Pooley, A.C. and C. Gans. 1976. The Nile crocodile. *Scientific American* 234:114-124.

Pulliam, H.R. and G.C. Millikan. 1982. "Social organization in the non-reproductive season." pp. 45-87. In: D.S. Farner and J.R. King, eds. *Avian biology.* Vol. 6. New York:Academic Press.

Pulliam, H.R. and T. Caraco, 1984. "Living in groups: Is there an optimal group size?" pp. 122-147. In: J.R. Krebs and N.B. Davies, eds. *Behavioural ecology: An evolutionary approach.* Sunderland, Mass.: Sinauer Association.

Rand, A.L. 1954. Social feeding bahvior of birds. *Fieldiana Zoology* 36:1-71.

Readshaw, J.L. 1973. "The numerical response of predators and prey density." In: Hughes, Ed., *Quantitative Evaluation of Natural Enemy Effectiveness. Journal of Applied Biology* 10:342-351.

Real, L. and T. Caraco. 1986. Risk and foraging in stochastic environments. *Annual Review of Ecology and Systematics* 17:371-390.

Reed, C.F., P.N. Witt and M.B. Scarboro. 1969. The orb web during the life of *Argiope aurantia* (Lucas). *Developmental Psychobiology* 3:251-265.

Reed, C.F., P.N. Witt, M. Scarboro and D. Peakall, 1970. Experience and the orb web. *Developmental Psychobiology* 3:251-265.

Riechert, S.E. 1974. Thoughts on the ecological significance of spiders. *Bioscience* 24:352-356.

Riechert, S.E. 1978. Energy-based territoriality in populations of the desert spider, *Agelenopsis aperta* (Gertsch). *Symposia of the Zoological Society of London* 42:211-222.

Riechert, S.E. 1981. The consequences of being territorial: Spiders, a case study. *American Naturalist* 117:872-892.

Riechert, S.E. 1982. Spider interaction strategies: communication vs. coercion. pp. 281-315. In: P.N. Witt and J.S. Rovner, eds. *Spider communication, mechanisms and ecological significance.* Princeton University Press. 440 pp.

Riechert, S.E. 1986. Spider fights as a test of evolutionary game theory. *American Science* 74:604-610.

Riechert, S.E., R. Roeloffs and A.C. Echternacht. 1986. The ecology of the cooperative spider *Agelena consociata* in equatorial Africa (Araneae:Agelenidae). *Journal of Arachnology* 14:175-192.

Rovner, J.S., G.A. Higashi and R.F. Foelix. 1973. Maternal behavior in wolf spiders: the role of abdominal hairs. *Science* 182:1153-1155.

Rypstra, A.L. 1979. Foraging flocks of spiders: A study of aggregate behavior in *Crytophora citricola* Forskal (Araneae:Araneidae) in West Africa. *Behavioral Ecology and Sociobiology* 5:291-300.

Rypstra, A.L. 1983. The importance of food and space in limiting web-spider densities: a test using field enclosures. *Oecologia* 59:312-316.

Rypstra, A.L. 1985. Aggregations of *Nephila clavipes* (L.) (Araneae, Araneidae) relation to prey availability. *Journal of Arachnology* 13:71-78.

Rypstra, A.L. 1986. High prey abundance and a reduction in cannibalism: The first step to sociality in spiders (Arachnida). *Journal of Arachnology* 14:193-200.

Rypstra, A.L. In press. Foraging success of web spiders: a comparison of aggregations and solitary individuals.

Schoener, T.W. and D.H. Janzen. 1968. Some notes on tropical vs. temperate insect size patterns. *American Naturalist* 101:207-224.

Shear, W.A. 1970. The evolution of social phenomena in spiders. *British Arachnology Society* 1:65-76.

Smith, D.R. 1983a. Ecological costs and benefits of communal behavior in presocial spider. *Behavioral Ecology and Sociobiology* 13:107-114.

Smith D.R. 1983b. Reproductive success of solitary and communal *Philoponella oweni* (Araneae:Uloboridae). *Behavioral Ecology and Sociobiology* 11:149-154.

Smith, D.R. 1986. Population genetics of *Anelosimus eximius* (Araneae: Theridiidae). *Journal of Arachnology* 14:201-218.

Smith-Trail, D.R. 1982. Evolution of Sociality in spiders - the costs and benefits of communal behavior in *Philoponella oweni* (Araneae: Uloboridae). Ph.D. thesis. New York:Cornell University.

Suter, R.B. 1985. Intersexual competition for food in the bowl and doily spider, *Frontinella pyramitela* (Araneae:Linyphiidae). *Journal of Arachnology* 13:61-70.

Thompson, W.A., J. Vertinsky and J.R. Krebs. 1974. The survival value of flocking in birds: A simulation model. *Journal of Animal Ecology* 43:785-820.

Tietjen, W.J. 1986a. Effects of colony size on web structure and behavior of the social spider *Mallos gregalis* (Araneae:Dictynidae). *Journal of Arachnology* 14:145-158.

Tietjen, W.J. 1986b. Social spider webs, with special reference to the webs of *Mallos gregalis*. pp. 172-206. In: W.A. Shear, ed. *Spiders: Webs, behavior and evolution*. California:Stanford University Press.

Turnbull, A.L. 1962. Quantitative studies of the food of *Linyphia triangularis* (Clerck) (Araneae, Linyphiidae). *Canadian Entomology* 96:568-579.

Turnbull, A.L. 1964. The search for prey by a web-building spider *Achaearanea tepidariorum* (C.L. Koch) (Araneae:Theridiidae). *Canadian Entomology* 96:568-579.

Uetz, G.W. 1985. Ecology and behavior of *Metepeira spinipes* (Araneae: Araneidae), a colonial web-building spider from Mexico. *National Geographic Research Report* 1985:597-609.

Uetz, G.W. 1986a. (Ed.) Symposium: Social behavior in spiders. *Journal of Arachnology* 14:145-281.

Uetz, G.W. 1986b. Web-building and prey capture in communal orb weavers. pp. 207-231. In: W.A. Shear, ed. *Spiders: Webs, behavior, and evolution*. California:Stanford University Press.

Uetz, G.W. and J.W. Burgess. 1979. Habitat structure and colonial behavior in *Metepeira spinipes* (Araneae:Araneidae), an orb weaving spider from Mexico. *Journal of Arachnology* 7:121-128.

Uetz, G.W., T.C. Kane and G.E. Stratton. 1982. Variation in the social grouping tendency of a communal web-building spider. *Science* 217:547-549

Uetz, G.W. and K.R. Cangialosi. 1986. Genetic differences in social behavior and spacing in populations of *Metepeira Spinipes* Pickard-Cambridge (Araneae:Araneidae), a communal/territorial orb weaver. *Journal of Arachnology* 14:159-174.

Uetz, G.W., T.C. Kane, G.E. Stratton and M.J. Benton. In press. "Environmental and genetic influences on the social grouping tendency of a communal spider." In: M.A. Huettel, ed. *Evolutionary genetics of invertebrate behavior*. Plenum Press.

Vollrath, F. 1982. Colony foundation in a social spider. *Zeitschrift für Tierpsychologie* 60:313-324.

Vollrath, F. 1985. Web spider's dilemma: A risky move or site dependent growth. *Oecologia* 68:69-72.

Vollrath, F. 1986a. Eusociality and extraordinary sex ratios in the spider *Anelosimus eximius* (Araneae:Theridiidae). *Behavioral Ecology and Sociobiology* 18:283-287.

Vollrath, F. 1986b. Environment, reproduction, and the sex ratios of the social spider *Anelosimus eximius* (Araneae:Theridiidae). *Journal of Arachnology* 14:267-281.

Vollrath, F. and D. Rohde-Arndt. 1983. Prey capture and feeding in the social spider *Anelosimus eximius*. *Zeitschrift für Tierpsychologie* 61:334-340.

Ward, P. and A. Zahavi. 1972. The importance of certain assemblages of birds as "information centres" for food finding. *Ibis* 119:517-534.

Wilson, E.O. 1971. *The insect societies.* Cambridge, Mass.:Belknap Press of Harvard University Press. 548 pp.

Wilson, E.O. 1975. *Sociobiology – The new synthesis.* Cambridge, Mass.: Harvard University Press. 697 pp.

Wise, D.H. 1975. Food limitation of the spider *Linyphia marginata*: Experimental field studies. *Ecology* 56:637-646.

Wise, D.H. 1979. Effects of an experimental increase in prey abundance upon the reproductive rates of two orb-weaving spider species (Araneae:Araneidae). *Oecologia* 41:289-300.

Wise, D.H. 1984. "The role of competition in spider communities: insights from field experiments with a model organism." pp. 42-53. In: D.R. Strong, Jr., D. Simberloff, L.G. Abele and A.B. Thistle, eds. *Ecological communities: Conceptual issues and the evidence.* New Jersey:Princeton University Press.

Witt, P.N. and C.F. Reed. 1965. *Spider-web building. Science* 149:1190-1197.

Witt, P.N., C.F. Reed and D.B. Peakall. 1968. *A spider's web. Problems in regulatory biology.* Berlin:Springer Verlag.

Witt, P.N. and J.S. Rover. 1982. *Spider communication: Mechanisms and ecological significance.* New Jersey:Princeton University Press. 440 pp.

Wittenberger, J.F. 1981. *Animal social behavior.* Boston, Mass.:Duxbury Press. 722 pp.

Chapter 16

RESOURCE INHERITANCE IN SOCIAL
EVOLUTION FROM TERMITES TO MAN

Timothy G. Myles

Department of Entomology
University of Arizona, Tucson 85721

I.   INTRODUCTION

A.   Resource Inheritance: An Ecological Basis
     for Sociality

This paper explores a common theme underlying social evolution in
many animal taxa. The basic idea is that the potential for resource inheritance provides an ecological basis for the origin and elaboration of
many social adaptations. This idea is not new. In fact, it is a traditional
concept in studies of cooperative breeding birds (Brown 1978; Woolfenden and Fitzpatrick 1984) and has many applications among philopatric
cooperative mammals (Emlen 1984; Waser and Jones 1983). The present
paper attempts to generalize the application of resource inheritance to
an even broader range of animals. A detailed examination is made of
the role of resource inheritance in the origin and evolution of termite
castes and sociality and the origin of hymenopteran sociality. I have
even risked some tentative speculations of its role in human social evolution. I have attempted to put resource inheritance in sharper focus by
contrasting it with its natural alternative which is dispersal. Also, I
have examined some of the basic diversifications of social behavior
which arise from competition over resource inheritance. Lastly, by gen-
eralizing resources to include social rank I show that dispersal/rank promotion tradeoffs are common.

I argue that ecological considerations often override relatedness
biases in determining why individuals disperse or help. In the lower
grades of sociality a stronger incentive for staying (than to help kin) is
that staying may eventually enable an offspring to reproduce without
dispersing. Moreover, staying permits one to be in a position to inherit a
reproductive situation sometimes far superior to any a solitary disperser
could obtain. This is because the activity of social groups often results
in the accumulation of enormously valuable material assets in the form
of nests, food reserves, territories, helper populations, burrow systems,

etc. Thus it may often be more relevant to consider the tradeoffs between dangerous dispersal, disinheritance of social wealth, and immediate reproduction versus secure staying, helping, and possible deferred reproduction with inheritance of socially accumulated assets, i.e., the "dispersal/inheritance tradeoff." Facultative adaptations for inheritance should be common in animals that have risky dispersal and valuable potential inheritance.

The payoff to helpers by inheritance must be contrasted with the payoff via kinship. While kinship may provide a genetic basis for certain aspects of sociality, resource inheritance provides a basis for the original clumping and centering on resources of social groups. Whereas kinship only helps to illuminate the subsocial route, resource inheritance provides a common basis for social evolution by both the subsocial and parasocial routes. Kin selection is often invoked to explain generalized tendencies for social beneficence; however, resource heritability provides incentives for a more specific expression of beneficence as a component of a long-term life history strategy. The sort of beneficence expected under resource inheritance is equivocating, condition dependent, temporary, and calculated. The calculations are not relatedness dependent (since this typically is a predictable constant) so much as dependent on the probability of successful dispersal which varies with individuals and circumstances. Resource heritability offers incentives for selfish and manipulative tendencies as well as beneficence. Resource inheritance tactics are well integrated, coadapted, behavioral-developmental programs consisting of both beneficent and maleficent interactions. Thus the "in-fighting" aspects of sociality are encompassed.

B.   Adaptations for Resource Inheritance

Resource inheritance can provide the selective advantage underlying a wide range of social behaviors including deferred dispersal or facultative nondispersal (expressed as wing polymorphism in insects), delayed or accelerated maturation, parental appeasement behaviors (filial subordination) which may take various forms such as abstention from courtship, mating, or various kinds of temporary helping (paying rent to live at home). Delayed dispersal may commonly be expressed as a prolonged adolescent or subadult period. Competition over resource inheritance can account for dominance contests, hierarchies among subordinate helpers, sibling rivalry, sibling manipulation, and even siblicidal behavior and the evolution of fighting or cannibalistic morphs (e.g., Cainisms, see Hrdy and Hausfater 1984). Such competition can lead to a differentiation of the sexes in terms of philopatry or dispersal distance and consequently to the structuring of matrilines or patrilines (Greenwood 1980). Competition may manifest itself as manipulation which may bias individuals to express cooperative default tactics. Competition to inherit should affect the conditions under which recruits are permitted to join the social group. The existence of competition to inherit should be expressed as a persistent subtle jockeying for dominance among subordinates in the presence of dominants and/or as sudden and violent power struggles to assume the inheritance during interregna when the reproductively dominant individuals falter or die. The ambivalence of subordinates to commit themselves irrevocably and wholeheartedly to helping, and the ceaseless internal strife and in-fighting which characterize the lives of social individuals reflect long-term tactics to move up within the system, inherit, and reproduce.

C. Resource Inheritance: A Minority Tactic

Opportunities for inheritance are generally few in comparison to the number of opportunities for dispersers; however, the inheritance option may often be less risky and/or more rewarding. Thus, the inheritance tactic is usually expressed facultatively and opportunistically. In most species nondispersal is necessarily employed by only a minority of the progeny; but, a series of replacements or supercedures, group fission, resource fragmentation, or budding off into adjacent areas can create numerous inheritance opportunities. Even so, these opportunities may be relatively few. However, since most animals with stable populations are in a zero-sum game, the few progeny that do mature as secure inheritors may be an important bet hedge against the many that attempt risky dispersal.

D. An Inheritance Tactic May Initiate Kin Overlap

Resource inheritance is widespread across solitary as well as social taxa. Within socioclines (phylogenetically related series of social species) opportunistic resource inheritance often occurs in the solitary ancestors and the lower grades of presocial and semisocial species (Waser and Jones 1983). The benefits of social membership in the lower grade social species appear, in large measure, to derive from the fair probability of resource inheritance relative to the low probability of resource acquisition and successful reproductive establishment by dispersal. Thus the potential for resource inheritance appears to be an important factor biasing on dispersal, and as such, an important prime mover of social evolution.

A secondary advantage which arises from a nondispersive tactic for inheritance is that it provides opportunities to assist kin. When parental behaviors are already in existence they imply a capacity of offspring to benefit from them, thus the benefits of alloparental behavior can be immediately enjoyed and reinforce the value of nondispersal. If parental care is not a preadaptation then kin helping by the nondisperser may evolve de novo after the overlap of generations is established and at that point reinforce nondispersal.

The lateralization of parental care as alloparental care by siblings gives nondispersing subordinate helpers an ideal indirect fitness outlet. This is because of the relatedness parity of siblings and offspring under monogamy and the fact that parent or sib caring can be undertaken without incurring dispersal, resource acquisition, and mating costs (establishment costs) which are normal prerequisites to offering parental care to offspring. It is virtually always much easier to alloparent sibs than to parent offspring and the returns in terms of promoted gene copies are the same if parents are monogamous. Thus subsocial alloparenting by nondispersers is not surprising but such behavior should not be construed as the impetus for nondispersal.

Further advantages of staying at home may be the mutualistic or reciprocal possibilities for foraging, defense, sharing effort or sharing body warmth. Whether the direct but belated reproductive advantage or the indirect kin care advantage or the mutualistic advantage or the sums of these are causal of the initial delayed dispersal and deferred reproduction is debatable and open to investigation in each instance. However, I suggest that the belated reproductive advantage is necessary and sufficient in the *origin* of delay in most instances. Mutualistic and nepotistic benefits are initially of minor importance and are secondary developments.

E.   Dispersal/Nondispersal Tradeoffs

When nondispersal does not entail reproductive subordination then one expects that nondispersal results in selfish, mutualistic, or reciprocal benefits which exceed the lost reproductive potential of dispersal. When nondispersal does entail reproductive subordination then the payoff for the subordinate may be either in terms of helping kin (indirect inclusive fitness component) or in terms of delayed reproduction or the sum of the helping and delayed reproduction components. The delayed reproduction may result from delayed dispersal under better conditions or it may result from eventual reproduction without dispersal. Delayed reproduction without dispersal may occur in one of three basic ways: (1) by usurping the dominant's position, (2) by replacing a dominant that dies, or (3) by budding off and reproducing at the periphery of the original group. Thus, the potential compensations for nondispersal are (1) the continuation of ongoing advantages associated with the present situation (familiarity, security, mutualistic or reciprocal interactions), (2) the helping of kin, (3) the improved chances or abilities for future dispersal and reproduction, and (4) the future opportunities for reproduction without dispersal, or some combination of these. Unique risks and unique average potential payoffs are associated with dispersal and each component of nondispersal. Furthermore, these risks and payoffs vary with conditions which may be assessable by the individual.

F.   Potential Benefits of Delayed Dispersal

1.   May permit dispersal at a more favorable time
   a.   environmental conditions for dispersal may improve later
   b.   a vacancy in the environment may open up later
   c.   may be stronger and more fit to disperse later with age
   d.   may overcome some inadequacy for current dispersal, for example, due to manipulation, malnourishment, or illness if wait until later
2.   May provide future opportunity to reproduce without dispersal
   a.   eliminates cost of dispersal

b.   may eliminate or reduce cost of mating
c.   eliminates other orientation, start-up and establishment costs
3.   May enable resource inheritance and enhanced reproduction
a.   nest materials, constructs, excavations, fortifications, territory
b.   cached food stockpiles, unused food, renewable resources
c.   immature and subordinate mature helpers
d.   "dear enemy" neighbors
e.   familiarity to environs
4.   Sets up or prolongs opportunities for interaction
a.   opportunities for selfishness, manipulation, exploitation
b.   opportunities for reciprocation or mutualism
c.   opportunities for nepotistic (altruistic) kin helping

G.   Autocatalytic Nature of Dispersal/Inheritance Tradeoffs

As animals defer dispersal the group enlarges in size which increases competition for inheritance and lowers the potential benefits from inheritance.   This is an immediate check on continued group enlargement.   However, if those individuals act as helpers, the helpers themselves become valuable heritable resources which then increase the value of nondispersal.   If the nondispersers make improvements to the nest or territory then this also heightens the value of the potential inheritance.   Some general examples of such improvements are nests, defensive fortifications, areas cleared of noxious plants or animals, escape routes, tunnel systems, collected materials, stored food, the enriching or disinfectant properties of excretory products, nest weatherproofing and camouflaging, etc.   Immature and mature helpers could be heritable and may serve as alloparents: nursing, feeding, hunting, babysitting, alarm sounding, and defending.   There are innumerable ways in which hangers-on at the nest can be serviceable.   The helpers may also accumulate food in caches, such as paralyzed insect prey in wasps; honey, pollen, and wax in some bees (Michener 1974); or prey, seeds, or honey in some ants.   In some ants the larger individuals may even serve as grossly distended storage vessels for honey--a unique form of lardercaching by individuals termed "repletes" (Rissing 1984).   Seeds are the most commonly stored food material by rodents and birds; enormous quantities of acorns are stored in "granaries" by acorn woodpeckers.   Excavated tubers are left uneaten as a stored food by colonies of naked mole rats.   Some social canids and other carnivores may scatter-cache buried carcasses (Smith and Reichman 1984; Sherry 1985).   It must be mentioned, however, that intraspecific competition over food hoards can also select for solitary behavior (Andersson and Krebs 1978; Nel 1975).
Dispersal may become progressively costly as neighboring territories become occupied by groups of territory-guarding nondispersers. When territory defense is aided by helpers, such population structuring could spread over a wider range of the suitable habitat.   This would have

the effect of dividing the entire landscape into well-policed private property. Floater harassment would reduce the ability to shop around for vacancies because of constant persecution for trespassing. As the safe shopping range constricts, the value of inheritance is further enhanced.

II.   TERMITE SOCIAL EVOLUTION

A.   Misconceptions about Termite Sociality

Part of the problem of understanding termite social evolution arises from misconceptions about the nature of termite sociality. For example, we find the following statement in Robert Trivers' otherwise admirable textbook: "All termites have a non-reproductive caste, which has often evolved into more than one form, usually by the addition of a soldier caste" (Trivers 1985). This is a critically erroneous premise. Lower termites do not have a permanently non-reproductive worker caste and the soldiers have not evolved from non-reproductive workers. In the Isoptera soldiers evolved before true workers appeared; termite soldiers are definitely not a derivative worker subcaste (Wheeler 1928; Hare 1937). Pseudoworkers (pseudergates) and true workers evolved independently of soldiers and apparently on separate occasions in several different taxa (see Watson and Sewell 1985; Noirot 1982, 1985a; Myles and Nutting in press). By starting with erroneous premises Trivers is led to ask inappropriate questions: "The key question is why there is so little reproductive conflict in the termites. Why have workers so completely abandoned personal reproduction and what exactly is the significance of secondary reproductives?" As will be discussed below, tremendous reproductive competition does exist within termite colonies (see also Ruppli 1969; Zimmerman 1983a,b; Myles and Chang 1984). Trivers should not be singled out for this criticism because statements asserting or implying the sterility of workers in all termites are fairly common. In fact, Trivers' query as to the significance of secondary reproductives is insightful and this paper deals largely with that question.

B.   Hamilton's and Bartz's Relatedness Asymmetry Models

Of the 27 orders of insects only two have eusocial members: the bees, ants and wasps in the Hymenoptera and the termites in the Isoptera. There are two prominent theoretical models to account for social evolution in these orders: Hamilton's (1964) haplodiploidy-based theory for the Hymenoptera, and Bartz' (1979) inbreeding-based theory for the Isoptera. These models are heuristically valuable but certain crucial assumptions may make them unrealistic (Lin and Michener 1972; West-Eberhard 1975, 1978; Page 1986; Myles and Nutting in press), at least as explanations of social origins. Both hypotheses emphasize relatedness asymmetries in which siblings are more closely related than potential offspring. Both models represent dispersal and helping as alternatives rather than as components of an overall strategy. The models assume that the ability of individuals to invest in siblings and offspring is about equal and therefore the determining factor is the degree of relatedness. Both purport to explain the origin of sociality due to an altruistic reproductive sacrifice in which workers gain more genetically by helping than by reproducing. Thus both models leap to an explanation of worker sterility rather than first explaining the phenomenon of temporary helping.

This paper challenges the general relevance of relatedness asymmetry models to early social evolution and suggests that the original payoff to temporary helpers is mainly delayed philopatric reproduction and not an overcompensating indirect component arising from altruism toward extraordinarily related siblings. The phylogeny of insect sociality as indicated by socioclines of extant species consistently shows that a decision to forego reproduction in order to invest fully in sterile helping is rarely the initiating event or prime mover of sociality. In presocial and primitively eusocial species the workers almost always retain significant reproductive potential (wasps, West-Eberhard 1978 and Lester and Selander 1981; bees, Michener 1975, 1977; ants, Ward 1983; termites, see below). Thus the initiating event in insect social evolution must be a tradeoff of early dispersal for a tactic of temporary helping with a certain probability for delayed reproduction. I contend that the important initiating conditions are not a relatedness asymmetry so much as a relatedness parity of sibs and offspring and opportunities for delayed reproduction, usually with resource inheritance. With these conditions *some* individuals are able to improve their inclusive fitness and reproductive value, or their "inclusive reproductive value" (see West-Eberhard 1981) by serving as temporary helpers. Furthermore, the individuals that fall into this category are generally disadvantaged by circumstances, manipulated, young, or otherwise inherently inferior so that helping is often a second-best or default tactic rather than the tactic of first choice.

My studies of wood-dwelling termites indicate that they meet the conditions of risky dispersal and valuable heritable resources. In the more primitive species a major component to fitness of nondispersers is the potential for reproduction as neotenics following orphaning. Inheritance of the nest by neotenic replacement reproductives seems to offer a large incentive for nondispersal and therefore would seem to have played a critical role in the origin of termite social behavior.

C. Dispersal-Related Alate Mortality in Termites

In termites the entire lifetime dispersal of an individual occurs during a single pre-reproductive flight after which the winged alate breaks off its wings (dealates) and settles into an essentially sedentary reproductive life. Exceptions to this general pattern are extremely rare, but there are reports of "sociotomy" in a few species of higher termites in which primary reproductives have been seen marching from their nests with a retinue of soldiers and workers to establish colonies elsewhere (Harris 1958; Nutting 1969). There is also the possibility, in species that nest in the soil, of dispersal through the soil followed by neotenic reproduction, a process known as budding. This is common in *Mastotermes* and the Rhinotermitidae (Watson and Abbey 1985; Lenz and Barrett 1982).

On the whole, however, dispersal is carried out by alates and is limited to a very short window of time, probably lasting from a few minutes to an hour or two. Experiments with *Pterotermes occidentis* showed that most individuals required at least 6 minutes of flight before they exhibited dealating behavior but individuals became flight exhausted between 30 and 70 minutes (Myles unpub.). In most species flights are triggered by species-specific seasonal and diurnal meteorological cues (Nutting 1969). This results in the synchronization of alate emergence from different nests in an area facilitating outbreeding and in some species also acting as a defensive mechanism by predator satiation (Nutting 1979). During this short period alates are subject to enormous mortality arising from both abiotic factors and predation. Alates are feeble fliers and easily captured on the wing by bats and various birds. They are not known to possess any mechanical or chemical defenses; on the contrary, they are known to be rich in fats (up to 56 percent) and are avidly sought by many predators (Nutting and Haverty 1976). They are taken in enormous numbers by frogs, toads, geckos, lizards, bats, birds, rodents, scorpions, spiders, centipedes, ants, and other predatory insects (Deligne et al. 1981). Some species are also subject to uncommonly intense abiotic selective forces. *Gnathamitermes perplexus* stages its flights during driving summer thundershowers and I have seen many languishing in the mud or drowned in puddles.

The vulnerability of alates is further exacerbated by the necessity of conducting during this brief period a number of vital behavioral tasks, such as locating a mate, dealating, courting (which may entail elaborate tandem running), locating a suitable nest site, digging in, and sealing the copularium. I can think of no other insects that are committed to carrying out their entire lifetime dispersal and courtship within such a narrow window of intense vulnerability. The age-specific mortality rate for dispersing alates, q_x, I believe is virtually unparalleled for adult animals, but may find parallels among the eggs and immature stages of pelagic organisms. Indeed dispersing alates are analogous to ejaculated gametes when one considers the colony as a sessile superorganism. Actual attempts to measure alate mortality have only been conducted for one higher termite species, *Odontotermes assmuthi*, for which it was estimated that more than 99.5 percent of alates were killed before colony foundation and another 88 percent of the incipient colonies failed within three years. Causes of colony failure in order of declining importance were fungus, nematodes, dipteran larvae, desertion of partner, and dehydration. Thus, the probability of surviving to produce alate offspring for dispersing alates was less than .0006 or 1 in 1,666 (Basalingappa 1970). It is likely that lower termites with less conspicuous emergences, such as *Pterotermes* and *Zootermopsis*, which release alates a few at a time throughout the night over an extended summer flight period, suffer substantially less mortality (Nutting 1966a). Even so, an estimate of 95 percent mortality would probably be conservative and mortality at this stage of the life cycle is undoubtedly higher than at any other time.

D. Attributes of Wood from a Termite's Perspective

Relative to the food items utilized by most insects, dead branches and trees are enormous single items. The nutritive value of such items greatly exceeds the requirements of any individual and therefore wood items are *shareable* resources.

Wood consists largely of cellulose, a polysaccharide of high caloric value, but it is a curious fact of life on earth that very few animals are able to digest it. Fungi are major decomposers of wood. Although fungi may compete with termites for the full utilization of wood, the action of fungi often preconditions wood for termite use in several ways: (1) by creating soft spots as points of entry, (2) by chemical degradation changing the nutrient quality of the wood, and (3) by structurally weakening the wood and hastening its fragmentation. Because wood is indigestible there is little animal competition for it and because fungi tend to improve its availability rather than consume it, it remains available for a prolonged period of time. In short, wood is a highly *durable* resource. The durability of wood may often exceed the 1-2 year generation time of termites by a factor of about five to 10. Thus, not only is the resource ample enough to be shared by many of one generation it may also be durable enough to be shared by a lineage over many generations.

Another feature of wood is the fact that much of it dies in an upright position. Standing dead branches are subject to termite infestation and subsequently must eventually break off the tree or fall over and break up further on impact with the ground. The cylindrical shape of branches tends over time to distend colonies linearly which increases the probability of colony fragmentation. Thus wood is a *fragmenting* resource.

Though rich, it comes in finite, *exhaustible*, nonrenewable chunks. Exhaustibility makes it worth monopolizing. Wood is *monopolizable* because galleries are confluent and can be patrolled so that potential usurpers can be detected and eliminated.

Wood is excavatable and noncollapsible, thus affording security and environmental buffering. While providing safety it also tends to enforce monogamy and proximity of interactants.

E. Consequences of Endoxylophagy and the Divergence of Cryptocercids and Protoispoterans

Endoxylophagy, the living in and eating of wood, clearly evolved early in the phylogeny of the order. The ancestor of termites was a cockroach-like detritivore, more ancient than any living cockroach. The most primitive roaches, the cryptocercids, evidently belong to the same monophyletic stem of endoxylophagous ancestors that gave rise to the termites because they share an almost identical gut fauna for digesting wood (Nutting 1956; McKittrick 1965). But the cryptocercids cannot be direct ancestors of termites because they are wingless as adults, while all termite adults are winged. The common ancestor presumably lived in damp forest leaf litter. Relative to leaf litter, logs and branches are large food items and relatively patchy in their dispersion. Thus the ancestors shifted from a sparse to a bonanza-type resource, rich but dispersed. What happened to these early founders of the endoxylophagous adaptive zone?

Interestingly, they seem to have diverged in opposite directions along the r-K continuum. The cryptocercids, represented by three extant species, are K-selection extremists. They are morphologically robust, have a long generation time of 3-6(?) years, have phenomenally low fecundity (average brood size ca. 20), are probably or nearly semelparous, and are subsocial with biparental care. Adult cryptocercids evidently invest heavily in the prolonged feeding of offspring and in the defense of offspring, mate, and residency of burrow. Their parental investment per offspring is huge compared to termites and provides an example of extreme parental investment in insects. The nest may even be endowed to nearly mature offspring. Their postreproductive adults have evolved into a functionally sterile defensive caste (a sort of menopause in females). Both parents are behaviorally specialized to respond to intrusion and morphologically specialized with enlarged pronota to block tunnels (Seelinger and Seelinger 1983; Nalepa 1984). Perhaps because of these specialization of the adults in defense, the immature nymphs are unable to render significant alloparental filial care to younger siblings. This could be one reason why the cryptocercids have failed to evolve cooperative brood care. A possible explanation of the fact that 90 percent of parents have only one brood (Nalepa 1984) may be that the initial brood would be hostile to a younger brood because of nest inheritance competition. For example, cannibalism by older siblings of younger broods occurs in Dermaptera (Vancassel 1984).

The adult cryptocercid evolved apterism by obligate neoteny so dispersal is cursorial, occurs at the preadult or adult stage, and presumably entails low mortality compared to termites. In termites, because of the risks of flight dispersal between wood items, the mature alate stage experiences high mortality. But, because of the security afforded by living inside wood, termite immatures are relatively protected from environmental perturbations, predators and starvation. Thus, whereas most insects tend toward type II or III survivorship curves (Price 1975) with heaviest mortality in young stages, primitive wood-inhabiting termites conform to a type I pattern. Thus, the critical difference between cryptocercids and termites is that, in order to take advantage of the potential for nest inheritance, termites evolved an apterous morph *facultatively* (neotenic reproductives) and retained the winged morph for dispersal. I speculate that because there is high dispersal-related mortality of dispersing alates, termites fell under a strong r-selective regimen. By sacrificing individual robustness for greater numbers, termites became feeble, evolved high fecundity, iteroparous reproduction, and care of parents by offspring rather than the cryptocercid pattern of care of offspring by parents. For these reasons the cryptocercids probably do not typify the ancestral social biology of pre-eusocial termites as suggested

by Seelinger and Seelinger (1983) and Nalepa (1984). Rather, they serve
to show what prototermites did not do in terms of parental investment.
I feel that the fundamental divergence of these groups warrants separate
ordinal status contrary to the proposed inclusion of both in the Dictyoptera.

F.   Neotenic Reproductives: The First Physical
     Caste in Termites

There is a remarkable convergence among wood- and cambium-
feeding insects for the evolution of wing polymorphism, i.e., macropter-
ous dispersers and apterous, non-dispersive morphs.   Hamilton (1978)
lists seven orders and 24 genera of insects as independently evolved
cases of flight polymorphism of insects living in dead trees.   Taylor
(1978) pointed out the possible similarity of the ecological conditions
favoring the evolution of flight polymorphism in wood-living ptiliid
beetles to the incipient stages of caste polymorphism in termites. However, she mistakenly interpreted the apterous reproductive morph as
analogous to a protoworker rather than as an analogy to the termite neotenic reproductive.
The neotenic reproductive caste of termites is an apterous reproductive form exactly analogous to the above examples.   Although
records do not exist for most species, the neotenic caste probably occurs
in all lower termites.   A survey of neotenic reproductives in the order
revealed a preponderance of records in species of the lower families
(Fig. 1).   However, they are numerically most abundant in two of the
lower soil-inhabiting families: Mastotermitidae and Rhinotermitidae.
They are less commonly known from species of the higher termites and
are suspected of having been entirely lost in the termitid subfamilies
Macrotermitinae and Apicotermitinae (Myles and Nutting in press).   An
alternative replacement mechanism by adultoids and microimagoes
occurs in several termitids (Sieber 1985; Roisin and Pasteels 1985).   This
comparative evidence supports the view that the neotenic caste was a
primitive feature of termite sociality and that their loss is a derivative
condition.   The primitive nature of neotenics along with the analogy of
neotenics to apterous reproductive forms in other wood-living insects
makes a strong argument in favor of neotenics as the first physical termite caste.   I propose that neotenic reproduction is a pleisiomorphy
within the Isoptera which originated prior to inception of protoworkers
and protosoldiers.   I argue below that not only did neoteny appear before
other forms of caste differentiation, but that the origin of neotenics
preadapted prototermites for subsequent caste and social evolution.
This explanation of caste evolution contrasts with the ergonomic theory
by which caste proliferation is assumed to have been selected to opti-
mize colony efficiency through a division of labor by colony level selection (Oster and Wilson 1978).   Since neotenics are reproductive there can
be no question that there was a direct component to individual fitness in
their origin.

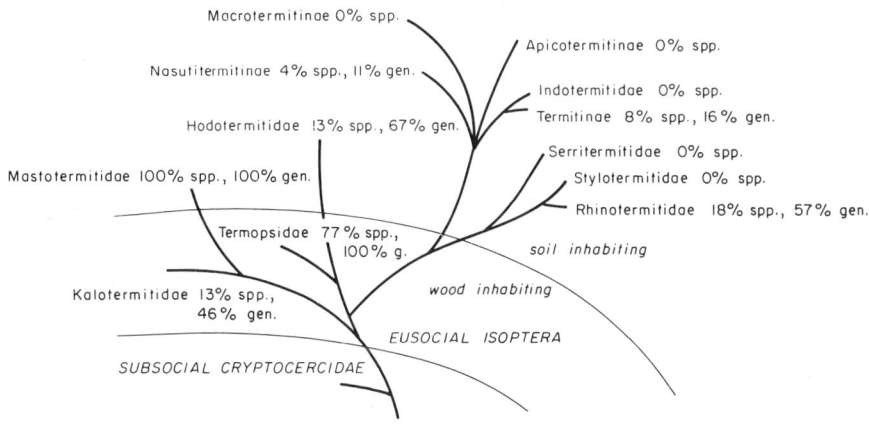

Fig. 1.   Dendrogram of the Cryptocercidae and Isoptera showing families and termitid subfamilies with percentages of species and genera with known records of the occurrence of the neotenic caste.   Note that although present knowledge is incomplete and disproportional for different groups, the available data indicate a preponderance of neotenic occurrence in the lower groups (phylogenetic relationships adapted from McKittrick 1965; Sands 1972; Roonwal 1975).

G.   Nest Inheritance by Neotenic Reproductives: An Alternative Resource-Acquisition Tactic

A distinctive feature of termites as compared to some other wood-inhabiting insects with flying and apterous morphs, such as *Micromalthus*, ptiliids, and zorapterans, is that the founding pair are long lived, highly fecund, and have a despotic monopoly on reproduction.   Unlike these other groups, in termites the primary pair inhibit maturation of the apterous reproductive morph.   Hence, neotenics in wood-inhabiting termites (Termopsidae and Kalotermitidae) almost always develop as replacement reproductives following orphaning.

Non-dispersive neotenic transformation is an alternative to alate maturation and dispersal.   It is, in effect, an alternative individual fitness-maximizing, resource-acquisition tactic.   Alternative tactics have mainly been studied in the context of mating (Austad and Howard 1984; Austad 1984).   While "sneaking" is the principal alternative to "courting" among mating tactics, I suggest that "resource inheritance" is the principal alternative to "independent establishment" among resource acquisition tactics.   Whie the alate acquires resources by investing in wings and risking flight, the neotenic employs an inheritance tactic.   The alate tactic is one of extraordinary risk with potential large gain.   The neotenic tactic is one of less risk but also less potential gain.

Neotenics receive the accumulated and residual assets of their late parents.   In wood-inhabiting lower termites the inherited log or log fragment can be thought of as a residual asset whereas the galleries and remaining population can be thought of as an accumulated asset. The probability of orphaning events and the value of log fragments depend on the host plant tree species and a variety of ecological factors peculiar to each species.

In soil-inhabiting termites neotenics frequently develop as supplementaries before orphaning occurs. They may still be considered heirs. Soil-dwelling termites may have diffuse nests or discrete mounds. In those with diffuse nests the population tends to move around opportunistically foraging where wood may be found, and their territories are flexible. New colonies are frequently founded by neotenics developing in a rich resource patch at the periphery of the foraging territory. This is called budding and is common in *Mastotermes*, rhinotermitids and some termitids. Other subterranean species form distinct hypogeal or epigeal mounds or arboreal nests and hold relatively stable foraging territories. Some of these, such as the Hodotermitidae, Psammotermitinae, some Apicotermitine, and some Nasutitermitinae construct polycalic nests (with several isolated subnests) which may facilitate fragmentation or isolation of subnests and provide inheritance opportunities (Roisin and Pasteels 1986a). Others, such as *Cubitermes* expand their nests by stacking on new chambers (Noirot 1977) which may also be susceptible to fragmentation. Certain arboreal nesting *Nasutitermes* are subject to periodic vertebrate predation and nest-fall and consequently retain short-winged microimagos as a reserve source for adultoid replacements in the event of orphaning (Roisin and Pasteels 1986b). The massive nests, cultivated fungus combs, extensive runway systems, and associated territories and worker populations of mound-building *Macrotermes* represent a considerable inheritance (Darlington 1982) and a succession of heirs in such nests seems possible but is only beginning to be investigated (Sieber 1985).

H.   Risk Associated with Neotenic Transformation

I have studied neotenic reproduction in a dry-wood termite, *Pterotermes occidentis*, regarded on morphological grounds as among the most primitive species of the Kalotermitidae (Krishna 1961). Unlike other types of caste differentiation, neotenics develop in excess of colony needs and then are eliminated by lethal combat between full sibs of the same sex (see also Ruppli 1969; Myles and Chang 1984). In these flights neotenics first inspect each other by anal licking. This is followed by head banging and snapping at the antennae. Finally attacks to the legs and flanks take place. This combat appears to be the major cause of mortality for neotenics. Such regicidal battles decimated orphaned populations by as much as 40 percent in *Neotermes jouteli* (Nagin 1972), 65 percent in *Cryptotermes brevis*, 25 percent in *Kalotermes flavicollis* (Lenz et al. 1985), and by an average of 11.4 percent in *Pterotermes occidentis* in artificially orphaned log fragments. It seems that either the threshold for elimination is higher or that elimination is delayed or condition-dependent in the Mastotermitidae, Termopsidae and Rhinotermitidae in which multiple functional neotenics may occur together (Lenz 1985). Since the probability of neotenic survival depends on the number of siblings of the same sex transforming to neotenics, survival probability is a function of transformation rates and is inversely related to the size of the orphaned group. The relative neotenic tendencies of stages multiplied by the stage ratios summed over all stages supply an estimate of the percentage molting to neotenics in a group. For *Pterotermes* this works out to about 10 percent, which is in close agreement with the average observed in artificially orphaned logs, 11.4 percent. Since *Pterotermes* has a 50:50 sex ratio, and since male and female neotenics develop at the same rate, the total number of neotenics divided by two gives the number of each sex. Since only one of each sex survives, the reciprocal of the number of each sex gives the probability of surviving. Thus, the probability of surviving neotenic combat in an average mature colony of 550 is $1/(\Sigma 550 \times R_x \times N_x) = 0.036$, where R_x is the stage ratio and N_x is the tendency for neotenic molting (Myles dissertation in prep.).

I.   Neoteny Lowers Threshold for Protoworker
     Differentiation

The potential for nest inheritance and neotenic reproduction profoundly alters and complicates an individual termite's life history options. Age-specific and condition-dependent switches should evolve which weigh the probable fitnesses via alation against neoteny. This gives individuals ways of optimizing total lifetime reproductive success under variable conditions. A common age-specific feature of lower termites is that the onset of wingpad development is delayed until the fifth or sixth instar. This puts off even a tentative commitment to alate development as long as possible. Also, the conditions which determine pseudergate (=pseudoworker) differentiation in lower termites are quite variable and suggestive of polyphyletic origins. For example, under conditions in which dispersal is likely to be unsuccessful, such as drought, a decision to abort wingpad development might effectively forestall alation and increase the probability of future neotenic reproduction. This may account for the mechanism of "wingpad abscission" which initiates pseudergate differentiation in the primitive genus *Zootermopsis* (Myles unpub.) (Fig. 2). Furthermore, the capacity for nestmates to make switches lowers the cost of manipulating them. This could explain the origin of manipulative wingpad and leg biting as mechanisms which determine pseudergate differentiation in kalotermitids (Zimmerman 1983a,b; Myles in press) (Fig. 2).

It is noteworthy that pseudergates in several different lower termite families have tendencies for neotenic transformation that are equal or usually greater than such tendencies by larvae (non-wingpadded) and nymphs (wingpadded immatures). In the three lower termite families, Termopsidae, Kalotermitidae, and Mastotermitidae, the worker-like forms are more prone to neotenic transformation than are the nymphs. *Zootermopsis* pseudergates derived from wingpad-aborted nymphs have a

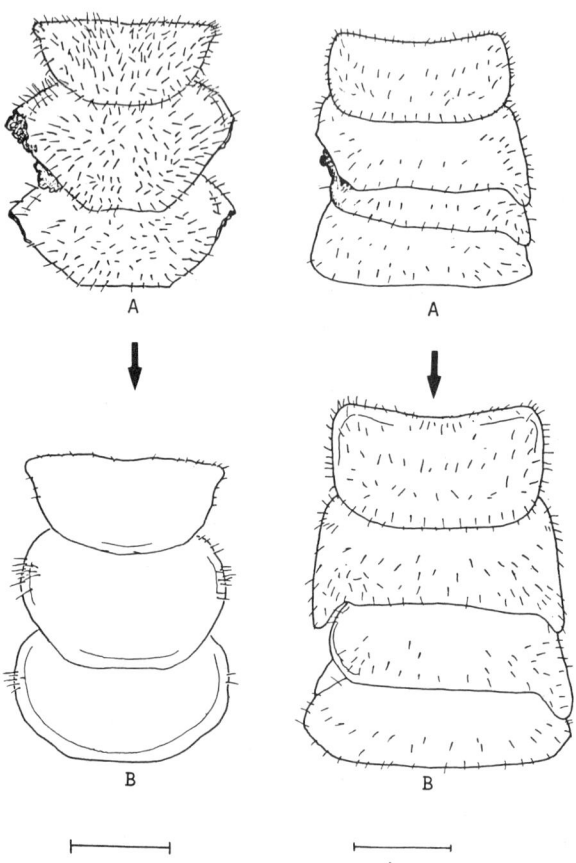

Fig. 2. Two mechanisms of pseudergate determination. Left side: Zootermopsis nevadensis (Termopsidae), A. nymphal thorax with wingpad abscission scars, B. pseudergate with rounded thoraces and altered pattern of setal distribution. Right side: Pterotermes occidentis (Kalotermitidae), A. nymph with nestmate-inflicted wingpad bite scar, B. pseudergate with asymmetrical wingpads and setal pattern not altered. Arrows indicate a molt.

heightened potential for neotenic transformation (Light and Illg 1945) as do pseudergates derived from wingpad-bitten nymphs in kalotermitids (Zimmerman 1983b; Sewell and Watson 1981). For *Mastotermes* see Watson et al. (1977). Interestingly, the opposite is true with nymphoid neotenics, more common than ergatoids in the more advanced, subterranean, lower termite families, Hodotermitidae (Mednikova 1977) and Rhinotermitidae (Buchli 1956) and for most higher termites, Termitidae (Noirot 1956). Thus, it is realistic to think of *lower* termite pseudergates not only as temporary helpers but also as very hopeful, neotenic, nest inheritors.

Apparently pseudergate differentiation promotes the chance of neotenic reproduction. The claim that pseudergates are a more or less sterile caste is therefore misleading. The relevant question is whether the pseudergate-neotenic and alate tactics are of equal or unequal fitness, isogignous or allogignous, in the terminology of Austad (1984).

J.   Reproductive Value of a Hopefully Neotenic
     Pseudergate May Approximate that of Dispersing
     Alate in *P. occidentis*

Pterotermes occidentis occurs throughout the Sonoran Desert of northwestern Mexico and southern Arizona (Nutting 1966b). It colonizes mainly standing dead branches of living palo verde trees of the genus *Cercidium*. For dead branches to be habitable by *Pterotermes* they must first be worked by various wood-boring beetles. The exit holes of the beetle serve as entrance holes for the founding alates. The weight distribution of suitable dead palo verde branches ranges from about 113 g to over 13,608 g with a mean of 1,542 g. Based on measurements of cross sections, I estimated maximum consumption by *Pterotermes* at around 80 percent and subtracted 20 percent for competition with wood-boring beetles and other termites and arrived at an estimate of 60 percent availability. Based on studies of wood consumption rates and instar durations I determined that the production of one alate costs about 295 mg of wood. Thus the average branch should yield, 1,542 x 0.6/0.295 = 3,136 alates. I assumed that a neotenic might expect half of the original resource after a replacement event (=1568 alates). I calculated the probability of alate establishment as the product of alate dispersal survival, 0.05 (conservative guesstimate), and the probability of surviving the critical first year, 0.35 (based on demographic records), 0.05 x 0.35 = 0.0175. This contrasts with the probability of neotenic establishment, 0.036 (see above).

Multiplying the probabilities of establishment by resource use (expressed as alates) gave figures for the relative reproductive value of the late and neotenic tactics: alate F_a = 0.0175 x 3136 = 55; neotenic (ave.) F_n = .0363 x 1568 = 57. Since the probability of neotenic establishment is very nearly twice that of alate establishment and since neotenic resource use was estimated at half that of alates the resulting figures are remarkably close. Comparable values for neotenics derived from specific stages depend on the stage-specific neotenic tendencies and percentage of the stage in the population. These figures are: pseudergates 0.25 and 21%; nymphs 0.9 and 35%; and larvae 0.04 and 41%. Thus for pseudergates to neotenics F_{p-n} = 57 xc 1.05 = 60; for nymphs to neotenics f_{n-n} = 57 x 0.62 = 35; and for larvae to neotenics F_{1-n} = 57 x 0.33 = 19.

These calculations provide a provisional demonstration that the pseudergate-neotenic developmental route may pay off roughly as well as direct nymphal-alate development and dispersal. I caution however that many assumptions and weak estimates underlie these calculations and much more precise knowledge of the costs of dispersal and neotenic competition are needed. Additional factors, such as the likely differential fighting success of neotenics derived from different stages, need to be factored in.

K.   Termite Protosoldiers: Inheritance Contest Specialists?

Competition over inheritance has produced some shockingly malevolent features of termite biology.   As noted above, in many species orphaning is followed by a flush of neotenic molting followed by violent intracolonial battles until a single neotenic of each sex remains.   The intensity of this social competition over inheritance suggests that it may have been a potent selective force capable of molding further adaptations.   I have proposed (Myles 1986) that the termite soldier may have originated as a selfish fighting morph.   In the ecologically most primitive, rotten-wood termites of the family Termopsidae, reproductive soldiers tend to develop in orphaned colonies. This suggests that soldiers originally evolved mandibular weaponry in the context of siblicidal neotenic fighting.   Ironically, structures which may have originally evolved for siblicide may have secondarily been modified for sibling defense. Lethal power struggles over nest inheritance are also found in other animals; for example, bumble bees, honey bees and naked mole rats. Analogous intraspecific cannibalistic morphs or siblicidal adaptations ("Cainisms," Hrdy and Hausfater 1984) have evolved in a wide variety of animals including *Amoeba*, flagellates, several genera of ciliates, *Asplancha* rotifers, *Scaphiopus* tadpoles, *Ambystoma* salamanders, *Dugesia* planarians, several fish, and sea gulls (see refs. in Polis 1981). In the life cycle of certain species of parasitic Hymenoptera belonging to the families Ichneumonidae, Trigonalidae, Platygasteridae, Diapriidae, and Serphidae the larvae undergo a temporary transformation into a bizarre fighting form that kills and eats its brothers and sisters occupying the same host insect (Wilson 1971b; Clausen 1940). Dimorphism with a distinct fighting morph among adult males occurs in many insects including agaonid fig wasps, torymid fig wasp parasites, the parasitoid wasps *Melittobia*, *Telenomus* and *Synagris*, the mites *Tetranychus*, *Caloglyphus*, and *Rhinoseius*, *Holothrips* thrips, the scolytid beetle *Ozopemon brownei*, the ants *Hypoponera* and *Cardiocondyla*, and the bees *Chilalictus* and *Evylaeus* (Hamilton 1979).   Siblicidal brood reduction is found in many birds, such as boobies, pelicans, egrets, oystercatchers, owls, egrets, skuas, eagles, hawks, and kittiwakes (Mock 1985). Examples of adaptive kin killing among mammals can be found in Hrdy and Hausfater (1984). In view of such widespread examples the proposal that termite soldiers originated as sibling fighters is less surprising. This "selfish soldier" hypothesis is also supported by a variety of circumstantial facts: (1) the existence of reproductive soldiers in the lowest family of termites which develop under conditions of orphaning and in a few cases have become established as the sole reproductives of their sex (Myles 1986); (2) the unusual appropriateness of termopsid soldier morphology for this function whereas the derivative soldiers in other families have evolved more muscular guards or chemical defenses (Deligne et al. 1981), and (3) the evolution of termite soldiers prior to ants which are termites' only significant social predators (Deligne et al. 1981; Carpenter and Hermann 1981). Phylogenetic comparisons in the order show that true nonreproductive soldiers evolved before true workers. Thus, if soldiers were initially selfish siblicidal morphs they must have been co-opted by kin or colony-level selection for their benesocial function as colony defenders at an early stage. Finally, it must be acknowledged that the origin of soldiers as sterile defensive morphs remains a viable hypothesis with good comparative examples of sterile soldiers evolving before workers in aphids (Aoki 1982) and a hymenopoteran parasite (Cruz 1981).

L.   Sequence of Termite Caste Origins

Although many invertebrates evolved coloniality in the oceans before them, termites were the first fully eusocial terrestrial organisms, appearing in the early Mesozoic, ca. 200 million y.b.p. Thus, they flourished during the era of the dinosaurs and have existed perhaps twice as long as the earliest social Hymenoptera (Carpenter and Hermann 1981).

Termite caste evolution can be envisioned to have proceeded in three steps. The single most important event initiating termite social evolution was the origin of the neotenic caste. Neotenics evolved because of dispersal/inheritance tradeoffs brought about by endoxylophagy. Flight dispersal to suitable new logs is dangerous while log inheritance is valuable. Neotenics are, in effect, opportunistic nest inheritors.

Second, the capacity for facultative neotenic development dramatically reduced the cost of delayed dispersal. This reduced cost was probably the prime impetus for the subsequent evolution of mechanisms for delayed dispersal and the origin of pseudergates. Several different mechanisms now appear to operate in determining pseudergate differentiation in lower termites including wingpad abscission under unfavorable conditions (*Zootermopsis*) and nestmate manipulation by wingpad and leg biting (Kalotermitidae). This second stage was also a period of intracolonial competition ("social selection," West-Eberhard 1979, 1981, 1983) over who should inherit. Neotenic combat may have led to the evolution of a specialized morph with hypertrophied mandibles, the so-called reproductive soldier. Thus, in the early stages of caste evolution all end points of the developmental pathways were reproductive (Fig. 3). This is essentially the stage at which rotten wood termites, Termopsidae, are at today except for the addition of sterile soldiers.

It is possible that both the protoworkers and protosoldier strategies were selected by reinforcing kin and individual selection from the outset since both strategies can involve selfish and altruistic components. It seems likely that neither strategy could have been possible without the selfish component. Direct reproduction by each caste in the early stages of their evolution might also have been important in the initial establishment of the genetic programs of differentiation. The pseudergate-neotenic and reproductive soldier ploys may be possible concurrently because each has special risks and benefits (see below).

The third stage has occurred between the lower and higher termites in which we have the emergence of a distinct neuter line with sterile true workers and soldiers. Distinct worker lines of development from the first or second instar and not initiated by any visible manipulation are found in several phylogenetically separated termite taxa, the Masto-termitidae, Hodotermitidae, and one or more times in the Rhinotermi-tid-Termitidae group. Each of these taxa coincide with independent cases of invasion of the soil habitat from a presumed ancestral condition of living in wood (Fig. 1). Thus it seems that canalization of worker development is brought about by selection forces associated with a sub-terranean habit and the need for foragers.

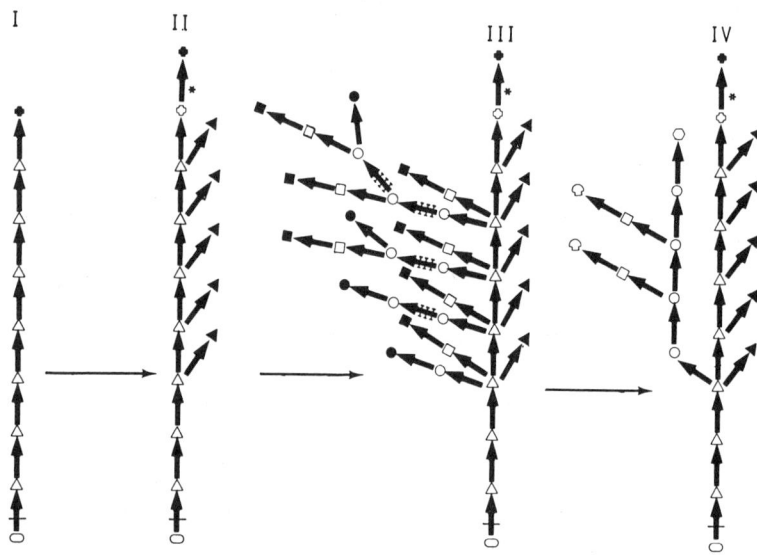

Fig. 3. Proposed sequence of caste origins in the Isoptera. I: simple development of roach-like ancestor; II: prototermites with facultative neotenic development; III: generalized hypothetical lower termite with all four basic physical castes and all developmental end-points reproductive; IV: generalized higher termite with an early branching distinct neuter line and sterile terminal castes. Arrows = molts, crossed arrows = indeterminant sequence of moles, asterisked arrows = dealation and sexual maturity, slashed arrows = eclusion from egg, oval = egg, open triangles = progressive immature instars, closed triangles = neotenic reproductive derived from progressive instar, open cross = alate, closed cross = dealate primary reproductive, open circle = pseudergate or immature worker, closed circle = neotenic derived from pseudergate, open square = presoldier, closed square = reproductive soldier, open hexagon = true worker, open mushroom = true soldier. Open symbols represent sexually immature or nonreproductive stages, closed symbols represent reproductive stages.

M. Termite Castes as Alternative Tactics for Maximizing Individual Fitness

1. **Alates** are subject to enormous dispersal and colony foundation-related mortality, but utilize major log portions and avoid neotenic costs (high risk, high direct reproductive gain tactic).

2. **Neotenics** are subject to mortality related to neotenic combat, but can avoid alate costs and inheritance of nest with helpers reduces colony foundation costs (moderate risk, moderate direct gain tactic).

3. **Pseudergates** avoid the large immediate costs associated with alate development due to unfavorable conditions or manipulation injuries but have an increased potential for neotenic transformation and retain potential for reversion to alate development. Pseudergates experience a prolonged opportunity to accumulate indirect fitness benefits through helping (a temporary default tactic with low risk and moderate indirect gain and potential deferred direct gain).

4. **True workers** are canalized from an early stage into a specialized form well suited for worker functions and with little chance of neotenic reproduction and no chance of transformation to the alate. Their personal fitness is largely to entirely indirect. Thus true workers are probably not individual tactics but evolved by reinforcing kin and colony-level selection for the function of foraging in the subterranean adaptive zone.

5. **Reproductive soldiers** avoid alate costs and have superior fighting ability in neotenic combat, but the presoldier stage is vulnerable and costly in terms of lost time during the highly critical period immediately following orphaning. Thus to be effective the presoldier molt may have to anticipate the orphaning event and if orphaning does not follow the direct fitness gamble is lost which may induce the presoldier to molt to a nonreproductive soldier.

6. **True soldiers** are subject to large risks but these very risks provide opportunities to render lifesaving protection, for which they have a heightened ability, to the colony (high risk, very high indirect gain tactic). Reinforcing kin and colony-level selection involved.

III. OTHER SOCIAL ANIMALS

A. Dispersal/Inheritance Tradeoffs in the Hymenoptera

The Hymenoptera is one of four megaorders of holometabolous insects with over 100,000 described species and probably between 200,000 and 250,000 total species (Snelling 1981). Within the Aculeate branch of the Hymenoptera eusociality has evolved independently at least 12 times and lower grades of sociality on many more occasions (Wilson 1971a). This is the densest cluster of separate social origins in the entire biota. It is therefore appropriate to ask whether preadaptations for dispersal/inheritance tradeoffs are in evidence. In Hymenoptera, since foraging flights are a routine aspect of adult life, a dispersal flight does not represent the inordinate risk that it does in termites. However, in many Hymenoptera independent colony foundation is difficult due to the effectiveness of predators, parasites and conspecific and heterospecific nest parasites. Thus it is appropriate to consider the tradeoff between the probability of establishment versus the prospect of inheritance since dispersal per se is not an overwhelming deterrent, e.g., an establishment/inheritance tradeoff.

B. Socially Preadaptive Resource Heritability in the Aculeate Hymenoptera

A striking pattern exists among the solitary ancestors of social Hymenoptera. All are (1) brood nest constructing and (2) provisiong members of the Aculeata (Wheeler 1928; Spadbery 1973; Michener 1974; Iwata 1976; Eickwort 1981). Sociality has never advanced beyond a rudimentary level in the vast assemblage of the non-nesting, non-provisioning phytophagous Symphyta and parasitic Parasitica of the Hymenoptera despite their haplodiploidy. Brood nests are usually burrows in the ground or wood, or may be constructs of mud or plant fibers. In at least two instances the emergence of sociality seems to have loosely coincided with the origin of pendent paper nests, e.g., Polistinae (Hansell 1986) and *Microstigmus* (Mathews 1968). Unlike ground nests with many divergent tunnels and scattered cells, the clumped cells of a pendent nest are more easily monopolized. This permits the strongest individual to skew reproduction in its favor making the value of the dominance role and its inheritance greater.

Besides the nest itself, there are resources which are stockpiled in the nest. Larval cells are provisioned with prey, pollen, nectar, or other food. The Aculeates are the most skilled insects in the ability to forage, collect, and repeatedly return to a central nest where resources are hoarded (Eickwort 1981). Only a few other insects share this ability to make repeated forays and cache food in the nest. These include burrowing crickets, burrowing bugs, earwigs, some stink bugs and a few seed-eating and detrivorous beetles (Andersson 1984; Tallamy and Wood 1986). Carabid and silphid beetles also mass provision nests but the provisions are extremely perishable. In none of these cases are the amount and durability of the stored food sufficient to provide significant resource inheritance opportunities for adult offspring. The critical difference in the Hymenoptera is that nest excavations, prepared brood cells, and provisions are potentially heritable to adult offspring and of considerable value. One measure of the value of such property is the widespread cuckoo parasitism of nests within and between species and cleptoparasitism of provisions (Iwata 1976). These phenomena indicate the preadaptedness of provisioned nests to the evolution of within-family facultative inheritance. Alexander (1986) noted at least 17 cleptoparasitic lineages in the Aculeata and suggested that theories about the repeated origins of eusociality should also account for the multiple origins of cleptoparas::ism. The resource inheritance theory presented here does just that.

Actually, 17 origins of cleptoparasitism is probably an underestimate. Emery's Rule states that cleptoparasitic species are usually the most closely related species of the host which suggests that nearly each cleptoparasitic species is of recent and separate origin (see Wilson 1971a). I suggest that cleptoparasites are obligatory alternative *strategies* for resource takeover whereas castes may be facultative alternative *tactics* for resource inheritance.

In more advanced social groups nests become increasingly durable and valuable. In the social bees nests are constructed of wax secretions, collected plant resins, and collected mud. These building materials are known as propolis, cerumen, and batumen. There are also resin stores in some species. Colonies also accumulate large stores of food (honey, pollen, etc.) in food pots (Michener 1974). Thus not only is the nest an energetically expensive construct and a safe refuge but it is also the repository of useful secretions and collected reserves representing the accumulated input of many individuals. The adult population itself is also a heritable resource. The nest is a staffed bank of sorts. Although the durability of provisioned nests is sometimes not great, even at the primitive stages it may be at least sufficient to give a nondispersing or joining adult a significant short-term advantage. Unlike termites in which resource heritability is a consequence of the size and durability of wood, in the Aculeates resource heritability is a consequence of the fact that the nest is the focal point of construction activities and resource stockpiling.

It is interesting that the importance of a nest as a preadaptation to hymenopteran sociality has been ascribed to the facts that (1) a progressively provisioned nest allows for overlap of generations (Wheeler 1928; Evans 1958; Michener 1958; Mathews 1968), (2) shared entrance to burrow obviates digging of separate entrances through hard soil crust (Michener 1958), (3) the nest presents possibilities for mutualistic defense (Lin 1964; Michener 1974; Jeanne 1975), (4) the nest permits control of certain aspects of the physical environment and promotes homeostasis (Michener 1958), and (5) the nest provides stigmeric cues and indirect contact between individuals (Michener 1974). However, the preadaptive quality of the nest as a potentially monopolizable, heritable, staffed construct and site of past and future food hoarding has generally been underemphasized though not entirely overlooked. Michener (1974) noted the frequent reuse of old nests in the parasocial *Augochloropsis diversipennis, A. sparsilis, Pseudagapostemon divaricatus, Lasioglossum ohei, L. malachurum, Pseudaugochloropsis* spp., the primitively eusocial Allodapines, *Lasioglossum umbripenne, Bombus atratus,* and the higher eusocial bees. Plateaux-Quenu (1961) reviewed replacement reproduction in social insects, and these examples involve nest reuse. Andersson (1984) and Brockman (1984) list nest inheritance as one of several potential benefits to helpers of sociality while continuing to stress the importance of indirect fitness benefits of helping. Eickwort (1981) comes closest, giving main emphasis to resource inheritance, stating that "subsocial behavior based on construction and provisioning of nests, plus re-use by a second generation, is perhaps the most critical prerequisite for eusocial behavior." Thus although nest reuse has been documented and acknowledged it has rarely been considered in terms of its importance to the fitness of potential inheritors, as an option to dispersal, and as an alternative resource acquisition tactic. Although the "hopeful reproductive" nature of "workers" and cofoundresses has been emphasized by West-Eberhard (1978, 1981) the potential *enhancement* of reproduction due to inheritance of the nest and its contents has generally been insufficiently appreciated. The hopeful reproductive worker's tactic can be viewed as "hopeful inheritance." Socially amassed resources are the analogs of human wealth. The benefits of resource inheritance can be important both initially and on a sustained basis in promoting caste and social evolution.

C.   Hymenopteran Workers as Hopeful Inheritors

If nest inheritance is at least a component of the payoff to incipient workers, one would expect to find in the primitively social species of the major hymenopteran socioclines that the auxiliaries or worker-like gynes (either parasocial joiners or subsocial first brood) should reproduce without dispersing at least occasionally. Just such evidence of occasional or delayed philopatric reproduction by subordinates or "workers" is abundant and the literature on this subject is expanding quickly. In fact, the tendency to generalize about worker "sterility" is unwarranted. Lin and Michener (1972) state, "In almost all social Hymenoptera some or many of the workers lay eggs. We are commonly not dealing with nonreproductive caste, but with individuals that are in varying degrees less productive than the queens and in many cases are productive only under certain circumstances." They also suggest that parthenogenetically produced males may be a significant alternative reproductive strategy for unmated workers: "Since males commonly mate more than once, at least in primitively social halictines, a worker's genes may be widely distributed by her offspring."

A sampling of worker reproduction in the social Aculeates is provided below.

VESPOIDEA

Vespidae: Stenogstrinae, *Parischnogaster nigricans serrei* usurpation by joiners common and helping, replacement of mother, or subsequent independent foundation by first daughters may occur (Turillazzi 1982). *Liostenogaster flavolineata* females can rise through the ranks to take over dominant position, if subordinate they do more foraging, or may leave to found or join another nest (Hansell et al. 1982).

Polistinae: In *Polistes gallica* workers establish a social hierarchy based on age, and if the queen dies the top workers assume the role of the queen, laying eggs which produce only males (Evans 1958). In *P. metricus* when queen removed the oldest worker assumed the queen role (Dew 1982). Takeovers of nest by subordinate cofoundresses and selfish joiners in *P. exclamans* (MacCormack 1982). Succession of queens common in lifespan of *Polistes* spp. (Lester and Selander 1981).

Polybiini: In *Belonogaster* the females are all identical structurally and physiologically; all of them mate and lay eggs, and none can be called workers (Evans 1958).

FORMICOIDEA

Ponerinae: In *Rhytidoponera impressa* Type b worker-reproductive colonies are headed by mated workers which produce winged males and reproduce by colony fission with more mated workers; mated workers have also been recorded in the ponerine genera *Diacamma*, *Hypoponera*, *Opthalmopone*, and *Pachycondyla* ('*Bothroponera*'), and there is circumstantial evidence for worker mating in *Streblognathus*, *Dinoponera* and *Mystrium* (Ward 1983); likely that, under natural conditions, replacement of the queen by mated workers would be readily accomplished in *Rhytidoponera metallica* (Ward 1986). *Opthalmopone berthoudi* lacks a distinct reproductive caste, only inseminated ergatoids produce eggs (Peeters 1982). Ergatoid queens in *Leptogenys nitida* and *Megaponera foetens* (Peeters 1986). In *Diacamma rugosum* part of the worker population is fertilized and is serving as queens (Moffett 1986).

Myrmicinae: In *Myrmica* spp. dominance hierarchy among queens, and workers produce males, in queen-right culture, males mostly develop from the eggs of newly eclosed workers (Elmes 1982). Gamergates in many *Megalomyrmex* spp. (Brandao 1986). Workers reproduce by parthenogenesis and almost all individuals oviposit in *Pristomyrmex pungens* (Tsuji 1986). Male production significant by workers of slave ant *Harpagoxenus sublaevis* (Bourke 1986).

Formicinae: In *Cataglyphis cursor minim*, small, medium and large workers are all able to lay reproductive eggs and to produce by parthenogenesis, queens, males and new workers (Cagniant 1983). In *Oecophylla longinoda* workers lay many of the eggs that become males and all those that become queens (see Lin and Michener 1972).

APOIDEA

Halictidae: In *Augochloropsis* all females are fertilized and pollen collectors may cease foraging and become egg layers; in *Lasioglossum zephyrum* new females stay in the nest, enlarge and guard it, and make and provision new cells, some mate, and these may replace the overwintering female as egg layers; castes are not sharply defined and all intergradations from queen to worker are found in nearly every nest (Batra 1964). Mated and egg-laying worker in first brood in *Evylaeus laticeps*, *pauxillus*, *linearis*, *malachurus* (Packer and Knerer 1982). In *Dialictus*, *Evylaeus*, *Halictus*, *Augochlora*, *Augochlorella*, *Pereivapis*, *Pseudaugochloropsis*, and *Augochloropsis* workers typically can later become replacement queens if the reproductive dies and laying workers are common in older colonies (Eickwort and Kukuk 1986).

Anthophoridae: In allodapine bees one to five supernumerary mated or unmated adult females remain in the nest with the mother and may help or lay eggs (Michener 1971). In *Allodape exoloma* and *Braunsapis* *foveata* chance events such as the death of the queen may be the major determinants of reproductive potential (Mason 1986).

Apidae, Bombinae: In *Bombus terrestris* an elite group of daughter offspring may drive the queen from the comb after which one becomes the dominant egg layer (van Honk 1982).

Apinae: In *Melipona favosa* facultative worker ovipositions occur in queenright colonies--queenlessness causes workers to compete for laying opportunities (Sommeijer 1986). In a queenless colony of *Trigona* or *Apis*, male-producing eggs are laid by the workers. Commonly 10 percent or more of the workers in an *Apis* colony have become layers a week after dequeening. There are records of laying workers producing 19 and 32 eggs per day--enormous rates compared to solitary bees but low compared to queens. A false queen of *A. cerana* laid eggs at the rate of 82 per day, an extraordinarily high rate for a laying worker (Michener 1974).

SPECOIDEA

Microstigmus comes, facultative nondispersal of one or a few daughters which help provision and appear usually not to reproduce but occasional reproduction or inheritance seems possible (Mathews 1968; West-Eberhard 1977).

Primitive workers, ergatoids, ergatogynes, mated laying worker (=gamergates) (Peeters and Crewe 1985), replacement gynes, auxiliaries, and cooperative cofoundresses may all be alternative reproductive tactics evolved at least in part to capitalize on the nondispersive option of reproduction with nest inheritance in the Hymenoptera.

D. Dispersal/Inheritance Tradeoffs in
   Vertebrate Social Evolution

Nondispersing "helpers" are found in several cichlid fish that are territorial around shelter sites and their increased chance of parental replacement and territory inheritance seems a likely explanation (Taborsky and Limberger 1981).

The importance of resource inheritance to the evolution of avian communal breeding systems is well established and numerous examples have been catalogued (Rowley et al. 1974; Brown 1978; Emlen 1982, 1984; Woolfenden and Fitzpatrick 1984). For most birds the heritable resource is an all-purpose territory upon which the birds reside permanently. The best-studied example of territory inheritance is that of the Florida scrub jay, *Aphelocoma c. coerulescens* (Woolfenden and Fitzpatrick 1984). The acorn woodpecker, *Melanerpes formicivorus*, is an interesting case in which the value of the territory is accentuated by a large food cache, a tree with many bored holes used for storing acorns called the granary (Hannon et al. 1985). For most birds the nest itself is not valuable enough to warrant the evolution of a nondispersive inheritance tactic; however, the inheritance of associated subordinate helpers is crucial. In the green woodhoopoe helping enables individuals to establish bonds with younger flock mates who will then serve as helpers after the current helpers inherit the wood cavity nest and territory. Ligon (1983) calls such helper inheritance "obligate reciprocity." Similarly, in the pied kingfisher helpers unrelated to the breeding pair stand a good chance of replacing the breeder of the same sex and thereby inheriting a mate, nest, territory and step-offspring as helpers (Reyer 1984).

In mammalian studies resource inheritance falls under the keyword "philopatry" (Waser and Jones 1983). Species with "helpers at the den" are designated "communal breeders" (Emlen 1984; Hoogland 1981). Numerous studies have established that in mammals it is the younger, weaker, socially subordinate individuals that are forced to disperse and experience high rates of mortality (Christian 1970). Thus, whether an individual is philopatric or dispersive is often mediated by dominance contests which establish hierarchical relationships. The heirs apparent hold the beta position in the hierarchy. It is usually the females that are philopatric in mammals (Greenwood 1980).

The burrow systems of fossorial rodents appear to be particularly conducive to the evolution of facultative nondispersal and incipient sociality. Postweaning associations of parent and juveniles occur in the mongolian gerbil, *Meriones unguiculatus*, which are detrimental to parental reproduction but which benefit the juveniles by improving future chances for successful dispersal or replacement of parents in the natal nest (Ostermeyer and Elwood 1985). Nondispersal by nonbreeding female black-tailed prairie dogs also depresses the annual reproductive success of breeders but appears to compensate the nonbreeders by increasing future opportunities for takeovers of neighboring coterie territories in the natal ward (Hoogland 1981). The only eusocial mammals, the naked mole rats, live entirely underground in a system of extensive foraging tunnels running at root level with a deeper nest area. They form unique cooperative "digging chains" in order to forage for tubers. They exhibit a rudimentary caste system involving frequent workers, infrequent workers, nonworkers, a beta female, and a breeding female and have a social structure remarkably reminiscent of wood-inhabiting termites (Jarvis 1981). Territory inheritance appears to be important in the social evolution of the European badger (Linstrom 1986). In mammalian species that have cooperative foraging, such as elephants (Dublin 1983), coatis (Russell 1983), and dwarf mongoose (Rood 1978), or cooperative hunting, such as wolves, coyotes, jackals, hyenas, Cape hunting dogs, and Indian wild dogs (Eisenberg 1981), group membership for subordinates may result in reproductive suppression (see Wasser and Barash 1983); but, the possibility of replacing the dominant and inheriting the cooperating subordinate population may provide sufficient fitness compensation. Rank inheritance is illustrated by gorillas in which nondispersing sons of dominant males may inherit leadership and reproductive prerogatives after the father dies (Harcourt et al. 1976). More examples can be found in Emlen (1984).

E.   Dispersal/Inheritance Tradeoffs in Human
     Social Behavior

The notion that property inheritance influences human sociality is
not new. Rouseau in *On the Origin of Inequality* wrote:

No individual was ever acknowledged as the father of many, till his
sons and daughters remained settled around him. The goods of the
father, of which he is really the master, are the ties which keep his
children in dependence, and he may bestow on them, if he pleases,
no share of his property, unless they merit it by constant deference
of his will.

Darwin also commented on the subject in *The Descent of Man.*

Man accumulates property and bequeaths it to his children so that
the children of the rich have an advantage over the poor in the race
for success, independent of bodily or mental superiority. On the
other hand, the children of parents who are short-lived, and are
therefore on an average deficient in health and vigour, come into
their property sooner than other children and will be likely to marry
earlier, and leave a larger number of offspring to inherit their inferior constitutions.

However, in recent textbooks on animal and human social evolution the
importance of resource inheritance has received little attention (Wilson
1980; Barash 1982; Trivers 1985). Although anthropologists have investigated the relationship between inheritance and certain aspects of human
sociality (e.g. Hartung 1976), sociobiologists have not to my knowledge
connected this with the animal analogues to resource inheritance.

F.   Man and Resources

The relationship of man to resources is unusual in many respects.
We are undoubtedly the most materialistic of creatures. Even the most
primitive peoples possess articles of many kinds: clothes, blankets,
adornments, tools, weapons, cooking and eating utensils, bags, pots,
domestic animals, and preserved food stores. Obviously many of these
possessions are durable and thus transmissible across generations.   In
addition to these durable goods, all human societies are territorial
(Wilson 1980); even nomadic tribes follow hereditary routes.   It seems
likely that all human societies have customs regulating the inheritance
of possessions and property.
     Humans are the only animals to have evolved a plasticity to re-
sources that enables them to inhabit the Arctic, Amazon and Kalahari.
Thsi flexibility results from our talent to learn and to culturally transmit
knowledge. We have an incredible diversity of ways for exploiting resources to make a living. We are also presumably the only animals with
cultural mechanisms for policing and maintaining ownership rights even
sometimes in the face of great inequalities.   The transmission across
generations of resources via dowries, trust funds, and wills have obvious
effects on the life history decisions of recipients, on marriage customs

(Goody 1969) and dispersal patterns. The fact that most humans maintain lifelong contact with their parents is probably peculiar to humans and may be partly to ensure inheritance (Alexander 1974). Our unusual resourcefulness, materialism, resource holding power, and sense of ownership make it likely that resource transfers may be important aspects of our biology relative to other animals.

Cultural inventions such as money, banking, and taxation have obvious effects on the control of inheritance and indicate that the cultural evolution of resource inheritance tactics continues to be dynamic up to the present and thus may influence modern social structure.

Stages in the resource inheritance are difficult to specify because of the labile nature of man's relationship with his environment and resources. However, we can imagine various resource extraction occupations associated with family land tenure which characterized progressive stages of human evolution. All extant hunter-gatherer societies can be characterized by some kind of land tenure rights or usufruct such as collecting, hunting, trapping, or fishing grounds, or access to water sources. We can also imagine a sequence in terms of the evolution of the home base. The home base may have progressed from sleeping trees, caves, improved caves, makeshift shelters or tents to permanent constructed dwellings to permanent settlements.

The transfer across generations of titles, authority, power, and community connections is undoubtedly nonrandom with respect to relatedness. The potential of inheriting these things is likely to have important bearing on individual life history decisions such as whether to, when to, and how to disperse, cooperate, and compete. It seems probable that throughout much of human evolution the intelligence to assess the potential for such kinds of resource inheritance and the acumen to maneuver accordingly would have been strongly correlated with Darwinian fitness. The aggregate consequences of these decisions, in turn, would have profoundly affected demographic structure and the nature of social interaction. The generational transfer of wealth, privilege, power and resources is reflected by the multiple independent and seemingly near-universal origin of dynastic power structures in the early stages of political structuring. Nepotistic transfers of political power and wealth or the means of production obviously persist as predominant features of modern human sociality in many cultures.

A final category consists of the accumulated knowledge of elders. This perhaps was an especially important resource among humans prior to the invention of writing which provides a superior way of preserving and transfering accumulated knowledge.

Thus there are moveable possessions, homesites, land tenure rights, hereditary rank privileges, and knowledge, all of which may be considered as heritable resources which should influence decisions about dispersal, subordination, within and between group competition and cooperation.   The subject greatly exceeds the scope of this paper; however, a few examples of land tenure inheritance follow.

G. Land Tenure or Use Privileges: A Birthright

It is likely that hunting may have provided mutualistic advantages in early human sociality as in other group hunting social carnivores, such as canids (Wilson 1980). To leave the group would dangerously subject one to predation and reduce one's ability to enjoy meat from large game kills. By staying in the group one would eventually move up in the leadership hierarchy. Turnover rates of group members may have been high due to hunting accidents, disease, etc. One would also inherit tools, hides, knowledge of hunting tactics, and access to the hunting ground and local game populations from hunting companions.

Wherever people have derived a living off the land the specific resources upon which the livelihood is based form the basis for territorial allotments or restricted use privileges (usufructs), and these territories or privileges are guarded by a kinship group through which inheritance passed. Examples from extant hunter-gatherer societies substantiate this. Systems of hereditary land tenure and their consequences on such major sociological parameters as matriarchy vs. patriarchy, matrilinearity vs. patrilinearity, uxorilocality vs. virilocality, monogamy vs. polygamy are too abundant and complex to review here but a few examples can be given. Plots for shallot cultivation are demarcated by the clan once every generation in certain native populations of southeastern Ghana (Nukunya 1973). Cultivated land for subsistence farming is traditionally acquired only through inheritance in the Gwemba Tonga (Scudder 1975). Rights to pinyon groves, trapping districts, and fishing sites are inherited bilaterally in the Paiute Indians of Nevada and California (Williams and Hunn 1982). Virilocal alignments of certain Alaskan Indian tribes such as the Tareomiut are related to male inheritance of the "family trapping territory," but this is absent in Eskimo boat-using tribes such as the Nunamiut, in which boat inheritance devolves to those who can best make use of it (Riches 1982). The !Kung bushmen have subdivided the Kalahari into hereditary landrights called n!ores. Each n!ore has enough food and water to sustain at least one band throughout its seasonal rounds in the year. All persons inherit one n!ore from their mother and one from their father. They do not inherit the land itself, but rather the right to exploit the resources of the land along with others who inherit similar rights (Wiessner 1982). Studies of the Nata River Basarwa of Botswana revealed that lineal inheritance patterns became more common in part because of the increased value of goods owned by households. These changes occurred over a few generations due to the sedentarism and resulting prosperity of the group compared to neighboring groups (Hitchcock 1982). A fascinating case of land tenure is found in the Okiek of Kenya. Lineage territories, called konoito, are actually rights to exclusive use for the purpose of gathering honey. Each lineage konoito is divided into as many as two dozen smaller areas called korets. Korets are the single most valuable commodity that an Okiek man will inherit from his father. Korets are almost as important socially as they are economically because they are used to satisfy a wide range of social indebtedness (Blackburn 1982).

IV.   SUMMARY

A.   Heritable Resources and Inheritance Tactics

The principle of a dispersal/inheritance tradeoff appears to apply to the origin of social tendencies by the offspring of founding pairs in many animals both invertebrate and vertebrate. Brown (1983) has formulated a similar concept which he terms the "dispersal benefit ratio," Z_x. In this paper I have used the phrase "resource inheritance" in a very broad sense. Heritable resources span a wide gamut from tangibles such as nests, burrow systems, unused food reserves, accumulated food or material caches, territories, adjacent territories, territorial improvements, subordinate nestmates or younger siblings, hunting or foraging group members, potential mates, etc., or intangibles such as rank, experience, familiarity, or transmitted knowledge. This definition of inheritance encompasses all advantages that individuals can come into as a consequence of delayed dispersal. When a tactic's payoff is mainly in indirect fitness then I classify it as nepotistic (= apparent altruism) rather than as an inheritance tactic. When the payoff for a tactic is delayed, based on nondispersal, and mainly in terms of direct fitness, then I classify it as an inheritance tactic. Resource heritability is the essential unifying quality underlying a diversity of explanations for helping, such as learning to parent, generational reciprocity, nest or territory usurpation, habitat saturation and shortage of mates (see Hoogland 1981; Emlen 1982, 1984; Ligon 1983).

There are several ways of inheriting: waiting to be orphaned (replacement), usurping (supercedure), coming into reproductive rank (promotion), or budding off the parental nest or territory (fission). Nest or resource fragmentation, as occurs in termites, is another method. The subordinate or waiting phase of an inheritance tactic is especially amenable to further socialization because the prolonged associations and overlap of generations provide opportunities for manipulation, helpfulness, and the development of psychological ties which become useful in the future.

B.   Sequence of Selective Processes in Social Macroevolution

Vehrencamp (1979) has devised indices by which to estimate the relative importance of individual, kin and group selection in the maintenance of sociality for any particular species. Another way to consider selection processes is in terms of their sequence of action at various stages of social evolution (Brown 1974, 1983). Most controversy concerns the precise role of indirect fitness. Brown's (1983) four parameters, D, d, I, and i (present and future, direct and indirect fitness respectively) are most useful in helping to clarify the components of fitness being selected. Below I outline a modified scheme involving a sequence of tradeoffs in the components of fitness.

K-Selection on Parents
   (1) A tradeoff of fecundity for greater parental investment (sacrifice of r for greater K), maximization of offspring survival, establishment of presociality, e.g., cryptocercid woodroaches.

Selection on Offspring
   (2) A dispersal/resource inheritance tradeoff of immediate reproduction for greater future reproduction (sacrifice of D for greater d)
resulting from situations in which dispersal is constrained and/or potential for future reproduction is enhanced by resource inheritance. Origin
of the first facultative developmental switch as a stay-and-inherit tactic. Maximization of total lifetime situational direct fitness. Establishment of overlap of brood generations and subsociality.
   (3) A "filial/parental-care cost asymmetry" (in which filial care is
less costly due to the bypass of start-up and procreation costs associated
with parental care) promotes the evolution of alloparental care by older
siblings. An indirect component to fitness, I, is added to the future
direct component, d, resulting in the transformation of the stay-and-
inherit tactic to a help-and-inherit tactic. Inclusive fitness maximization with both d and I important but with d > I. Establishment of cooperative brood care and incipient eusociality.
   (4) Parental and/or sibling manipulation to exploit the indirect care-
giving component of the help-and-inherit tactic. Manipulation dimin-
ishes the potential for future reproduction forcing the manipulated to
compensate by prolonged or intensified helping. A gradual manipulation-
compensation process may eventually lead to a tactic in which I > d.
Opportunities for specialized care giving such as foraging, or increasing
reliance on care giving by super beneficiaries may also tip the balance in
favor of I.
   (5) At some point, early physiological determination may result in
total I exceeding the sum of I + d. This step transforms the help-and-
inherit tactic into an essentially sterile specialized worker caste herald-
ing the ultrasocial threshold at which point a society may begin to take
on superorganismal character. Thereafter the evolutionary unit or "individual" becomes the colony (Hull 1980) which evolves to maximize "ergonomic efficiency" (Oster and Wilson 1978) with caste numbers and proportions producing an "adaptive demography" (Wilson 1985) much like the
evolution of cytodifferentiation in a multicellular organism. Soldiers
can evolve by specialization of sterile workers.

Alternative Routes to Sterility
   (1a) A first-instar sterile soldier may evolve in sibling clones, I off-
sets lost D because relatedness is perfect and predation risk is high, e.g.,
some aphids (Aoki 1982) and a parasitoid (Cruz 1981).
   (3b) Competition among siblings for inheritance may result in the
evolution of siblicidal behaviors in which gains in D compensate for
losses in I, e.g., siblicidal neotenic replacement reproductives in dry-
wood termites.

(4b) Siblicidal behaviors may be followed by the evolution of morphological specializations which enhance fighting ability, e.g., possibly the reproductive soldiers in Termopsidae.

(5b). Adaptations for selfish fighting may preadapt for the ability to function as a specialized defender morph (soldier caste) in which I compensates for lost D.

An interesting feature of the (1-5) and (3a-5a) secenarios is that direct fitness is relinquished gradually and via intermediate stages in which delayed or accelerated direct fitness is of greater importance than indirect fitness.   In sequence (1-5) a eusocial threshold is established with incipient differentiation of caste before the indirect component compensates for lost direct fitness at which point "helpers" become "workers."   Termite pseudergates are similar to "helpers at the nest or den" in birds and mammals.   Like them it appears that the largest component to total fitness lies in future direct fitness, d rather than in present or future indirect components $I + i$.

As sequences (1a) and (3b-5b) indicate, it may be easier for a fighter or defender morph to switch to total indirect fitness and become a soldier than it is for a helper to become a strictly worker morph.   This is because the value of defense may be greater to kin and the nature of the morphological change may constrain reproduction.   For example, soldiers evolved before workers in termites and have evolved in the absence of workers in some aphids (Aoki 1982) and a parasitoid (Cruz 1981).   The ability of Aculete hymenopteran females to defend with the sting may be a factor explaining the tendency for advanced caste differentiation in this group (large I compensating for lost $D + d$).

C.   Relatedness Patterns and Situational Dynamics

It is the contention of this paper that kin selection and relatedness asymmetries were not important in the origin of the social insects and most social vertebrates.   A more fundamental initiating mechanism across social taxa is the existence of a dispersal/resource inheritance tradeoff promoting the evolution of a tactic for nondispersal which tends to result in enhanced delayed reproduction.   Competition over inheritance continues to have strong influences on selection of social tactics.

Relatedness asymmetries due to haplodiploidy in the Hymenoptera and alternating inbreeding and outbreeding in the Isoptera do not now appear to be the important prime movers of insect sociality they were once thought to be.   The downfall of haplodiploidy resulted from reviews by Michener (1974) and West-Eberhard (1978) showing that many primitively social bees and wasps start colonies with multiple unrelated foundresses, a review by Page (1986) showing multiple insemination to be common in both lower and higher social Hymenoptera, and rebuttals to Trivers and Hare (1976) indicting that the 3:1 sex ratio is not well supported (Andersson 1984).   The downfall of Bartz's inbreeding hypothesis for termites resulted from a review by Myles and Nutting (in press) showing that primary reproductives are long lived and do produce large numbers of alate offspring, too much outbreeding occurs in most wood-inhabiting termites, inadequate alternation of inbreeding and outbreeding may occur in groups that do have high inbreeding, and groups with high levels of inbreeding constitute only about 9 percent of termite species and are not correlated with especially pronounced altruism or sociality, e.g., *Heterotermes* and *Reticulitermes*. Therefore, asymmetries of relatedness in which individuals are more related to siblings than to potential offspring probably do not account for the origin of insect castes. Nevertheless, a basic degree of within-family relatedness, achieved in termites because of monogamy, is important because it gives socially victimized (manipulated), socially inferior (dominated), or ecologically disadvantaged individuals alternative ways (tactics) for promoting situational inclusive fitness (temporary kin helping). Thus kinship and indirect fitness definitely have a place in insect social evolution even if relatedness asymmetries do not.

For a time, attention to "altruism" and the magic numbers of kinship asymmetries and sex investment ratios distracted attention from possibly more important factors. It now appears that delayed direct fitness, d, is more important in the *origin* of sociality than indirect fitness, I. Furthermore, the preadaptations promoting the origin of the nondispersal/helper tactic are ecological and behavioral factors which generate situational variation between individuals rather than genetical factors. The reductionism of sociality to such genetical features as haplodiploidy, inbreeding patterns, chromosome number, and translocation complexes has given way to greater emphasis in understanding the condition-specific and age-specific situational complexity of individuals and how alternative tactics can improve situational fitness.

The new orientation to alternative reproductive tactics brings out the importance to individual fitness of neotenic reproduction in termites, worker reproduction in hymenopterans, and delayed reproduction in vertebrate social groups. With this there has been increasing attention to the ecological and behavioral factors which influence the cost of alternative behaviors. For example, the important ecological determinants of sociality in termites are predation causing poor survivorship of dispersing alates, and the durability and large size of logs which makes them heritable resources. Behavioral factors include sibling and parental manipulation by wingpad and leg biting which determines pseudoworker differentiation, pseudoworkers' higher tendency than nymphs or larvae to transform to replacement neotenic reproductives, and neotenic reproductives' engagement in violent siblicidal takeover battles.

D. Alternative Tactics and Situational Fitness

A large literature is starting to appear in animal behavior under the term "alternative tactics" (Austad 1984; Dominey 1984). This needs to connect with an older literature in insect life history in which similar phenomena were designated as polyphenism, polyphasy, or environmentally induced polymorphism (Kennedy 1961; Clark 1976). One of the advantages of insects is that they often express tactics in morphologically discrete forms whereas vertebrates express tactics more cryptically as behavioral plasticity. Selfish phenotypes can evolve by intraspecific co-evolutionary arms races (Dawkins and Krebs 1979) (=social selection, West-Eberhard 1979, 1981). Each selfish phenotype tends to broaden or expand the situational fitness of a genome containing several genetic programs for a conditional overall strategy. Thus the genome creates the selective environment in which the genome subsequently evolves; phenotypes coevolve against each other within one genome.

It may be generally useful for understanding the evolution of pheno-typic plasticity to define increases in "situational fitness" as the addition of alternative tactics to the behavioral repertoire. A general feature of social evolution may be the proliferation of tactics with the maximiza-tion of situational fitness. This is manifest as polymorphism in inverte-brates and as behavioral plasticity and intelligence in invertebrates. A fruitful approach to the study of adaptive phenotypic variation may be to identify situations and then consider the fitness improvements possi-ble by responses involving facultative switches. The term "situation" is used here in a restricted sense to refer to intermittently recurring con-tingencies that may strongly affect the fitness outlook of individuals but to which not all individuals are subjected. Examples would be the order of birth, sibling or nest composition at egg or pupal eclosion, parental presence or absence, receipt of manipulative injuries from nestmates, asymmetrical competitive encounters, breeding vacancies, etc. Since situations involve complexities of group composition and environmental constraint, appropriate response may often depend on subtle powers of situation assessment and kin recognition. All situational adaptations are additions to the pre-existing phenotypic repertoire; thus, they expand the genome's overall phenotypic capability and may not have to be se-lected at the expense of present adaptational capabilities. Also, the conditions for selection do not require that a new tactic exceed the prior fitness value of old tactics but that they provide fitness improvements for individuals that are ecologically disadvantaged or socially inferior. Alternative tactics probably yield lower reproductive success and inclu-sive fitness than the primary tactic. Perhaps behavior may generally be considered situational adaptation.

REFERENCES

Alexander, B. 1986. Eusociality and parasitism in the Aculeate Hymen-optera. *Proceedings of the 10th International Congress of the Inter-national Union for the Study of Social Insects,* Munchen, p. 126.
Alexander, R.D. 1974. The evolution of social behavior. *Annual Review of Ecology and Systematics* 4:325-383.
Andersson, M. 1984. The evolution of eusociality. *Annual Review of Ecology and Systematics* 15:165-190.
Andersson, M. and J. Krebs. 1978. On the evolution of hoarding behavior. *Animal Behaviour* 26:707-711.

Aoki, S. 1982. "Soldiers and altruistic dispersal in aphids." pp. 154-158. In: M.D. Breed, C.D. Michener and H.E. Evans, eds. *The biology of social insects.* Proceedings 9th Congress International Union of the Study of Social Insects. Boulder:Westview Press.

Austad, S.N. 1984. A classification of alternative reproductive behaviors and methods for field-testing ESS models. *American Zoologist* 24:309-319.

Austad, S.N. and R.D. Howard. 1984. Introduction to the symposium: Alternative reproduction tactics. *American Zoologist* 24:307-308.

Barash, D.P. 1982. *Sociobiology and behavior.* 2nd ed. New York: Elsevier.

Bartz, S.H. 1979. Evolution of eusociality in termites. *Proceedings of the National Academy of Science, USA* 76:5764-5768.

Basalingappa, S. 1970. Environmental hazards to reproductives of *Odontotermes assmuthi* Holmgren. *Ind. Zoology* 1:45-50.

Batra, S.W.T. 1964. Behavior of the social bee *Lasioglossum zephyrum,* within the nest. *Insectes Sociaux* 11:159-186.

Blackburn, R.H. 1982. "In the land of milk and honey: Okiek adaptations to their forest and neighbours." pp. 283-292. In: E. Leacock and R. Lee, eds. *Politics and history in band society.* Cambridge:Cambridge University Press.

Bourke, A. 1986. Alternative reproductive strategies in workers of the slave-making ant *Harpogoxenus sublaevis. Proceedings of the 10th International Congress of the International Union for the Study of Social Insects, Munchen.* p. 89.

Brandao, C.R.F. 1986. Queenlessness in Megalomyrmex (Formicidae: Myrmicinae), with a discussion on the effects of the loss of the true queens in ants. Proceedings of the 10th International Congress of the International Union for the Study of Social Insects, Munchen. p. 47.

Brockmann, H.J. 1984. "The evolution of social behavior in insects." In: J.R. Krebs and N.B. Davies, eds. *Behavioural ecology, an evolutionary approach.* Sunderland:Sinauer.

Brown, J.L. 1974. Alternative routes to sociality in jays--with a theory for the evolution of altruism and communal breeding. *American Zoologist* 14:63-80.

Brown, J.L. 1978. Avian communal breeding systems. *Annual Review of Ecology and Systematics* 9:123-155.

Brown, J.L. 1983. Cooperation--A biologist's dilemma. *Advances in the Study of Behavior* 13:1-39.

Buchli, H. 1956. Die neotenie bei *Reticulitermes. Insectes Sociaux* 3:131-143.

Cagniant, H. 1983. La parthenogenese thelytoque et arrhenotoque des ouvrieres de la fourmi *Cataglyphis cursor* Fonscolombe (Hymenoptera:Formicidae). Etude biometrique des ouvrieres et de leurs potentialites reproductices. *Insectes Sociaux* 30:241-254.

Carpenter, F.M. and H.R. Hermann. 1981. Antiquity of sociality in insects. pp. 81-90. In: H.R. Hermann, ed. *Social insects.* Vol. 1. New York:Academic Press.

Christian, J.J. 1970. Social subordination, population density, and mammalian evolution. *Science* 168:84-90.

Clark, W.C. 1976. The environment and the genotype in polymorphism. *Zoological Journal of the Linnean Society* 58:255-262.

Clausen, 1940. *Entomophagous insects*. New York:McGraw-Hill.

Cruz, Y.P. 1981. A sterile defender morph in a polyembryonic hymenopterous parasite. *Nature* 294:446-447.

Darlington, J.P.E.C. 1982. The underground passages and storage pits used in foraging by a nest of the termite *Macrotermes michaelseni* in Kajiado, Kenya. *Journal of Zoology (London)* 198:237-247.

Dawkins, R. and J.R. Krebs. 1979. Arms races between and within species. *Proceedings of the Royal Society of London, B* 205:489-511.

Deligne, J., A. Quennedey and M.S. Blum. 1981. "The enemies and defense mechanisms of termites." pp. 1-76. In: H.R. Hermann, ed. *Social insects*. Vol. 2. New York:Academic Press.

Dew, H.E. 1982. "Effects of queen removal on intracolony dynamics of *Polistes metricus*." p. 217. In: M.D. Breed, C.D. Michener and H.E. Evans, eds. *The biology of social insects*. Proceedings 9th Congress International Union of the Study of Social Insects. Boulder:Westview Press.

Dominey, W.J. 1984. Alternative mating tactics and evolutionarily stable strategies. *American Zoologist* 24:385-396.

Dublin, H.T. 1983. "Cooperation and reproductive competition among female African elephants." pp. 291-313. In: S.K. Wasser, ed. *Social behavior of female vertebrates*. New York:Academic Press.

Eickwort, G.C. 1981. "Presocial insects." pp. 199-281. In: H.R. Herman, ed. *Social insects*. Vol. 2. New York: Academic Press.

Eickwort, G.C. and P.F. Kukuk. 1986. Reproductive castes in primitively social bees (especially Halictidae). Proceedings of the 10th International Congress of the International Union for the Study of Social Insects, Munchen. p. 87.

Eisenberg, J.F. 1981. *The mammalian radiations*. University of Chicago Press.

Elmes, G.W. 1982. "Intracolonial competition in ants, with special reference to the genus *Myrmica*." pp. 212-216. In: M.D. Breed, C.D. Michener and H.E. Evans, eds. *The biology of social insects*. Proceedings 9th Congress International Union of the Study of Social Insects. Boulder:Westview Press.

Emlen, S.T. 1982. The evolution of helping. I. An ecological constraints model. *American Naturalist* 119:29-39.

Emlen, S.T. 1984. "Cooperative breeding in birds and mammals." pp. 305-339. In: J.R. Krebs and N.B. Davies, eds. *Behavioural ecology, an evolutionary approach*. Sunderland:Sinauer.

Evans, H.E. 1958. The evolution of social life in wasps. *Proceedings of the 10th International Congress of Entomology* 2:449-457.

Goody. 1969. Inheritance, property and marriage in Africa and Eurasia. *Sociology* 3:55-76.

Greenwood, P.J. 1980. Mating systems, philopatry and dispersal in birds and mammals. *Animal Behaviour* 28:1140-1162.

Hamilton, W.D. 1964. The genetical evolution of social behavior, I and II. *Journal of Theoretical Biology* 7:1-16, 17-52.

Hamilton, W.D. 1978. "Evolution and diversity under bark." pp. 154-175. In: L.A. Mound and N. Waloff, eds. *Diversity of insect faunas.* London:Royal Entomological Society, Symposium 9.

Hamilton, W.D. 1979. "Wingless and fighting males in fig wasps and other insects." pp. 167-220. In: D.W. Tinkle, ed. *Sexual selection and reproductive competition.* New York:Academic Press.

Hannon, S.J., R.L. Mumme, W.D. Koenig and F.A. Pitelka. 1985. Replacement of breeders and within-group conflict in the cooperatively breeding acorn woodpecker. *Behavioral Ecology and Sociobiology* 17:303-312.

Hansell, M. 1986. The evolution of paper technology in the social Vespidae. Proceedings of the 10th International Congress of the International Union for the Study of Social Insects, Munchen. p. 127.

Hansell, M.H., C. Samuel and J.I. Furtado. 1982. *"Liostenogaster flavolineta:* Social life in the small colonies of an Asian tropical wasp." pp. 192-195. In: M.D. Breed, C.D. Michener and H.E. Evans, eds. *The biology of social insects.* Proceedings 9th Congress International Union of the Study of Social Insects. Boulder:Westview Press.

Harcourt, A.H., K.S. Stewart and D. Fossey. 1976. Mate emigration and female transfer in wild mountain gorilla. *Nature* 263:226-227.

Hare, L. 1937. Termite phylogeny as evidenced by soldier mandible development. *Annals of the Entomological Society of America* 30:459-486.

Harris, W.V. 1958. Colony formation in the Isoptera. *Proceedings of the 10th International Congress of Entomology* 2: 435-439.

Hartung, J. 1976. On natural selection and the inheritance of wealth. *Current Anthropology* 17:607-622.

Hitchcock, R.K. 1982. "Patterns of sedentism among the Basarwa of eastern Botswana." pp. 223-267. In: E. Leacock and R. Lee, eds. *Politics and history in band society.* Cambridge:Cambridge University Press.

Hoogland, J.L. 1981. "Nepotism and cooperative breeding in the black-tailed prairie dog (Sciuridae:*Cynomys ludovicianus).*" pp. 283-310. In: R.D. Alexander and D.W. Tinkle, eds. *Natural selection and social behavior.* New York:Chiron Press.

Hrdy, S.B. and G. Hausfater. 1984. *Infanticide: Comparative and evolutionary perspectives.* New York:Aldine.

Hull, D.L. 1980. Individuality and selection. *Annual Review of Ecology and Systematics* 11:311-332.

Iwata, K. 1976. *Evolution of instinct: Comparative ethology of Hymenoptera.* New Delhi:Amerind Publications.

Jarvis, J.U.M. 1981. Eusociality in a mammal: Cooperative breeding in naked mole-rat colonies. *Science* 212:517-573.

Jeanne, R.L. 1975. The adaptiveness of social wasp nest architecture. *Quarterly Review of Biology* 50:267-287.

Kennedy, J.S. 1961. *Insect polymorphism.* Symp. 1. Royal Entomological Society of London.

Krishna, K. 1961. A generic revision and phylogenetic study of the family Kalotermitidae (Isoptera). *Bulletin of the American Museum of Natural History* 122:303-408.

Lenz, M. 1985. "Is inter- and intraspecific variability of lower termite neotenic numbers due to adaptive thresholds for neotenic elimination?--Considerations from studies on *Porotermes adamsoni* (Froggatt) (Isoptera:Termopsidae)." pp. 125-146. In: J.A.L. Watson, B.M. Okot-Kotber and Ch. Noirot, eds. *Caste differentiation in social insects* (Current Themes in Tropical Sciences. Vol. 3). Oxford, Pergamon Press.

Lenz, M. and R.A. Barrett. 1982. Neotenic formation in field colonies of *Coptotermes lacteus* (Froggatt) in Australia, with comments on the roles of neotenics in the genus *Coptotermes*. *Sociobiology* 7:47-60.

Lenz, M., R.A. Barrett and E.R. Williams. 1985. "Reproductive strategies in *Cryptotermes*: Neotenic production in indigenous and "tramp" species in Australia (Isoptera: Kalotermitidae)." pp. 147-162. In: J.A.L. Watson, B.M. Okot-Kotber and Ch. Noirot, eds. *Caste differentiation in social insects* (Current Themes in Tropical Sciences. Vol. 3). Oxford, Pergamon Press.

Lester, L.J. and R.K. Selander. 1981. Genetic relatedness and the social organization of *Polistes* colonies. *American Naturalist* 117:146-166.

Light, S.F. and P.L. Illg. 1945. Rate and extent of development of neotenic reproductives in groups of nymphs of the termite genus *Zootermopsis*. *University of California Publications in Zoology* 53:1-40.

Ligon, J.D. 1983. Cooperation and reciprocity in avian social systems. *American Naturalist* 121:366-384.

Lin, N. 1964. Increased parasitic pressure as a major factor in the evolution of social behavior in halictine bees. *Insectes Sociaux* 11:187-192.

Lin, N. and C.D. Michener. 1972. Evolution of sociality in insects. *Quarterly Review of Biology* 47:131-159.

Lindstrom, E. 1986. Territory inheritance and the evolution of group-living in carnivores. *Animal Behaviour* 34:1825-1835,

MacCormack. 1982. "Foundress associations and early colony failure in *Polistes exclamans*." p. 219. In: M.D. Breed, C.D. Michener and H.E. Evans, eds. *The biology of social insects*. Proceedings 9th Congress International Union of the Study of Social Insects. Boulder:Westview Press.

Mason, C.A. 1986. Social interactions and division of labour in two species of primitively eusocial allodapine bees: *Allodape exoloma* and *Braunsapis foveata*. Proceedings of the 10th International Congress of the International Union for the Study of Social Insects, Munchen. p. 90.

Mathews, R.W. 1968. *Microstigmus comes*: Sociality in a specid wasp. *Science* 160:787-788.

McKittrick, F.A. 1965. A contribution to the understanding of cockroach-termite affinities. *Annals of the Entomological Society of America* 58:18-22.

Mednikova, T.K. 1977. Caste differentiation in the termite *Anacantho-termes ahngerianus* Jacobson (Isoptera:Hodotermitidae). *Proceedings IUSSI Congress.* 118-120.

Michener, C.D. 1958. The evolution of social behavior in bees. *Proceedings of the 10th International Congress of Entomology* 2:441-447.

Michener, C.D. 1971. Biologies of the African allodapine bees. *Bulletin of the American Museum of Natural History* 145:221-301.

Michener, C.D. 1974. *The social behavior of bees.* Cambridge:Harvard University Press. 404 pp.

Michener, C.D. 1977. "Aspects of the evolution of castes in primitively social insects." *Proceedings of the 8th International Congress IUSSI.* Wageningen. pp. 2-6.

Mock, D.W. 1985. Siblicidal brood reducton: The prey size hypothesis. *American Naturalist* 125:327-343.

Moffett, M.W. 1986. Evidence of workers serving as queens in the genus *Diacamma* (Hymenoptera:Formicidae). *Psyche* 93:151-152.

Myles, T.G. 1986. Reproductive soldiers in the Termopsidae (Isoptera). *Pan-Pacific Entomology* 62:293-299.

Myles, T.G. In press. Evidence of sibling and/or parental manipulation in three species of Hawaiian termites (Isoptera). *Proceedings Hawaiian Entomological Society.*

Myles, T.G. and F. Chang. 1984. The caste system and caste mechanisms of *Neotermes connexus* (Isoptera:Kalotermitidae). *Sociobiology* 9:163-321.

Myles, T.G. and W.L. Nutting. In press. Termite eusocial evolution: A re-examination of Bartz' hypothesis and assumptions. *Quarterly Review of Biology.*

Nagin, R. 1972. Caste determination in *Neotermes jouteli* (Banks). *Insectes Sociaux* 19:39-61.

Nalepa, C.A. 1984. Colony composition, protozoan transfer and some life history characteristics of the woodroach *Cryptocercus punctulatus* Scudder (Dictyoptera:Cryptocercidae). *Behavioral Ecology and Sociobiology* 14:273-279.

Nel, J.A.J. 1975. Aspects of the social ecology of some Kalahari rodents. *Zeitschrift für Tierpsychologie* 37:322-331.

Noirot, Ch. 1956. Les sexues de remplacement chez les termites superieurs (Termitidae). *Insectes Sociaux* 3:145-158.

Noirot, Ch. 1977. Nest construction and phylogeny in termites. *Proceedings of the 8th International Congress IUSSI.* Wageningen. pp. 177-180.

Noirot, Ch. 1982. La caste des ouvriers, element majeur du succes evolutif des Termites. *Rivista di Biologia* 75:157-192.

Nukunya, G.K. 1973. "Population pressure, land tenure, and agricultural development in southeastern Ghana." pp. 191-217. In: S. Tax, ed. *Population, ecology, and social evolution.* Hague:Mouton Publishing.

Nutting, W.L. 1956. Reciprocal transfaunations between the roach *Cryptocercus* and the termite *Zootermopsis. Biological Bulletin* 110:83-90.

Nutting, W.L. 1966a. The seasonal flights of *Pterotermes* and *Zootermopsis* in southern Arizona. Proceedings 2nd Workshop on Termite Research. Biloxi, Miss.:Building Research Advisory Board, National Academy of Science-National Research Council. Nov. 8-10, 1965. pp. 17-19.

Nutting, W.L. 1966b. Distribution and biology of the primitive dry-wood termite *Pterotermes occidentis* (Walker) (Kalotermitidae). *Psyche* 73:165-178.

Nutting, W.L. 1969. "Flight and colony foundation." pp. 233-282. In: K. Krishna and F.M. Weesner, eds. *Biology of termites*. Vol. 1. New York:Academic Press.

Nutting, W.L. 1979. Termite flight periods: Strategies for predator avoidance? *Sociobiology* 4:141-151.

Nutting, W.L. and M.I. Haverty. 1976. Seasonal production of alates by five species of termites in an Arizona desert grassland. *Sociobiology* 2:145-153.

Oster, G.F. and E.O. Wilson. 1978. *Caste and ecology in the social insects*. Princeton:Princeton University Press.

Ostermeyer, M.C. and R.W. Elwood. 1985. Helpers(?) at the nest in the Mongolian gerbils, *Meriones unguiculatus*. *Behaviour* 91:61-77.

Packer, L. and G. Knerer. 1982. "Social evolution in the *Evylaeus malachurus* group." p. 220. In: M.D. Breed, C.D. Michener and H.E. Evans, eds. *The biology of social insects*. Proceedings 9th Congress International Union of the Study of Social Insects. Boulder:Westview Press.

Page, R.E. 1986. Sperm utilization in social insects. *Annual Review of Entomology* 31:297-320.

Peeters, C. 1982. "The reproductive strategy of the ponerine *Ophthalmopone berthoudi*: An insight into the evolution of ant eusociality." p. 220. In: M.D. Breed, C.D. Michener and H.E. Evans, eds. *The biology of social insects*. Proceedings 9th Congress International Union of the Study of Social Insects. Boulder:Westview Press.

Peeters, C. 1986. The diversity of reproductive systems in ponerine and other primitive ants. Proceedings of the 10th International Congress of the International Union for the Study of Social Insects, Munchen. p. 85.

Peeters, C. and R. Crewe. 1985. Worker reproduction in the ponerine ant *Ophthalmopone berthoudi*: An alternative form of eusocial organization. *Behavioral Ecology and Sociobiology* 18:29-37.

Plateaux-Quenu, C. 1961. Les sexues de remplacement chez les insectes sociaux. *Ann. Biology*. Ser. 3. 37:177-216.

Polis, G.A. 1981. The evolution and dynamics of intraspecific predation. *Annual Review of Ecology and Systematics* 12:225-251.

Price, P.W. 1975. *Insect ecology*. New York:John Wiley & Sons. 514 pp.

Reyer, H. 1984. Investment and relatedness: A cost/benefit analysis of breeding and helping in the pied kingfisher (*Ceryle rudis*). *Animal Behaviour* 32:1163-1178.

Richards, O.W. 1961. "An introduction to the study of polymorphism in insects." pp. 1-10. In: J.S. Kennedy, ed. *Insect polymorphism*. Oxon:Classey.

Riches. 1982. *Northern nomadic hunter-gatherers*. London:Academic Press. 242 pp.

Rissing, S.W. 1984. Replete caste reproduction and allometry of workers in the honey ants *Myrmecocystus mexicanus* Wesmael (Hymenoptera:Formicidae). *Journal of the Kansas Entomological Society* 57:347-350.

Roisin, Y. and J.M. Pasteels. 1985. Imaginal polymorphism and polygyny in the Neo-Guinean termite *Nasutitermes princeps* (Desneux). *Insectes Sociaux* 32:140-157.

Roisin, Y. and J.M. Pasteels. 1986a. Reproductive mechanisms in termites: Polycalism and polygyny in *Nasutitermes polygynus* and *N. coastalis*. *Insectes Sociaux* 33:149-167.

Roisin, Y. and J.M. Pasteels. 1986b. Replacement of reproductives in *Nasutitermes princeps* (Desneux) (Isoptera:Termitidae). *Behavioral Ecology and Sociobiology* 18:437-442.

Rood, J.P. 1978. Dwarf mongoose helpers at the den. *Zeitschrift für Tierpsychologie* 48:277-287.

Roonwal, M.L. 1975. Phylogeny and status of termite families Stylotermitidae and Indotermitidae with three-segmented tarsi, and the evolution of tarsal segmentation in the Isoptera. *Biol. Zbl.* 94:27-43.

Rowley, I., L.G. Grimes, G.E. Woolfenden and A. Zahavi. 1974. Cooperative breeding in birds. *Proceedings 16th International Ornithological Congress*, Canberra. pp. 655-693.

Ruppli, E. 1969. Die elimination uberzahliger ersatzgeschlechtstiere bei der termite *Kalotermes flavicollis* (Fabr.). *Insectes Sociaux* 16:235-248.

Russell, J.K. 1983. "Altruism in coati bands: Nepotism or reciprocity." pp. 263-290. In: S.K. Wasser, ed. *Social behavior of female vertebrates*. New York:Academic Press.

Sands, W.A. 1972. The soldierless termites of Africa (Isoptera: Termitidae). *Bulletin of the British Museum of Natural History (Entomology)* Suppl. 18.

Scudder. 1975. *The ecology of the Gwemba Tonga*. Manchester University Press. 274 pp.

Seelinger, G. and U. Seelinger. 1983. On the social organization, alarm and fighting in the primitive cockroach *Cryptocercus punctulatus* Scudder. *Zeitschrift für Tierpsychologie* 61:315-333.

Sewell, J.J. and J.A.L. Watson. 1981. Developmental pathways in Australian species of *Kalotermes* Hagen (Isoptera). *Sociobiology* 6:243-323.

Sherry, D.F. 1985. Food storage by birds and mammals. *Advances in the Study of Behavior* 15:153-188.

Sieber, R. 1985. "Replacement of reproductives in Macrotermitinae (Isoptera:Termitidae)." pp. 201-208. In: J.A.L. Watson, B.M. Okot-Kotber and Ch. Noirot, eds. *Caste differentiation in social insects* (Current Themes in Tropical Sciences. Vol. 3). Oxford, Pergamon Press.

Smith, C.C. and O.J. Reichman. 1984. The evolution of food caching by birds and mammals. *Annual Review of Ecology and Systematics* 15:329-352.

Snelling, R.R. 1981. "Systematics of social hymenoptera." pp. 370-454. In: H.R. Hermann, ed. *Social insects*. Vol. 2. New York:Academic Press.

Sommeijer, M.J. 1986. An ethological analysis of intra-colonial competition in relation to reproductive dominance in the stingless bee *Melipona favosa*. Proceedings of the 10th International Congress of the International Union for the Study of Social Insects, Munchen.

Spradbery, J.P. 1973. *Wasps: An account of the biology and natural history of solitary and social wasps*. University of Seattle Press. 408 pp.

Taborsky, M. and D. Limberger. 1981. Helpers in fish. *Behavioral Ecology and Sociobiology* 8:143-145.

Tallamy, D.W. and T.K. Wood. 1981. Convergence patterns in subsocial insects. *Annual Review of Entomology* 31:369-390.

Taylor, V.A. 1978. A winged elite in a subcortical beetle as a model for a prototermite. *Nature* 276-73-75.

Trivers, R.L. 1985. *Social evolution*. Menlo Park:Benjamin/Cummings Publishing Company. 462 pp.

Trivers, R.L. and H. Hare. 1976. Haplodiploidy and the evolution of social insects. *Science* 191:249-263.

Tsuji, K. 1986. Social structure of the Japanese queenless ant, *Pristomyrmex pungens*. Proceedings of the 10th International Congress of the International Union for the Study of Social Insects, Munchen. p. 54.

Turillazzi, S. 1982. "Some aspects of the biology and social behavior of *Parischnogaster nigricans serrei* (Hymenoptera: Stenogastrinae)." p. 222. In: M.D. Breed, C.D. Michener and H.E. Evans, eds. *The biology of social insects*. Proceedings 9th Congress International Union of the Study of Social Insects. Boulder:Westview Press.

Vancassel, M. 1984. Plasticity and adaptivity radiation of dermapteran parental behavior: Results and perspectives. *Advances in the Study of Behavior* 14:51-80.

van Honk, C.G.J. 1982. "The social structure of *Bombus terrestris* colonies: A review." pp. 196-200. In: M.D. Breed, C.D. Michener and H.E. Evans, eds. *The biology of social insects*. Proceedings 9th Congress International Union of the Study of Social Insects. Boulder: Westview Press.

Vehrencamp, S.L. 1979. "The role of individual, kin, and group selection in the evolution of sociality." pp. 351-394. In: P. Marler and J.G. Vandenbergh, eds. *Handbook of behavior and communcation. Vol. 3. Social behavior and communication.*

Ward, P.S. 1983. Genetic relatedness and colony organization in a species complex of ponerine ants. *Behavioral Ecology and Sociobiology* 12:285-299.

Ward, P.S. 1986. Functional queens in the Australian greenhead ant, *Rhytidoponera metallica* (Hymenoptera:Formicidae). *Psyche* 93:1-12.

Waser, P.M. and W.T. Jones. 1983. Natal philopatry among solitary mammals. *Quarterly Review of Biology* 58:355-390.

Wasser, S.K. and D.P. Barash. 1983. Reproductive suppression among female mammals: Implications for biomedicine and sexual selection theory. *Quarterly Review of Biology* 58:513-538.

Watson, J.A.L. and J.J. Sewell. 1985. "Caste differentiation in *Mastotermes* and *Kalotermes*: Which is primitive?" pp. 27-40. In: J.A.L. Watson, B.M. Okot-Kotber and Ch. Noirot, eds. *Caste differentiation in social insects* (Current Themes in Tropical Sciences. Vol. 3). Oxford, Pergamon Press.

Watson, J.A.L. and H.M. Abbey. 1985. Development of neotenics in *Mastotermes darwiniensis* Froggatt: An alternative strategy. pp. 107-124. In: J.A.L. Watson, B.M. Okot-Kotber and Ch. Noirot, eds. *Caste differentiation in social insects* (Current Themes in Tropical Sciences. Vol. 3). Oxford, Pergamon Press.

Watson, J.A.L., E.C. Metcalf and J.J. Sewell. 1977. A reexamination of the development of castes in *Mastotermes darwiniensis* Froggatt (Isoptera). *Australian Journal of Zoology* 25:25-42.

West-Eberhard, M.J. 1975. The evolution of social behavior by kin selection. *Quarterly Review of Biology* 50:1-33.

West-Eberhard, M.J. 1977. Morphology and behavior in the taxonomy of *Microstigmus* wasps. Proceedings of the 8th International Congress IUSSI, Wageningen. pp. 123-125.

West-Eberhard, M.J. 1978. Polygyny and the evolution of social behavior in wasps. *Journal of the Kansas Entomological Society* 51:832-856.

West-Eberhard, M.J. 1979. Sexual selection, social competition, and evolution. *Proceedings of the American Philosophical Society* 123:222-234.

West-Eberhard, M.J. 1981. Intragroup selection and the evolution of insect societies. pp. 3-17. In: R.D. Alexander and D.W. Tinkle, eds. *Natural selection and social behavior*. New York:Chiron Press.

West-Eberhard, M.J. 1983. Sexual selection, social competition, and speciation. *Quarterly Review of Biology* 58:155-183.

Wheeler. 1928. *The social insects*. New York:Harcourt Brace. 378 pp.

Wiessner. 1982. Risk reciprocity and social influences on !Kung San economics. pp. 61-84. In: E. Leacock and R. Lee, eds. *Politics and history in band society*. Cambridge University Press.

Williams, N.M. and E.S. Hunn. 1982. *Resource managers: North American and Australian hunter gatherers*. Boulder:Westview Press. 267 pp.

Wilson, E.O. 1971a. *The insect societies*. Cambridge:Harvard University press. 548 pp.

Wilson, E.O. 1971b. Competitive and aggressive behavior. pp. 181-217. In: J.F. Eisenberg and W.S. Dillion, eds. *Man and beast: Comparative social behavior.* Washington, D.C.:Smithsonian Institution Press. 401 pp.

Wilson, E.O. 1980. *Sociobiology: The abridged edition.* Cambridge, Mass.:Belknap of Harvard University Press. 366 pp.

Wilson, E.O. 1985. The sociogenesis of insect colonies. *Science* 228:1489-1495.

Woolfenden, G.E. and J.W. Fitzpatrick. 1984. The Florida scrub jay: Demography of a cooperative-breeding bird. *Monographs in Population Biology: 20.* Princeton:Princeton University Press. 406 pp.

Zimmerman, R.B. 1983a. Sibling manipulation and indirect fitness in termites. *Behavioral Ecology and Sociobiology* 12:143-145.

Zimmerman, R.B. 1983b. Individual and colony life history of *Pterotermes occidentis* and its relation to theories of termite evolution. Tucson:University of Arizona. Ph.D. dissertation. 170 pp.

INDEX